·dbook of Extemporaneous
· ation

Handbook of Extemporaneous Preparation

A guide to pharmaceutical compounding

Edited by
Mark Jackson BSc, MPhil, MRPharmS
Deputy Director, QCNW/Head of QA/QC, Liverpool Pharmacy Practice Unit, Liverpool, UK

Andrew Lowey DPharm, MRPharmS
Clinical Pharmacy Manager, Leeds Teaching Hospitals, Leeds, UK

On behalf of
The NHS Pharmaceutical Quality Assurance Committee

Published by the Pharmaceutical Press
1 Lambeth High Street, London SE1 7JN, UK

Pharmaceutical Press is the publishing division of the Royal Pharmaceutical Society of Great Britain

First published 2010
Reprinted 2011

Typeset by Thomson Digital, Noida, India
Printed in Great Britain by TJ International, Padstow, Cornwall

ISBN 978 0 85369 901 9

A catalogue record for this book is available from the British Library.

Contents

Preface

My first experience of extemporaneous dispensing was sitting in the corner of the dispensary after school watching my father being handed pieces of paper from customers written in an unfamiliar foreign language with strange hieroglyphics and indecipherable handwriting. He would then peer at the paper from under his glasses, consult a tattered little black book and proceed to select a variety of powders and liquids. Out would come the scales, the pestle and mortar, and he would proceed to make up a mystical potion which would be bottled up, labelled and gratefully received by the customer. The process had an air of mystery to it and I was intrigued. It was these early first-hand experiences of compounding that were my inspiration for a career in pharmacy.

Extemporaneous preparation or pharmaceutical compounding has historically been a core component of the pharmaceutical profession since its inauguration. However in the modern era, the large-scale manufacture of medicines in industry has led to the majority of medicines becoming commercially available. Now, when a patient has a special need for a custom-made product, the majority of pharmacy departments, quite rightly, outsource the service to a specialist company or hospital. Hence the need for pharmacists to retain compounding skills has diminished as it is no longer part of their daily work and consequently this has led to decline in expertise within the profession in this area.

However, there are still circumstances where custom-made products are required for patients and we need to ensure that we as a profession retain the skills to ensure this is done safely, whether to prepare it locally in the pharmacy or to establish the credentials of a third party to make it on our behalf. Unfortunately there have been some well-publicised incidents where due diligence has not been taken, resulting in patient harm.

Licensed medicines represent the 'gold standard' for quality, safety and efficacy. There are, however, circumstances in which there is no licensed medicine which fully meets the clinical needs of a particular patient or patients. In these circumstances it is sometimes necessary for the pharmacist to extemporaneously prepare a limited quantity of a custom-made product for a specific patient. Oral liquid medicines are commonly prepared extemporaneously because of a relative lack of licensed formulations for groups such as the young and the elderly who are unable to swallow tablets or capsules, or for whom the required dose is less than a single tablet or capsule.

It is widely recognised that the extemporaneous preparation of medicines carries significant risks. Pharmacists and pharmacy departments have a key role in ensuring that patients receive medicines of the appropriate quality whatever the source, whether dispensed, manufactured locally or procured from a third party. This book aims to provide an updated standard for extemporaneous preparation, taking into account previous NHS standards and regulatory guidance.

To give some historical background to the book, a working party of the NHS Pharmaceutical Quality Assurance Committee first produced the *Guide to the Preparation of Extemporaneous Products* in 2001. This guide provides detailed guidance to pharmacists relating to the extemporaneous preparation of medicines in accordance with a prescription, under the exemption conferred on pharmacists under Section 10 of the UK Medicines Act 1968.

In December 2005, the UK National Advisory Board for the Modernisation of NHS Hospital Medicines Manufacturing and Preparation Services commissioned a study into improving the quality and formulation of unlicensed, non-sterile oral medicines made by the NHS.

In April 2008, the Pharmaceutical Inspection Convention, Pharmaceutical Inspection Co-operation Scheme (PIC/S) published a *Guide to Good Practices for the Preparation of Medicinal Products in Healthcare Establishments*. This document sets out guidance for the preparation of medicines for human use normally performed by healthcare establishments for supply directly to patients. The UK Medicines and Healthcare products Regulatory Agency (MHRA) have stated that this guidance is not applicable to NHS hospitals working under a manufacturer's 'Specials' licence. However, it is applicable to products prepared under Section 10 exemption to the Medicines Act 1968, including extemporaneous dispensing.

The NHS Pharmaceutical Quality Assurance Committee reviewed this document and the authors of this book were tasked with updating the 2001 'Guide to the preparation of extemporaneous products' in line with the PIC/S guidance.

This book aims to provide the reader with comprehensive and relevant guidance about extemporaneous preparation that incorporates the principles of the PIC/S guidance document. It also incorporates the key findings and outputs from the UK National Advisory Board study, including a formulary of individual stability summaries for the top 50 most commonly extemporaneously prepared medicines in NHS hospitals. It will be adopted as the standard for extemporaneous dispensing for NHS patients. Although the standards set out in this book are primarily written for implementation in NHS hospitals, the principles should be equally applied across the profession.

The focus for the formulary section of the book is oral liquid medicines, due to the remit of the National Advisory Board study; however, the standards equally apply to other extemporaneously prepared dosage forms (e.g. creams, ointments, lotions).

The book is an important reference for any pharmacist, pharmacy technician or student involved with extemporaneous preparation. It also includes sections relating to clinical risk assessment and advice for procuring 'Specials' from manufacturers, and is an important reference for both clinical and procurement pharmacists.

The Editors would like to thank members of the Editorial Board for all their hard work and conscientiousness in preparing these standards, and the members of the UK NHS QA Committee and UK National Paediatric Pharmacists' Group (NPPG) for their help and advice. Our hope is that this book will keep these essential skills alive and that the art of compounding can be practised safely throughout the profession, offering inspiration to future generations.

Mark Jackson
Leeds, February, 2010

Further reading

Fenton-May V'I (2003). *Guide to the Preparation of Non Sterile Extemporaneous Products in NHS Hospitals*. Regional Pharmaceutical Quality Control Committee.

Pharmaceutical Inspection Convention, Pharmaceutical Inspection Co-operation Scheme (2008). *Guide to Good Practices for the Preparation of Medicinal Products in Healthcare Establishments*. Pharmaceutical Inspection Convention Pharmaceutical Inspection Co-operation Scheme. April 2008. www.picscheme.org (accessed April, 2006).

About the editors

Mark Jackson

Mark graduated from Brighton School of Pharmacy in 1990; he subsequently gained an MPhil from Bradford University and has completed the Pharmaceutical Technology and Quality Assurance (PTQA) postgraduate diploma at Leeds University.

Mark started his career as a clinical/QA pharmacist at Bradford Royal Infirmary before moving to New Zealand to take up the post of production manager at Dunedin Hospital. He returned to the UK in 2001 to take up the post of QA/QC manager at Leeds Teaching Hospitals NHS Trust and performs the role of regional QA specialist for Yorkshire and Humber. In March 2010 Mark moved to Liverpool Pharmacy Practice Unit to take up the post of Head of QA and Deputy Director for Quality Control North West.

Mark is a member of the National NHS Pharmaceutical Quality Assurance Committee and is chairman of the Working Party for Extemporaneous Preparation. He also acted as the project manager for the UK research project entitled 'Improving the Quality and Formulation of Unlicensed, Non-Sterile, Oral Medicines Prepared in the NHS', sponsored by the UK National Advisory Board.

Andrew Lowey

Andrew graduated from Bradford University in 1999 and went on to study for a clinical diploma at the University of Wales, Cardiff. Following this, he entered the DPharm doctoral degree programme at Bradford University in 2001, during which he practised as a hospital pharmacist in a variety of clinical and technical roles around Yorkshire.

While at Harrogate District Hospital, he won the Pharmaceutical Care Award for Innovation in Hospital Pharmacy (2002) for helping to develop a pharmacist-led cardiac risk clinic for patients with type 2 diabetes mellitus.

In 2004, he moved to the Quality Assurance Department in Leeds to carry out a doctoral thesis focusing on the quality of unlicensed, oral liquid medicines in the NHS. The funding for the underpinning project was provided by the UK National Advisory Board for the Modernisation of NHS Hospital Medicines Manufacturing and Preparation Services, and much of the final work is included in this book. While carrying out the project, Andrew also served on the British Pharmacopoeia Working Group for Unlicensed Medicines.

After graduating from the DPharm programme in 2007, he moved back to clinical pharmacy in 2008 to take up the post of clinical pharmacy manager for Leeds Teaching Hospitals NHS Trust. He has published several articles on clinical and technical aspects of pharmacy, and is passionate about promoting high-quality pharmacy research.

Membership of the editorial board

Acknowledgements

The editors would like to thank:

NHS Pharmaceutical QA Committee members
The Neonatal and Paediatric Pharmacists' Group (NPPG) Committee members
All the pharmacists at Leeds Teaching Hospitals NHS Trust
Chris Acomb, Clinical Pharmacy Manager, Leeds Teaching Hospitals NHS Trust
Dr Christine Alexander, Quality Assurance Manager, Tayside Pharmaceuticals
John Bane, Senior Pharmacist, Sheffield Children's Hospital
Richard Bateman, QA Regional Specialist Pharmacist, Guy's Hospital
Brian McBride, Royal Pharmaceutical Laboratory Service, Belfast City Hospital
Phil Bendell, Principal Pharmacist, Unit Manager, Torbay PMU, South Devon Healthcare
Caroline Brady, Senior Pharmacy Technician, County Durham and Darlington Acute Hospitals NHS Trust
Roger Brookes, Technical Services Manager, Royal Hallamshire Hospital (at the time of the project)
Professor Henry Chrystyn, Professor of Pharmacy Practice, Bradford University (at the time of the project)
Professor David Cousins, Head of Safe Medication Practice, National Patient Safety Agency
Peter Cowin, Deputy Quality Assurance Manager, Pharmacy, Charing Cross Hospital, Hammersmith Hospitals NHS Trust
Dr Diana Crowe, Principal Pharmacist, Regional Pharmaceutical Laboratory Service, Belfast City Hospital
Philip Dale, Paediatric Pharmacist, Royal Cornwall Hospital
Jackie Eastwood, Specialist Pharmacist (Gastroenterology), St. Mark's Hospital, London

V'Iain Fenton-May, Quality Controller, St Mary's Pharmaceutical Unit, Cardiff and Vale NHS Trust (at the time of the project)
Andy Fox, Principal Pharmacist, Southampton University Hospitals NHS Trust
Wayne Goddard, Laboratory Manager, Quality Control West Midlands, University Hospitals Birmingham NHS Foundation Trust
Dr Frank Haines-Nutt, Quality Controller, South Devon Healthcare, Long Rd, Paignton, Devon
David Harris, Principal Paediatric Pharmacist, Leicester Royal Infirmary
Dr Chris Hiller, Quality Control Department, Newcastle Upon Tyne Hospitals NHS Trust
Dr Denis Ireland, Deputy Director, Quality Control North West, Pharmacy Practice Unit, Liverpool
Sue Jarvis, Senior Pharmacist (PICU), Bristol Royal Hospital for Children
Professor Liz Kay, Clinical Director of Medicines Management and Pharmacy, Leeds Teaching Hospitals NHS Trust
Simon Keady, Principal Pharmacist, University College London Hospitals NHS Foundation Trust
Tasneem Khalid, Principal Pharmacist Clinical Services/Chair, Paediatric Oncology Pharmacist's Group, Central Manchester and Manchester Children's University Hospitals NHS Trust
Martin Knowles, Regional QA Specialist Pharmacist, London and the South-East
Lisa Lawrie, Pharmacy Technician, Bradford Teaching Hospitals NHS Foundation Trust
Robert Lowe, Director of Quality Assurance Specialist Services, East of England and Northamptonshire
Rowena McArtney, Senior Information Pharmacist, Welsh Medicines Information Centre
Liz Mellor, Lead Pharmacist for Clinical Governance, Leeds Teaching Hospitals NHS Trust
Peter Mulholland, Pharmacy Department, Southern General Hospital, Glasgow
Dr Trevor Munton, Regional QA Pharmacist, Blackberry Hill Hospital, Bristol
Tony Murphy, University College London Hospital
Jodi New, Quality Assurance Pharmacist, Calderdale and Huddersfield NHS Foundation Trust
Catherine Norris, Consultant Pharmacist, Harrogate and District NHS Foundation Trust
Penny North-Lewis, Paediatric Liver Pharmacist, Leeds Teaching Hospitals NHS Trust

Claire Norton, Lead Pharmacist for Clinical Trials and Psychiatry, West Midlands Medicines for Children Research Network Pharmacist Adviser, Birmingham Children's Hospital
Tony Nunn, Clinical Director of Pharmacy, Royal Liverpool Children's NHS Hospital
Hema Patel, University Hospital Lewisham
Reena Patel, Children's University Hospital, Dublin, Eire
Susan Phillips, Production Technician, Whittington Hospital NHS Trust Pharmacy Department
Bob Shaw, NHS Pharmacy Practice Unit, University of East Anglia
Tim Sizer, Assistant Course Director, PTQA, Leeds University (at the time of the project)
Jon Silcock, Lecturer in Pharmacy, School of Healthcare, University of Leeds
Julian Smith, Quality Controller, St Mary's Pharmaceutical Unit, Cardiff and Vale NHS Trust
Professor Peter Taylor, Director of Pharmacy, Airedale NHS Trust (at the time of the project)
John Timmins, Clinical Director of Pharmacy and Medicines Management, Sheffield Children's NHS Foundation Trust
Bob Tomlinson, Principal Pharmacist, Calderdale and Huddersfield NHS Foundation Trust
Dr Catherine Tuleu, School of Pharmacy, University of London
Dr Andrew Twitchell, Compounding Manager, Nova Laboratories Limited (at the time of the project)
Ross Walker, Pharmacist, United Bristol Healthcare Trust
Don Wallace, Regional QA Pharmacist, Belfast City Hospital
Angela Watkinson, Senior Technician for Extemporaneous Preparation, Leeds Teaching Hospitals (at the time of the project)
Susan Williamson, Senior Pharmacist, DIAL Information Service, Royal Liverpool Children's NHS Hospital
Louise Wraith, Non-Sterile Production Manager, Royal Free Hospital
Dr David Wright, Senior Lecturer in Pharmacy Practice, School of Chemical Sciences and Pharmacy, University of East Anglia
David Woods, School of Pharmacy, University of Otago, Dunedin, New Zealand

Standards

1

Introduction

1.1 Background

Pharmacists are responsible for ensuring that drug use is safe and effective. Wherever practicable, licensed medicines are used and represent the 'gold standard' for quality, safety and efficacy. There are, however, circumstances in which there is no licensed medicine that fully meets the clinical needs of a particular patient or patients. In these circumstances it is sometimes necessary for the pharmacist to extemporaneously prepare a limited quantity of a custom-made medicine for a specific patient. Oral liquid medicines are commonly extemporaneously prepared because of a relative lack of licensed formulations for groups such as the young or elderly who are unable to swallow tablets or capsules, or for whom the required dose is less than a single tablet/capsule.

Extemporaneous preparation remains one of the highest risk preparative activities carried out in the pharmacy, as the risks of unlicensed medicines are combined with inherent risks associated with the pharmaceutical compounding process (Marshall and Daly, 1996). In addition, extemporaneously prepared medicines are commonly given to some of the most vulnerable patients in hospitals and communities (e.g. neonates, children, elderly patients, stroke victims, patients with feeding tubes *in situ*). These vulnerable patients are often not capable of alerting carers or staff to any adverse drug events they may be experiencing.

There have been a number of reports of errors associated with the use of extemporaneously prepared medicines, resulting in serious harm to patients. The most notable incident in recent UK history was the 'peppermint water case', where the use of the wrong strength of chloroform water led to the death of a child (Anon, 1998). This case highlighted the issues of toxic ingredients and calculation errors, particularly where the strength of one or more ingredients is stated in a historical or non-standard fashion. Similar incidents have occurred in the USA, including the death of a child from a superpotent imipramine liquid, and another of a 5-year-old child who received a thousand-fold overdose of clonidine (Kaye, 2003; Kirsch, 2005).

Administration errors have also been reported, including five-fold, ten-fold, hundred-fold or thousand-fold mistakes in calculating or measuring doses. Some of these errors have been attributed to inadequate labelling (e.g. confusion between strengths expressed per mL or per 5 mL).

The relative lack of research and development work supporting these products is associated with the potential for formulation failure or poor dose uniformity, resulting in the risk of overdose or underdose. The lack of high-quality data to support historical formulae has been acknowledged by authors across the world (Stewart and Tucker, 1982; Crawford and Dombrowski, 1991; Woods, 1997).

Even where a given formulation has been shown to achieve suitable physical, chemical and microbiological stability, the bioavailability and palatability of the preparation may be unproven. Very few extemporaneous preparations are supported by any data to demonstrate a suitable absorption profile and/or bioequivalence with a licensed preparation. Other issues include concerns about inadequate access to equipment and materials needed to provide a safe extemporaneous dispensing service and the highest possible quality products (Fisher *et al.*, 1991; Davis, 1997).

All these risks have been potentiated by declining expertise in pharmaceutics and formulation within the pharmacy profession (Crawford and Dombrowski, 1991; Pappas *et al.*, 2002; Chowdhury *et al.*, 2003). Compounding of oral medicines is often delegated to junior or trainee staff, and there is commonly no quality assurance system in place to support practice (Pappas *et al.*, 2002). Until the publication of the *Guide to the Preparation of Non-Sterile Extemporaneous Products in National Health Service (NHS) Hospitals* in 2003 (Fenton-May *et al.*, 2003), there were no agreed, detailed standards for this area of extemporaneous preparation in the UK.

There is therefore an identified need to investigate and improve the quality of formulation of the oral medicines currently prepared, with a view to offering further guidance to pharmacists in order to minimise risks to patient safety.

This book has been written on behalf of the NHS Pharmaceutical Quality Assurance Committee, following the completion of a UK research project entitled 'Improving the quality and formulation of unlicensed, non-sterile, oral medicines prepared in the NHS'. The project was supported by Modernisation Board funding and directed by the NHS Pharmaceutical Quality Assurance Committee – Working Group for Extemporaneous Preparation. It aimed to consider patient needs with regard to non-sterile preparation and manufacture, and formed one part of a wider programme of research and development. The book includes an update of the *Guide to the Preparation of Non-Sterile Extemporaneous Products in National Health Service (NHS) Hospitals*, together with the 50 technical monographs generated from the research project.

1.2 The NHS Modernisation Agenda

The NHS Modernisation Agency (now superseded by the NHS Institute for Innovation and Improvement) was established in the UK in 2001 to support the NHS and its partner organisations in the task of modernising services and improving outcomes and experiences for patients. Key standards and frameworks were developed by the Department of Health, against which continuous improvements can be measured (Department of Health, 2004).

In 2000, concerns about risk management and the sustainability of NHS hospital manufacturing units led ministers to commission a UK-wide risk assessment of NHS medicines manufacture. A multidisciplinary Advisory Group was established. Their main recommendation was that 'the NHS Manufacturing Service should be restructured as a national service and the move to this national service should be facilitated by an over-arching Implementation Board' (Department of Health, 2003). The Modernisation Board generated an implementation plan for the service.

Integration into the patient safety agenda and creation of strong working links with the Medicines and Healthcare products Regulatory Agency (MHRA), the National Patient Safety Agenda (NPSA) and other key UK stakeholders was deemed essential, as was embedding NHS medicines manufacturing services in clinical practice to ensure that patients get the best from their medicines. The four central pillars of modernisation are:

- national coordination
- clinical governance
- working with partners in industry
- capital investment.

A key part of modernisation strategy is to review which products the NHS should continue to use, by which processes they should be made, and where and by whom they would most appropriately be made. For products for which continued use has been justified by clinical need, the review should also consider whether manufacture should continue in the NHS or whether outsourcing to non-NHS manufacturers might be more appropriate and release NHS capacity for other purposes and products. Collaboration with independent providers is essential if this is to be achieved.

The ultimate goal is to ensure that the NHS has a rational approach to provision and cost-effective use of unlicensed medicines by ensuring that:

- patients have ready access to all the medicines they need
- unlicensed medicines are used only when there is no licensed medicine which meets the clinical needs of a particular patient or group of patients
- the NHS makes and uses only a limited range of clinically essential unlicensed medicines

- all medicines are of the highest possible quality
- NHS medicine manufacturing capacity is used cost effectively and is prioritised for clinically essential products which only the NHS can make.

In order to facilitate this process, mutually agreed guidelines for working with commercial partners are necessary in order to make sure that all available manufacturing capacity is used to best effect.

1.3 Clinical governance and quality assurance

Clinical governance is defined by the Department of Health (England) as 'the system through which NHS organisations are accountable for continuously improving the quality of their services and safeguarding high standards of care, by creating an environment in which clinical excellence will flourish'. Clinical governance must be applied to all aspects of medicines management.

The principles of clinical governance and quality assurance should underpin all modernisation strategies and the modernisation strategy highlighted that pharmacy quality assurance services play a key role in the provision of manufacturing and aseptic services preparation through:

- assessment of quality and risk potential of purchased medicines
- coordination of minor product defect reports and communication of drug alerts
- audit and advising on standards
- stability testing to support allocated shelf-lives
- research, often in relation to new product development
- technical information and advice
- quality assessment and testing of devices and dressings.

The relationship between the environment in which medicines are prepared, the level of quality assurance applied to the processes involved and the level of residual risk to patients treated with those medicines is shown in Figure 1.1.

Preparation in clinical areas and extemporaneous preparation in the pharmacy represent the lowest levels of quality assurance and the highest risk, whereas manufacture of licensed medicines provides the most robust assurance of quality, safety and efficacy. Manufacture of unlicensed medicines under a manufacturer's 'Specials' licence (MS) represents a level of intermediate overall quality and risk. The key difference between licensed medicines and 'Specials' is that all licensed medicines are supported by robustly assessed evidence for the pharmaceutical quality of the formulation and for safety and efficacy in clinical use, whereas the evidence for most 'Specials' is much weaker and has not been subjected to regulatory assessment.

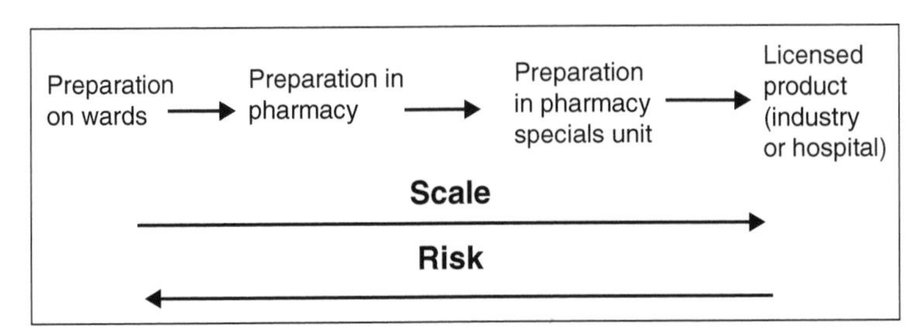

Figure 1.1 The progression of risk. (From Alison Beaney, personal communication, January 2006.)

1.4 Rationalisation and standardisation

A key objective for clinical governance of unlicensed medicines is to ensure that use of products prepared in clinical areas or by extemporaneous preparation in the pharmacy is limited in favour of products manufactured under a manufacturer's 'Specials' licence (MS) or, ideally, a full manufacturer's licence (ML). In order to achieve this, it is essential that key decisions about product choice and presentation are made in conjunction with expert clinicians and are subject to peer review. This process may need to involve professional bodies such as the relevant Royal Colleges and other key stakeholders. In the UK, senior doctors, pharmacists and other allied healthcare professionals sit on Drug and Therapeutics Committees. These committees also play (or should play) a key role in evaluating the evidence base to support the use of unlicensed medicines as well as licensed medicines.

In the UK, unlicensed non-sterile, oral, liquid (and other) medicines may be made by 'Specials' manufacturers under an MS, or can be extemporaneously prepared under the supervision of a pharmacist. Extemporaneous dispensing is carried out under the exemptions allowed by Section 10 of the 1968 Medicines Act. The modernisation agenda aims to improve the extemporaneous dispensing practice and identify opportunities to replace preparation in clinical areas or extemporaneously in the pharmacy by preparation under an MS or, where possible, an ML. It is essential to involve colleagues in 'Specials' units and in the pharmaceutical industry to move preparations along the quality and risk progression in this way.

Rationalisation of the product inventory and standardisation of the remaining product range offers safety and quality benefits:

- Practitioners and patients become more familiar with the products available: greater familiarity reduces risks of prescribing, dispensing and administration errors.

- Opportunities are created for product development to improve formulations, enhance stability and shelf-life, improve patient acceptability and improve availability.
- Opportunities for batch manufacture and economies of scale may lead to cheaper products as well as higher quality.

Historical UK legal restrictions on advertising named unlicensed medicines have severely restricted the amount of information readily available to inform decision-making by practitioners trying to identify and source 'Specials'. We hope that the data, formulae and supporting information contained in this handbook, together with the continued use of the NHS 'Pro-File' database, will help procurement and clinical colleagues. We also hope it will help those working in 'Specials' units (both NHS and non-NHS) and the pharmaceutical industry to identify products, particularly oral liquid medicines, for which the potential market is large enough to make production as licensed products commercially viable.

1.5 Summary

After preparation in clinical areas, extemporaneous preparation represents the preparative activity posing the highest risk to patient safety and subject to the lowest level of quality assurance. An update of the existing standards together with the standardisation and rationalisation of current products and formulae will help to improve product quality, decrease the overall risk to patient safety, and will support the progressive transfer of preparation of many products to batch manufacture in a controlled environment.

1.6 References

Anon (1998). Baby dies after peppermint water prescription for colic. *Pharm J* 260: 768.

Chowdhury T, Taylor KMG, Harding G (2003). Teaching of extemporaneous preparation in UK Schools of Pharmacy. *Pharm Educ* 3: 229–236.

Crawford SY, Dombrowski SR (1991). Extemporaneous compounding activities and the associated needs of pharmacists. *Am J Hosp Pharm* 48: 1205–1210.

Davis NM (1997). Do you really need glacial acetic acid on your shelf? *Hosp Pharm* 32: 611.

Department of Health (2003). *Modernising the NHS Hospital Medicines Manufacturing Service. Implementation Plan: Allocation of capital funding for Financial Years 2004/05 and 2005/06.* www.dh.gov.uk (accessed 18 August 2007).

Department of Health (2004). *Standards for Better Health.* Updated April 2006. www.dh.gov.uk/en/Publicationsandstatistics/Publications/PublicationsPolicyAndGuidance/DH_4086665 (accessed 21 March 2007).

Fenton-May V, ed. (2003). *Guide to the Preparation of Non-Sterile Extemporaneous Products in NHS Hospitals.* London: NHS Quality Assurance Committee.

Fisher CM, Corrigan OI, Henman MC (1991). Quality of pharmaceutical services in community pharmacies. *J Soc Adm Pharm* 8: 175–176.

Kaye T (2003). The quandary of compounding for MCOs: administrative costs, risks, and waste. *Managed Care*, 42–46.

Kirsch L (2005). Extemporaneous quality. *J Pharm Sci Technol* 59: 1–2.

Marshall IW, Daly MJ (1996). Risk management in hospital pharmacy. *Pharm Manage* 12: 44–47.

Pappas A, McPherson R, Stewart K (2002). Extemporaneous dispensing: Whatever happened to it? A survey of Australian general practitioners *J Pharm Res* 32: 310–314.

Stewart PJ, Tucker IG (1982). A survey of current extemporaneously prepared paediatric formulations. *Aust J Hosp Pharm* 12: 64–68.

Woods DJ (1997). Extemporaneous formulations – problems and solutions. *Paediatr Perinat Drug Ther* 1: 25–29.

2

Risk management

2.1 Introduction

The extemporaneous preparation of medicines is associated with a number of potential risks to patients, healthcare staff and their organisation. These all need to be carefully considered in determining the best treatment option; they then need to be minimised when the use of this category of medicine is necessary. A risk assessment should be performed before making a decision to extemporaneously prepare a medicine in line with the local unlicensed medicines policy. This process should be underpinned with a procedure in place and records of risk assessments should be maintained on file.

This section gives guidance on the risks associated with extemporaneous preparation, the assessment and management of these risks and alternative options to extemporaneous preparation.

2.2 Legal background and organisational risks

Medicines legislation requires that medicinal products are licensed before they are marketed in the UK. Accordingly no medicinal product may be placed on the market without a marketing authorisation.

The marketing authorisation provides assurance of the safety and efficacy of the drug in relation to a specified use, which has been reviewed and accepted by an official expert body. It also defines the legal status of the product and assures its quality. A marketing authorisation specifies the clinical condition(s), dose(s), routes of administration, and packaging for the particular preparation, all of which are detailed in the Summary of Product Characteristics (SPC).

Extemporaneously prepared medicines are unlicensed medicines and are not subject to these regulatory safeguards. Therefore neither prescribers nor pharmacists can make the same assumptions of quality, safety and efficacy about these products as they do for licensed medicines.

It should also be noted that the extemporaneous preparation of medicines from licensed starting materials (e.g. tablets, capsules, injections) also removes these regulatory safeguards unless specifically covered in the SPC. It is therefore an area of pharmaceutical activity which carries potentially increased risk to the patient, the supervising pharmacist and any other healthcare professionals involved in preparation and/or administration.

The pharmacist responsible for preparing or procuring an extemporaneously prepared medicine should therefore take responsibility for ensuring that the medicine is of suitable quality, and is safe and efficacious. A failure to do so puts both the pharmacist and organisation at risk in terms of both civil liability (negligence, breach of contract: Sale and Supply of Goods Act 1974) and criminal liability (Medicines Act 1968, Health and Safety at Work Act 1974, Corporate Manslaughter and Corporate Homicide Act 2007, Consumer Protection Act 1987). The pharmacist should also ensure that the prescriber is aware of the unlicensed status of the medicine and any associated risks with its use.

Extemporaneous preparation should therefore only be considered when an equivalent licensed product is unavailable or is unsuitable for use and if the use can be clearly justified clinically and pharmaceutically. Consideration should be given to all alternatives before choosing this option.

However, it is recognised that some patients may have special clinical needs that cannot be met by licensed medicinal products or by a viable alternative option. In these circumstances it would be inappropriate to curtail the patient's treatment, as this would have a detrimental effect on their condition. Whenever carrying out a risk assessment, the risks of not treating the patient should also be considered and be at the forefront of the decision-making process.

2.3 Alternatives to extemporaneous preparation

There are a number of alternative options that should be carefully considered as part of a patient-specific clinical risk assessment before opting for extemporaneous preparation. Each option has its own associated merits and risks, and the best option will vary according to specific circumstances surrounding both the patient's condition and the urgency of commencing the treatment.

2.3.1 Therapeutic substitution

The use of a licensed medicine from the same therapeutic classification should be considered and may provide a better clinical option than the use of an extemporaneously prepared medicine which has limited data to support its formulation and stability. However, the decision to switch to a different medicine should also take into account the condition of the patient and the relative toxicity of the drug. For example, if a patient is stabilised on a medicine with a narrow therapeutic index, it may have a more detrimental impact on the patient's well-being to switch to a different, but therapeutically equivalent drug, than use a medicine that has been extemporaneously prepared against a validated formulation. However, in either case, the patient should be closely monitored following the change to their treatment to ensure their condition remains under control.

The use of a less potent steroid rather than diluting a potent agent is an example where a therapeutic alternative may eliminate the need for an extemporaneous preparation.

The use of an alternative route of administration, for example use of the rectal rather than the oral route, could also be considered if an appropriate formulation is available.

2.3.2 Procurement options

2.3.2.1 Use of an imported product

The importation of a suitable product that carries a marketing authorisation in its country of origin should be considered. However, it must be noted that the presence of a non-UK marketing authorisation confers no legal status on the medicine in the UK and that importation can only take place through a company holding a Wholesale Dealer (Import) Licence.

The preparation selected should be licensed for use in a country with equivalent or similar licensing arrangements and regulatory standards to the UK (e.g. EU, Canada, Australasia). This will provide the requesting pharmacist with assurance that the quality, safety and efficacy of the medicine have been reviewed by a competent regulatory authority. However, care should be taken to ensure that the medicine has been licensed for use in the country of origin and placed on the market there, rather than being manufactured solely for export.

The procuring pharmacist should also ensure that the company used for importation has adequate quality systems in place to ensure that the medicine comes from a reputable source; that counterfeit detection measures are in place; and that the cold chain (where appropriate) is maintained to the point of delivery (see Chapter 11 for more details).

From a clinical perspective, the procuring pharmacist needs to be aware that if the medicine is being used outside of its intended purpose, the safety and efficacy review may not apply to their specific clinical indication. Therefore it is important that the procuring pharmacist reviews the SPC and patient information leaflet (PIL) to ensure that they are appropriate for the intended use and provide alternative guidance if necessary.

Patient information, user guidance and the label must include comprehensive, relevant information in English. When importing a product from a non-English-speaking country, provision must be made to ensure that the product is labelled appropriately and that sufficient guidance is provided to the clinician and patient to ensure safe use.

When importing borderline substances such as vitamins and food supplements, preference should be given to procuring products that have been marketed as medicines wherever possible. Where this is not possible, a quality assessment should be carried out to ensure that the product is free from transmissible spongiform encephalopathies (TSE) as a minimum.

2.3.2.2 Use of a `Special' manufactured in a MHRA-licensed unit

The commissioning of a suitable preparation from a licensed 'Specials' manufacturer within the UK should be considered. The benefit of purchasing a 'Special' is that the product should be made to a validated formula with supporting stability data in accordance with the principles of good manufacturing practice (GMP). Licensed 'Specials' units are regularly inspected by the Medicines and Healthcare products Regulatory Agency (MHRA) to ensure these principles are upheld. However, the purchasing pharmacist will still need to review the supporting documentation (e.g. specification, Certificate of Analysis/Conformity, TSE statement) to assess whether the product is of appropriate quality.

Information on Specials manufacturers is available in the British National Formulary (BNF).

Guidance relating to the procurement of extemporaneously prepared patient-specific doses from 'Specials' manufacturers can be found in Chapter 11.

Further guidance for assessing the quality of both imports and 'Specials' can be found in the NHS Pharmaceutical Quality Assurance Committee guidance document 'Guidance for the purchase and supply of unlicensed medicinal products' (NHS Pharmaceutical Quality Assurance Committee, 2004). Advice can be sought from medicines information centres, regional quality assurance specialists, licensed importers of medicines and individual 'Specials' units.

2.3.3 Practical options

2.3.3.1 Use of soluble or dispersible tablets

Soluble or dispersible tablets may be a useful and convenient alternative to preparation of liquid extemporaneous products. Some tablets can be dispersed, even if this is not within the terms of their marketing authorisation (licence). Most tablets will disperse in a small volume of water (~ 10 mL) within a few minutes. This practice presents fewer health and safety risks than crushing tablets, which can expose the carer to potentially harmful dusts via inhalation.

When dispersing tablets, the dose should be prepared and administered immediately, as stability cannot be guaranteed. It should be noted that slow or modified release preparations should not be used in this manner.

Care should be taken, however, if part doses are required. The practice of taking aliquots from insoluble, dispersed tablets for smaller doses presents a significant risk of dose inaccuracy. This is because water has no suspending properties, commonly resulting in aggregation and sedimentation of the drug, leading to poor dosage accuracy. For this reason, tablet dispersion may not be a practical option in paediatrics where the required doses are frequently fractions of the lowest available strength tablet.

2.3.3.2 *Cutting tablets*

The use of tablet cutters can sometimes provide an acceptable option, especially when tablets are effectively scored and designed to help in the administration of part doses. However, tablets cannot be cut with great accuracy of dose and research suggests that the variability may range from 50% to 150% of the desired dose even when using commercially available tablet cutters (Breukreutz *et al.*, 1999; Teng *et al.*, 2002).

2.3.3.3 *Use of a preparation intended for a different route*

The use of a suitable preparation intended for a different route of administration can sometimes be a practical alternative; for example the use of an injection solution orally, or an oral solution rectally. However, this practice has its own inherent risks and the pharmacist should ensure that the presentation used will be absorbed by this route and that it will be tolerated by the patient.

When using an injection by the oral route, consideration should be given to the possibility of rapid absorption and elevated peak levels, the potential for rapid drug degradation due to exposure to gastric acid and problems with first-pass metabolism. The pH of an injection should also be considered, as extremes of pH can adversely affect the gastric mucosa.

Some consideration should also be given to other excipients in the formulation such as propylene glycol and ethanol, which may be problematic if large volumes of the injection are required to provide the dose.

2.4 Risks associated with extemporaneous preparation

The technical and clinical risks associated with extemporaneously prepared medicines are considered below.

2.4.1 Formulation failure

All formulae used for extemporaneous preparation should be validated and have supporting stability data. Suitable sources include pharmacopoeial formulations, industry-generated expert reports and published papers.

However, it is recognised that there is a lack of standardised formulae available, leading to a plethora of different approaches and formulations being used which are commonly not peer reviewed or published. There are a number of risks associated with the use of non-standard formulations that need to be considered before taking this option.

Formulation failure can occur when a formulation has not been adequately validated, potentially resulting in either under- or overdose and associated toxicity or therapeutic failure. If a poorly formulated medicine that lacks dose uniformity is used, both underdosing and overdosing may occur during a course of treatment.

The causes of formulation failure are numerous and can be complex, including physical incompatibilities, drug/excipient binding issues and drug degradation. Generally, as the complexity of the formulation increases so does the risk of problems occurring. Formulations should therefore be kept as simple as possible for these reasons.

Oral liquids are usually formulated as either a suspension or solution. Solutions have the benefit of ensuring uniformity of dose, but drugs are more susceptible to degradation in solution than in the solid state and this should be considered when preparing a solution.

An insoluble drug suspended in a suitable vehicle may be less susceptible to drug degradation, but may settle out of the suspension over time, leading to sedimentation and caking. In this state, there will be a higher concentration of drug at the bottom of the bottle than at the top. If taken, this will result in the patient being underdosed at the beginning and overdosed towards the end of a treatment course. In order to ensure uniformity of dose, these formulations need to be shaken properly before use and patients need to be adequately counselled.

The majority of liquid formulations are prepared for children where small doses are required. In a number of cases, even suspended 'insoluble' drugs will be partially soluble at these concentrations and therefore it is important to review drug stability data and solution kinetics when assessing the formulation.

It should also be noted when using tablets as starting materials in the preparation of oral liquids that many of the excipients will be insoluble, even if the drug is soluble. These excipients can bind some of the drug and therefore it is prudent to use a suspending agent as the drug vehicle to ensure uniformity of dose. For this reason, filtration of this type of preparation should not be carried out.

2.4.2 Microbial contamination

Microbial contamination can pose a significant risk to immunocompromised patients, while by-products of microbial degradation can lead to physical or chemical changes in the preparation. Microbial growth can lead to spoilage, affecting product appearance and producing foul odours.

The choice of preservative for a formulation needs to take into account a number of factors including pH, physical compatibility and the intended patient group. Unpreserved preparations should be stored in a refrigerator and assigned a short shelf-life to limit microbial growth. A maximum shelf-life of 7 days at 2–8°C should be assigned to unpreserved oral liquid preparations unless sufficient validation work has been carried out to support an extended shelf-life.

2.4.3 Calculation errors

Calculation errors pose the greatest risk of causing serious patient harm and the greater the complexity of calculation required, the higher the risk of an error. Formulations should be kept as simple as possible and all calculations should be independently checked and documented on a worksheet.

Common calculation errors associated with extemporaneous preparation include errors when converting units from one to another (e.g. milligrams to micrograms, conversions from weight in volume to millimoles).

Problems can also arise when doses can be prescribed as free base or salt, leading to potential calculation errors when making and administering preparations (e.g. two-fold errors if caffeine citrate is confused with caffeine base).

Care should be taken when diluting concentrates; calculation errors have been known to lead to 1000-fold overdoses (Kirsch, 2005).

Decimal point errors are commonplace and extra vigilance is needed to ensure that documentation is clear (especially worksheets and formulations) and that products are labelled without using decimal points wherever possible (e.g. 0.5 g should be labelled as 500 mg). Guidance relating to reducing the risk of medication errors can be found in Chapter 6.

Errors have also occurred when unfamiliar terminology is used to describe the strength of solutions and this was highlighted in the 'peppermint water' case (Anon, 1998) where concentrated chloroform water was used instead of double strength chloroform water, resulting in the death of a child.

2.4.4 Starting materials

The use of some historical formulae carries the associated risk of using ingredients that are no longer suitable. For example, chloroform has now been recognised as a class III potential carcinogen and is present in a number of old BP monographs (CHIP3 Regulations, 2002).

The toxicity of some ingredients is age-specific and they may be inappropriate for children, and some ingredients are unsuitable for certain religious groups (e.g. phenobarbital elixir (BNF) contains 38% alcohol). Alcohol has been linked to CNS-depressant and hypoglycaemic effects (Woods, 1997). Care should also be taken with the use of cariogenic sugars (e.g. sucrose) in paediatric formulations as it has been associated with dental cavities. It is therefore important to list all such excipients on the product label so that end-users are made aware of their presence in the formulation

All starting materials, particularly those of animal origin (e.g. gelatin) should be certified free from TSE.

2.4.5 Patient acceptability issues

Consideration should be given to the palatability and presentation of oral liquid medicines as there is a good argument that taste is crucial to achieving good compliance in children, especially for the treatment of longstanding conditions such as in cardiology.

2.4.6 Health and safety risks

The risks to the operator should also be considered. A Control of Substances Hazardous to Health (COSHH) risk assessment should be carried out and any risks should be identified and carefully evaluated before undertaking an extemporaneous preparation. (Note: once performed, this assessment does not have to be repeated each time the preparation is made, provided the assessment is up to date and available on the premises.)

When handling hazardous products, units should be equipped with suitable containment devices and systems should be put in place to eliminate the risk of cross-contamination.

2.4.7 Therapeutic risks and clinical consequences

When identifying the potential clinical consequences of a formulation failure or calculation error associated with an extemporaneously prepared medicine, it is important to review both the inherent properties of the drug and the patient's clinical condition as part of the risk assessment.

Any inaccuracy of dosing associated with medicines that have a narrow therapeutic index can lead to significant morbidity, whether due to under-dosing leading to treatment failure or overdosing leading to toxicity. By contrast, any inaccuracy of dosing associated with drugs with a wide therapeutic index may have little or no impact on the therapy.

Patients with certain clinical conditions or from vulnerable patient groups may be at greater risk of morbidity than others and therefore it is important that the risk assessment takes into account the patient-specific circumstances rather than being solely a drug-based risk assessment.

Where there is significant risk of morbidity associated with a non-standard or complex formulation, all alternative options should be explored and extemporaneous preparation should be seen as a last resort.

2.4.8 Associated clinical risk factors

The majority of patients receiving extemporaneously prepared products, in particular oral liquid medicines, tend to be from vulnerable patient groups (e.g. neonates, children, stroke victims) who are either unaware of ill-effects associated with their treatment or who cannot communicate with their clinician. Coupled with this, extemporaneous preparations may not be routinely identified as high-risk therapies by pharmacists and therefore such treatments are not commonly given the level of scrutiny and close monitoring they require. Therefore when embarking on the use of an extemporaneously prepared product, the pharmacist should ensure that systems are in place to monitor the effectiveness of the therapy.

Pharmacists should regard patients receiving extemporaneous preparations as at increased risk and regularly review their condition to ensure the treatment is effective. Any issues should be documented and reported to the manufacturer (and MHRA if necessary for serious adverse events – see MHRA website for guidance) as part of an ongoing pharmacovigilance programme.

2.5 Managing the risks

The following checklists may provide a helpful summary guide to the risk management of patients requiring an extemporaneously prepared medicine.

2.5.1 Clinical risk reduction

- Identify extemporaneous preparations as high-risk therapy.
- Carry out a risk assessment.
- Consider alternative therapies.
- Review all available evidence to support the use of the preparation.
- Evaluate drug toxicity – consider therapeutic index.
- Monitor patient for clinical effect, toxicity and adverse drug reactions (ADRs).
- Document any problems and successful treatments for future reference.

2.5.2 Technical risk reduction

2.5.2.1 Formulation

- Use standard, validated formulae where possible (e.g. pharmacopoeia, expert report (industry generated), published papers).
- Evaluate data using first principles (in-house expert review by suitable qualified personnel, e.g. QC department).
- Gather information on or evidence of effective use from other units or pharmaceutical companies.
- Use information resources (e.g. Pharmaceutical Codex, Compounding Interest Group, NHS QA website – QAinfozone, Pharminfotech, Paddock Laboratories).
- If no formula is available, keep it simple using readily available, pharmaceutical-grade starting materials and standard vehicles.
- Restrict the shelf-life to limit degradation and spoilage (maximum of 28 days if preserved, 7 days if unpreserved).

2.5.2.2 Preparation

- Ensure extemporaneous dispensing facilities and practices comply with this guidance and are subject to systems of audit and self-inspection.
- Use QA-approved worksheets and procedures.
- Ensure facilities and equipment are appropriate and validated/calibrated.
- Ensure all operatives are appropriately trained.
- Use licensed or approved (e.g. QC-tested) starting materials.
- Perform COSHH assessment on both the starting materials and the preparation process.

2.5.2.3 Risk matrix

The risk matrix in Figure 2.1 may be helpful in risk evaluation.

<table>
<tr><th>Overal risk assessment</th><th colspan="2">Low</th><th>Medium</th><th colspan="2">High</th></tr>
<tr><td>Risks to quality</td><td>Validated formula and supporting stability data available
• Published papers
• Pharmacopoeia
• Developed by licensed manufacturer
Rating: Low</td><td>Formula available, but not validated. No supporting stability data
• Evaluation of formula and shelf-life from first principles by suitably experienced staff
• Experience of safe and effective use in NHS
Rating: Low</td><td>Formula available, but not validated
• No supporting stability data or evaluation
• Experience of safe and effective use in NHS
• Reduced shelf-life (max 7 days)
Rating: Medium</td><td>Formula available, but not validated
• No supporting stability data or evaluation
• No evidence of safe and effective use in NHS
Rating: High</td><td>No formula available
Rating: High</td></tr>
<tr><td>Risks to safety/efficacy</td><td>Low toxicity
Short-term use
Rating: Low</td><td>Wide therapeutic index
Short-term use
Rating: Low</td><td>Wide therapeutic index
Maintenance therapy
Vulnerable patient groups
Rating: Medium</td><td>Narrow therapeutic index
Short-term use
Bioavailability could be significantly changed by crushing tablet
Rating: High</td><td>Narrow therapeutic index
Maintenance therapy
Bioavailability could be significantly changed by crushing tablet
Rating: High</td></tr>
<tr><td>H & S risks</td><td colspan="2">Full supporting COSHH data
Control measures in place
Rating: Low</td><td colspan="3">Inadequate supporting COSHH data
No control measures in place
No COSHH assessment carried out
Rating: High</td></tr>
</table>

Figure 2.1 Risk assessment matrix. Low risk: Prepare worksheet and make in accordance with local SOPs. Use licensed or QC approved starting materials only. Medium risk: Make for short-term use only and monitor patient for clinical effect and ADRs. Consider outsourcing to a 'Specials' unit or alternative therapy for long-term use. High risk: Consider all alternatives before making – only make as last resort. Monitor patient closely for clinical effect, toxic effects and ADRs.

2.6 References

Anon (1998). Baby dies after peppermint water prescription for colic. *Pharm J* 260: 768.

Breitkreutz RT, Wessel T, Boos J (1999). Dosage forms for peroral drug administration to children. *Paediatr Perinatal Drug Ther* 3: 25–33.

CHIP3 Regulations (2002). Chemicals (Hazard Information and Packaging for Supply) Regulations. SI 2002/1689.

Kirsch L (2005). Extemporaneous quality. *J Pharm Sci Technol* 59(1): 1–2.

NHS Pharmaceutical Quality Assurance Committee (2004). Guidance for the purchase and supply of unlicensed medicinal products. Unpublished document available from regional quality assurance specialists in the UK or after registration from the NHS Pharmaceutical Quality Assurance Committee website: www.portal.nelm.nhs.uk/QA/default.aspx.

Teng J, Song CK, Williams RL, Polli JE (2002). Lack of medication dose uniformity in commonly split tablets. *J Am Pharm Assoc* 42: 195–199.

Woods DJ (1997). Extemporaneous formulations – problems and solutions. *Paediatr Perinatal Drug Ther* 1: 25–29.

3

Quality management

3.1 General issues

The assurance of quality of extemporaneously prepared pharmaceuticals is of prime importance. Patients rely on pharmacists providing medicines that are consistently safe, efficacious and of a quality fit for their intended use. To achieve this it is essential that a comprehensive pharmaceutical quality system is in place.

This quality system should comply where appropriate with published guidance (International Organization for Standardization, 2000, 2005, 2006, 2008) and should include all aspects of quality assurance (QA) and good preparation practice (GPP) as described in this guidance document. It should be fully documented (e.g. in a quality manual). Recorded monitoring of its effectiveness should be in place.

The pharmaceutical quality system (PQS) should:

- describe the quality policy
- state the scope of the PQS
- identify the PQS processes
- incorporate risk management principles (see Chapter 2)
- identify management responsibilities (see Chapter 4)
- include process performance and product quality monitoring
- include policy and procedures for corrective and preventive actions
- include policy and procedures for change management
- include policy and procedures for management review.

3.2 Quality assurance

Quality assurance is 'a wide ranging concept which covers all matters which individually or collectively influence the quality of a product. It is the total sum of the organised arrangements made with the object of ensuring that medicinal products are of the quality required for their intended use' (MHRA 2007). In the context of extemporaneous preparation this incorporates GPP and should ensure that:

- extemporaneously prepared medicines are formulated and prepared in compliance with current legal requirements and standards
- preparation and quality control arrangements are documented and in compliance with current GPP requirements
- all products prepared are of a quality suitable for their intended use
- products are released for patient use only by a pharmacist (see Chapter 9)
- documentation and records comply with recommendations in this guidance document.

3.3 Good preparation practice

Good preparation practice (GPP) is that part of QA which ensures that extemporaneous products are consistently prepared to the quality standards appropriate to their intended use. It includes extemporaneous preparation activities and appropriate quality control (QC) arrangements.

This guidance document gives detailed guidance on GPP requirements, including premises and equipment, documentation, compounding, quality control, work contracted out, complaints and recalls, audit, cleaning and hygiene.

3.4 Quality control

Quality control (QC) is that part of GPP concerned with sampling, specifications and testing. It is also associated with the organisation, documentation and release procedures which ensure that appropriate tests are carried out and that products are not released for use until their quality has been judged to be satisfactory.

Full details are given in Chapter 9.

3.5 Change management

Change management or change control is a formal system whereby proposed or actual changes that may affect the status of validated facilities, equipment, systems or processes are reviewed by appropriate and qualified representatives of each discipline that may be affected by the change.

The process should ensure that sufficient supporting data are provided to assess the impact of any change and the revised process will ensure that products continue to be consistently prepared to the quality standards appropriate to their intended use and meet their approved specification. The change control process should be documented and include any associated impacts or risks associated with the change, together with proposed actions for re-qualification or re-validation if necessary.

An example of a change control document can be found in Appendix 1.

3.6 Deviation management

A deviation is defined as a departure from an approved process, instruction or established standard. Deviations can be either planned or unplanned but should be routinely documented together with a risk assessment relating to any potential impact on the quality of the product.

All deviations should be formally assessed as part of the product approval procedure. Deviations are key performance indicators in the ongoing quality management process and should be regularly reviewed to identify if the system requires a formal change. An example of a deviation reporting form can be found in Appendix 2.

3.7 References

International Organization for Standardization (2000). ISO 9004: Quality management systems – Guidelines for performance improvements. International Organization for Standardization. Geneva. www.iso.org (accessed January 2009).

International Organization for Standardization (2005). ISO 9000: Quality management systems – Fundamentals and vocabulary. International Organization for Standardization. Geneva. www.iso.org (accessed January 2009).

International Organization for Standardization (2006). ISO 10014: Quality management – Guidelines for realizing financial and economic benefits. International Organization for Standardization. Geneva. www.iso.org (accessed January 2009).

International Organization for Standardization, (2008). ISO 9001: Quality management systems – Requirements. International Organization for Standardization. Geneva. www.iso.org (accessed January 2009).

MHRA (Medicines and Healthcare products Regulatory Agency) (2007). *Rules and Guidance for Pharmaceutical Manufacturers and Distributors*. London: Pharmaceutical Press.

4

Personnel and training

4.1 General issues

4.1.1 There must be a nominated Accountable Pharmacist (AP) who is responsible for the extemporaneous preparation service. Deputising arrangements should be in place for when the AP is absent.

4.1.2 The AP must be conversant with the following areas and able to apply them in practice:

- the Medicines Act in relation to unlicensed medicines and extemporaneous preparation
- current professional guidance and standards (e.g. *Handbook of Extemporaneous Preparation*; *Medicines Ethics and Practice* (RSPGB, 2009))
- good preparation practice
- assessment of risk and medication error potential
- quality assurance and quality control
- health and safety and Control of Substances Hazardous to Health (COSHH)
- theory and principles of chemical and pharmaceutical stability
- formulation.

4.1.3 Any pharmacist deputising for the AP must have the necessary level of knowledge and training or be clear about the limits of his or her responsibilities and have access to a pharmacist with the necessary knowledge and training.

4.1.4 Extemporaneous preparation must be supervised by a pharmacist. Under current UK legislation, products prepared extemporaneously under a Section 10 exemption from the Medicines Act 1968 are required to be released by a pharmacist who takes responsibility for the quality of the product. Further guidance on product release is given in Chapter 9. The duties and responsibilities of the supervising pharmacist must be formally documented and included in their training programme.

4.1.5 The pharmacist must ensure that the quality management system requirements described in Chapter 3 are in place.

4.1.6 The pharmacist must ensure that the staff, facilities and systems in place are capable on a day-to-day basis of providing an adequate quality service able to meet the needs of patients.

4.1.7 There should be an adequate number of competent personnel at all times.

4.1.8 There should be an organisation chart showing staff and management accountability.

4.1.9 The duties and responsibilities of all personnel, including deputies, should be clearly documented (e.g. in written job descriptions).

4.1.10 All staff involved in extemporaneous preparation must receive training Appropriate to their role. Typically this will:

(a) provide them with knowledge in good extemporaneous preparation practice, local practices including health and safety, formulation, expiry periods and quality assurance appropriate to the level of involvement, assessment of risk and medication error potential, the pharmacy, its products and services provided

(b) demonstrate competency in the necessary extemporaneous preparation skills and pharmaceutical calculations and dilutions.

4.1.11 A written training programme should be available and completion of training should be documented. This applies to all personnel working in the preparation area, including those not directly involved in preparation processes (e.g. cleaning staff).

4.1.12 Competence in extemporaneous preparation skills and knowledge of local practice should be assessed and documented.

4.1.13 During training, staff must be carefully supervised and checked and trainees must understand the limits of their responsibilities.

4.2 Hygiene

4.2.1 Standard operating procedures should be in place for hygiene, behaviour and clothing.

4.2.2 Great care must be taken at all times to ensure that products are not contaminated microbiologically or chemically. The reconstitution of antibiotic powders is a potential source of contamination for a pharmacy if not adequately controlled, as powder can displaced into the environment when adding water to the bottle. Good standards of cleaning, personal hygiene and protective clothing will assist and must be assessed regularly.

Attention to procedures, adequate preparation and storage space and prevention of overcrowding will help in the avoidance of contamination

4.2.3 Staff preparing extemporaneous products should wear clean clothing which both protects the product from unnecessary contamination and protects the operator from the ingredients and the product. Clothing (e.g. hat, white coat or plastic apron) should be reserved for extemporaneous dispensing and changed regularly and when soiled. Specialised clothing such as gloves and masks must be available and should be worn when the ingredients or product pose a hazard to staff handling them.

4.2.4 Eating and drinking must not be allowed in preparation areas.

4.2.5 Written procedures must include information to staff on the reporting of respiratory and other infections and skin lesions to the pharmacist in charge. The AP must assess the risk of contaminating products, taking advice if necessary. If the AP deems that the risks to either the product or staff member are unacceptable, the affected staff member must not prepare the product.

4.2.6 Handwash facilities with hot water must be available in toilet areas. Particular care should be taken to ensure handwashing occurs after eating, handling stock, visits to the toilet, etc.

4.3 Reference

RPSGB (Royal Pharmaceutical Society of Great Britain) (2009). *Medicines, Ethics and Practice 33: A guide for pharmacists and pharmacy technicians*. London: Pharmaceutical Press.

5

Premises and equipment

5.1 Premises

5.1.1 Premises and equipment should be suitable for the intended purposes, must not present any hazard to the quality of the product, and must minimise the risk of errors.

5.1.2 There should be a dedicated facility/room for extemporaneous preparation. However, for pharmacies preparing small quantities of a limited range of products for immediate dispensing then a designated area or temporarily designated area is permissible. The area must be clearly demarcated during preparation. Within UK hospitals, the Health Building Note 14-01 should be consulted for further advice.

5.1.3 Measures must be taken to reduce any risk of cross-contamination. Only one product should be handled at a time. Premises and equipment should be easy to clean. Following maintenance or repair, thorough cleaning and, where appropriate, disinfection should take place.

5.1.4 Environmental conditions, including temperature, humidity and lighting, must be adequate to permit safe and comfortable preparation of medicinal products to limit the degradation of ingredients.

5.1.5 Facilities should be laid out in a way to provide adequate space for the activities undertaken. Work should flow logically between working areas without the risk of product cross-contamination.

5.1.6 Where possible, weighing should be carried out in a dedicated area with a minimum of draughts and vibration.

5.1.7 Storage areas (including cold storage) should be defined, monitored and, where necessary, controlled. Monitoring results should be documented and assessed. Where conditions fall outside defined limits, adequate corrective actions should be taken and documented.

5.1.8 Before undertaking preparation of certain drugs, including antibiotics, sensitising agents and cytotoxics, a risk assessment should be carried out. This should include any hazards of the material handled in the operation, any control measures possible and the risks of cross-contamination of other products. Dedicated equipment is recommended for handling these substances. For further information, refer to the Control of Substances Hazardous to Health (COSHH) regulations.

5.1.9 All materials brought into the preparation area should be clean.

5.1.10 Dedicated garments, hair covers and gloves should be worn during preparation.

5.1.11 Quality control activities should be performed in a separate designated area.

5.1.12 Rest and refreshment rooms should be separate from all other areas.

5.2 Equipment

5.2.1 A policy for the segregation of specific equipment for certain product types, including cytotoxic drugs and external products, should be available; this should take into account local COSHH assessments.

5.2.2 Equipment should be calibrated and validated to demonstrate satisfactory performance. The advice of a local NHS Pharmaceutical Quality Assurance Service may be sought on appropriate methodology.

5.2.3 Measures should be calibrated for specific volumes and should have levelling lines.

5.2.4 Glassware should be inspected regularly, before and after each use, for chips and cracks, and replaced as necessary. Consideration should be given to the merits of other materials. Stainless steel of a suitable grade is a good material but care must be taken with some agents as it is susceptible to oxidation. Plastics are susceptible to surface scratches and stains.

5.2.5 Balances should be suitable for the weights being measured (50–250 mg Class 1, above 250 mg Class II). All balances brought into service since January 2003 must be labelled accordingly to indicate compliance with the European Directive 90/384/EEC. Balances purchased before this date should be assessed for suitability.

5.2.6 The accuracy of balances should be checked on a regular basis and records of this process maintained.

5.2.7 Protective equipment, including fume cupboards, should be regularly tested for performance and records maintained; it should be appropriate to the types of products handled. For powders, a bench-top filtered extract system may be appropriate; for solvents a fume cupboard, and for cytotoxic agents total containment is appropriate.

5.2.8 Faulty equipment should be taken out of use and labelled as such until repaired or removed.

5.2.9 Equipment should be stored in a clean and dry condition.

5.3 Cleaning

5.3.1 Areas and equipment should be cleaned effectively with a suitable detergent before and after use. Equipment should be rinsed with an appropriate grade of water after washing (potable or sterile water for irrigation/injection) and then dried. Critical surfaces should be effectively sanitised with a suitable agent such as 70% alcohol before use.

5.3.2 Sinks should be available for washing up of equipment, with separate facilities for handwashing. Sinks should be a suitable distance away from the preparation areas to reduce the risk of microbial contamination from water splashes. Washing and cleaning activities should not themselves be a source of contamination.

5.3.3 All equipment should be visibly clean prior to use. Ensure that all residues of cleaning agent have been removed.

5.3.4 Containers and lids should be checked to ensure that they are clean and dry before use.

5.3.5 Adequate pest control measures should be taken.

5.4 Further reading

Health and Safety Executive website: www.hse.gov.uk. (accessed 29 July 2009).

6

Documentation

6.1 General

All extemporaneous preparation should be adequately documented either electronically or in hard copy to prevent errors from spoken communication and to provide an audit trail.

Specifications, master documents and standard operating procedures (SOPs) should not be altered. New documents should be produced when necessary. Documents should be regularly reviewed at defined intervals. Superseded documents should be clearly identified as such and should be retained for at least 5 years (Eastern Pharmacy Network Senior Pharmacy Managers, 2003).

Documentation should comprise the following:

6.2 Risk assessments

6.2.1 The decision to prepare an extemporaneous product should be documented as described in Chapter 2.

6.3 Specifications

6.3.1 Simple specifications are required for starting materials which are not licensed, and for packaging. These should be appropriately authorised and dated. (An example is shown in Appendix 3.)

6.3.2 Specifications for starting materials and, where applicable, packaging materials should include:

- name (and reference to pharmacopoeia, where applicable)
- description
- instructions for sampling and testing
- qualitative and quantitative requirements, with acceptance limits
- requirements for storage, where applicable
- safe handling requirements, where applicable
- expiry period/retest date.

6.3.3 Specifications for finished products, if produced separately from the master document for regularly prepared extemporaneous products, should include:

- name of product
- description of dosage form and strength
- formula, including the quantity and grade of all starting materials
- packaging details
- storage conditions
- special handling precautions
- expiry period (unopened and in-use if appropriate) with reference to supporting stability data.

6.3.4 If not produced as a separate specification, this information should be included on the master document.

6.4 Product-specific instructions (more commonly known as master documents)

6.4.1 These give processing, packaging and release instructions. For products regularly prepared, master documents which combine the specification and product specific instructions should be produced and independently checked. Alternatively, a simple final product specification may be produced in addition to the master document. Master documents should be clear and detailed and should have a standardised style within any one pharmacy.

6.4.2 Simple stepwise instructions for preparation, along with any specific notes or cautions (e.g. COSHH requirements) should be included on the master document for each product.

6.4.3 The master document for each product should contain simple packaging instructions including:

- pack size
- master label
- description of all packaging materials (type, size, grade)
- simple stepwise instructions for packaging.

6.4.4 In the absence of a master document, general instructions should be available for each product type (e.g. capsules, ointments, etc.) and a master document prepared if the product is subsequently requested.

6.5 Worksheets and labels

6.5.1 A record should be kept of the key stages of processing, packaging and release of products to provide an audit trail of the quality relevant facts of the history of an extemporaneous product during preparation. This is normally achieved by completion of an individual worksheet, usually by photocopying the master document (see Appendix 4 for an example worksheet).

6.5.2 Worksheets should be checked for accuracy, preferably by an independent checker, prior to commencement of preparation.

6.5.3 Worksheets will vary for each pharmacy but should include:

- the name and formula of the product, and the source of the formula
- a unique identification number to enable traceability
- the manufacturer, batch numbers of each starting material (or QC reference number)
- the date of preparation
- the expiry date of the product (and time if appropriate)
- the initials of staff carrying out each stage of the preparation, packaging and checking procedures
- the signature of the pharmacist supervising the preparation process
- the signature of the pharmacist performing the release of the product (this will often be the supervising pharmacist as above)
- a record of the label used on the product
- the patient's name, where applicable
- a comments section for recording any unusual occurrences or observations.

All critical processes should be recorded.

6.5.4 Labels must comply with all statutory and professional requirements, and should include the following information:

- the name of the product, and where necessary the nature of the salt and any waters of crystallisation
- quantity and strength of active ingredients
- the pharmaceutical form
- the contents by weight, volume or number of dose units
- excipients of known effect (available on the European Medicines and European Agency (EMEA) website www.emea.europa.eu)
- method and route of administration, if necessary
- dispensing date
- expiry date (and time if applicable) expressed unambiguously
- a unique identifier linked to preparation records, e.g. batch number
- name of patient
- name and address of pharmacy

- 'keep out of sight and reach of children'
- statutory warnings required by SI 1994/3144 for particular actives; for example, aspirin, paracetamol.

The following, as applicable

- instructions for use, including any appropriate cautionary notices
- any special storage and handling requirements
- location of patient (e.g. ward).

6.5.5 To reduce the risk of medication error, the following key principles should be noted:

- Fractions of a milligram or microgram should not be used.
- Micrograms should be stated in full and not abbreviated.
- The dose should be stated in the usual dose volume (e.g. 5 mg in 5 mL, or possibly 1 mg in 1 mL for paediatrics).
- Any percentages should be stated accurately, for example % w/w, %, v/v.

Note that strength is often expressed in different ways. For Texample, 10 mg per mL is the same as 1% w/v, which is also sometimes termed a 1 in 100 dilution. Particular care should be taken with products which require dilution (e.g. concentrated chloroform water which is 40 times stronger than single strength chloroform water). These are high-risk products.

- Numbers of 1,000 or more should contain commas and the use of trailing or unnecessary zeros should be avoided (e.g. 5,000 mg should be stated as 5 g; 5.000 mg should be stated as 5 mg).
- As advised by the National Patient Safety Agency, a minimum of 12 point text, a mixture of upper and lower case, and a plain Sans Serif typeface should be used for labelling. Key information such as the generic name of the medicine and its strength should be easily readable, even in dim light (e.g. a ward at night). Particular care should be taken with look-alike and sound-alike drug names.
- Consideration should be given to the use of colour to emphasise key information (although not for colour coding). In all instances there should be adequate contrast between the text and the label background, however. User trials on readability are also often helpful.
- Negative labelling should be avoided (e.g. 'store below 25°C' is preferable to 'do not store above 25°C').

6.5.6 Any computerised systems used for worksheet or label generation should be password protected.

6.6 Standard operating procedures (SOPs)

6.6.1 Written in the imperative, SOPs provide instructions for the performance of standardised operations to ensure consistency of quality of the extemporaneous product.

6.6.2 There should be current approved SOPs for extemporaneous preparation. These must be written in the imperative and should include the following:

- facilities to be used
- pre-preparation activities (e.g. cleaning, balance calibration, approval of starting materials)
- use of documentation, including labelling
- authority of personnel for preparation, checking and final product approval
- post-preparation activities (e.g. storage and distribution)
- handling of complaints, defective products, etc.
- training of personnel, including hygiene requirements.

6.7 Other documentation

6.7.1 The formulation of non-pharmacopoeial products should be conveyed to other relevant practitioners by approved methods; for example, discharge letters, labels or other documentation.

6.7.2 Documentation covering training programmes and recording satisfactory compliance of personnel with such training should be available.

6.7.3 Additional documentation includes:

- logs for calibration of balances
- cleaning records
- audit observations and action plans
- equipment maintenance and calibration records
- error logs
- change control requests
- deviation reports
- training records
- complaints and recalls.

(This list is not exhaustive.)

6.8 Reference

Eastern Pharmacy Network Senior Pharmacy Managers (2003). Recommendations for the retention of pharmacy records. *Hosp Pharm* 10: 222–224.

7

Preparation

7.1 General

The preparation process includes all the practical procedures and processes required to extemporaneously prepare, package and label a medicine, ready for supply to a named patient.

Critical processes in the preparation of extemporaneous products should be identified using process mapping techniques. Critical processes include the following:

- prescription verification
- worksheet and label generation
- assembly of components
- weighing
- measurement of liquids
- grinding tablets into uniform powders
- mixing
- reconciliation
- packaging and labelling.

Appropriate control measures should be implemented at all critical points in the process and should include a procedural check (independent, wherever possible) by a suitably trained person as a minimum. These measures should be validated (in the case of reproducible or automated processes) or subject to competence assessment (in the case of operator-dependent processes).

7.2 Prescription verification

7.2.1 All prescriptions must be signed by an approved prescriber.

7.2.2 Prescriptions should be clear, unambiguous and accurate. The medicine(s) and dose(s) selected must be appropriate to meet agreed therapeutic objectives.

7.2.3 Pharmacists should ensure that medication errors do not occur from any cause, including as a result of prescribing, formulation or calculation errors. It is the responsibility of the authorising pharmacist to ensure that prescriptions are verified.

7.2.4 A written approved procedure should be in place for prescription verification which should include checks for the following where appropriate:

- full signature of authorised prescriber and date
- patient details (e.g. full name, hospital number, consultant, ward, date of birth, weight where appropriate).
- correct dose calculation
- administration details
- compatibility with prescribed constituents
- stability of formulation
- correct presentation for intended route of administration
- absence of contraindications, allergies or drug interactions.

7.2.5 Records must indicate who carried out the verification of each prescription.

7.2.6 These records should be retained for a period that is in accordance with legal requirements and with the local policy.

7.3 Documentation (also see Chapter 6)

7.3.1 The relevant master document should be selected and a product-specific worksheet and label should be prepared before starting preparation.

7.3.2 Where a calculation is required, it should be independently checked by an appropriate second person wherever possible.

7.3.3 The check should be performed prior to preparation to minimise the chance of product wastage.

7.4 Starting materials

7.4.1 All starting materials should be checked against the worksheet to ensure that the correct materials have been selected.

7.4.2 Ensure that all starting materials are either licensed or have been approved for use following a quality assessment (see Chapter 9).

7.4.3 The chemical and microbiological quality of any water used in preparations should be specified and monitored if taken from the mains or local water systems. Monitoring is not necessary if sterile bottled water for injections or irrigation is used. Sterile bottled water should be given a maximum shelf-life of 24 hours once opened, with its status clearly labelled.

7.4.4 Ensure that the starting materials are 'in date' at the time of production and for the duration of the product's shelf-life.

7.4.5 Check that containers are clean and fit for use.

7.5 Weighing

7.5.1 Before starting the process, ensure the weighing range of the balance is appropriate for the quantity of material to be weighed.

7.5.2 The balance should be levelled according to the levelling bubble (where appropriate) and should be protected from draughts and risk of other disturbance.

7.5.3 The balance should be checked regularly using appropriate weights and be re-calibrated if necessary. Records of checks and calibrations should be maintained.

7.5.4 Where possible the balance should be linked to a printer so that a permanent record of the weight measured can be retained with the worksheet.

7.5.5 Record accurately the batch number and expiry date of the starting materials on the worksheet.

7.5.6 Check that the balance is weighing the correct units of weight (micrograms, milligrams, grams, kilograms, etc.).

7.5.7 Before commencing weighing, the balance should be tared.

7.5.8 Where possible, a second person should check that the correct materials and weights of material have been used and measured. This should be recorded on the worksheet.

7.5.9 Common errors in weighing to look out for include:

- putting the decimal point in the wrong place (e.g. 14.4 g instead of 1.44 g)
- reversing the order of digits (e.g. 15680 g instead of 15860 g)
- misreading double digits 14556 g instead of 14566 g).

7.6 Measurement of liquids

7.6.1 The measuring of all liquid starting materials should be undertaken in appropriate and validated measuring vessels wherever possible. In some cases the use of unvalidated devices may be appropriate (e.g. syringes for small volumes).

7.6.2 The selected measure should be appropriate for the volume of liquid required.

7.6.3 Where possible, glass measures should be avoided to minimize the risk of undetected contamination with broken glass. When glass is used, it should be regularly inspected for chips and cracks and replaced as necessary. It is recognised that in some cases glass may be the container of choice (e.g. when measuring oils and solvents).

7.6.4 Where possible, there should be an independent check to ensure that the correct material and volume of material have been measured, especially where multiple measurements have been required in the process. This should be recorded on the worksheet.

7.7 Crushing tablets

7.7.1 When tablets are used as a starting material for making a suspension, care should be taken to ensure they are ground to a fine uniform powder. A lack of uniformity in particle size may have a detrimental effect on dose uniformity.

7.7.2 Tablets should be ground to a powder using a pestle and mortar, and a check of powder uniformity should be made and documented on the worksheet.

7.8 Mixing

7.8.1 Powders

Where two powders are to be mixed, the process should be undertaken in a way that will achieve homogeneity. This is commonly accomplished by triturating equal aliquots of the powders, mixing well between each addition, until the entire quantity of the bulk material has been incorporated.

7.8.2 Liquids

Miscible liquids will eventually mix completely by diffusion. To speed up the process the liquids should be mixed using a stirring rod or paddle. Avoid the use of glass stirring rods or stainless steel stirring rods in glass containers, as this can increase the risk of contamination with glass fragments.

Where powder is to be dissolved in or added to a liquid to make a solution or suspension, it should be added to the liquid slowly with continuous stirring to prevent it from aggregating. Once the powder is dissolved or evenly dispersed, the solution should be made up to the required volume.

7.8.3 Semi-solids

The final product should always be uniform in appearance and consistency.

Where a dispersion of particles or solution into a semi-solid is required (e.g. ointment, cream, paste), a small amount of the semi-solid should be mixed with the powder/liquid to form a stiff paste before incorporating the remainder of the vehicle, in small aliquots, to form a homogeneous product.

Where a mechanical mixer is used to assist in the preparation of a semi-solid product, consideration should be given to the optimum mixing period, as too short or too long a time may have detrimental effects on the final product.

During the mixing, the mixer should be periodically stopped to allow for scraping down of the paddles, sides and bottom of the bowl.

7.9 Packaging and labelling

7.9.1 The primary packaging for the product (e.g. bottle and closure) must be fit for purpose. It must adequately protect the product from the environment while being compatible with the product.

7.9.2 The physico-chemical properties of the drug and formulation should be considered when selecting the type of packaging.

- For medicines sensitive to light, a light-proof pack should be considered (e.g. opaque or amber bottle).
- When a medicine is sensitive to oxidation, a tight-fitting closure and impermeable pack is required.
- The potential for medicine loss by absorption and permeation, or leaching of additives from plastics needs to be considered for certain formulations (e.g. poorly soluble medicines formulated with solvents to improve drug solubility).

7.9.3 Generally glass can be considered to be inert and can be universally used as a container for the vast majority of formulations (with the exception of some concentrates, e.g. sodium bicarbonate).

7.9.4 Labels should be clear and legible, and must comply with all the statutory and professional requirements (see Chapter 6). The information on the label should reflect the product prepared, and should tie in with the master label on the worksheet.

7.9.5 Where possible the contents of the label should be independently checked against the worksheet and the prescription/requisition.

7.10 Waste

7.10.1 All waste should be disposed of in accordance with local guidelines on the handling and disposal of pharmaceutical waste (NHS Pharmaceutical Quality Assurance Committee, 2009).

Reference

NHS Pharmaceutical Quality Assurance Committee (2009). Guidelines on the handling and disposal of hospital pharmacy waste. Unpublished document available from regional quality assurance specialists in the UK or direct from the National QAWebsite: www.qainfozone.nhs.uk (accessed January 2009).

8

Formulation and stability

8.1 Introduction

Where there is a demonstrable need to undertake extemporaneous preparation, the accountable pharmacist must choose an appropriate formula. Ideally, all formulae used for extemporaneous dispensing should be validated and have supporting stability data. This evidence should be maintained on file for all preparations and reference sources should be documented on individual worksheets.

8.2 Finding and choosing an appropriate formulation

There are many potential sources of extemporaneous formulations. Typical references are listed in the further reading section, and include papers in peer-reviewed scientific journals, pharmacopoieal formulations, expert reports, and databases supported by commercial compounding companies.

As with clinical papers, the pharmacist must consider the quality of the evidence source and the overall appropriateness of the formulation. Comprehensive data should include information on physical and microbiological stability, as well as chemical stability testing. In rare cases, a given formulation may also be supported by successful clinical outcome data.

When choosing a formulation, it is imperative to consider the holistic needs of the prospective patient. This includes decisions regarding the form, strength, viscosity and excipient content of any given formula. For example, a formulation designed for use in adults may contain inappropriate excipients for use in a child, and a viscous suspension may not be suitable for administration via a nasogastric tube. Furthermore, historical formulations may contain starting materials which are no longer considered to be appropriate (e.g. chloroform).

These examples highlight the difference between the *quality* of a formulation and its *fitness-for-purpose*. The accountable pharmacist must be aware of the needs of the individual patient before making a decision on the most appropriate formulation; such responsibility cannot be delegated to a formulation scientist.

8.3 What if there isn't a validated, appropriate formulation?

Due to the paucity of data, many pharmacists faced with a request for an extemporaneous product may not be able to find a fully validated formula. When deciding on whether to formulate an extemporaneous product in these circumstances, the pharmacist must consider the risks of withholding treatment, as well as the risks inherent in extemporaneous formulation.

In this instance, advice may be available from NHS and commercial 'Specials' manufacturers, quality control departments and medicines information departments. Tertiary paediatric centres may also hold valuable information on possible formulae.

In the absence of published data, using first principles may be the only available option to evaluate a formulation. In these instances, pharmacists should consider obtaining advice from a quality assurance pharmacist or other personnel with expertise in formulation and stability testing.

Where supporting data are incomplete, the formulation should be kept as simple as possible, in order to minimise risks associated with manipulation and incompatibilities. Shelf-lives should be limited to as short as practically possible until further information is available. Formulations should be designed to include readily available starting materials and standard vehicles.

With regard to oral liquid formulations, insoluble drugs can often be suspended in universally available suspending agents for a short time, with little or no significant chemical degradation. Information regarding solubility can typically be found in the British Pharmacopoeia or Pharmaceutical Codex.

8.4 Hints on choosing a formulation – things to consider

8.4.1 Shelf-life

Given the risks associated with extemporaneous formulation, it is recommended that the shelf-life of any extemporaneous product should not exceed 28 days if preserved or 7 days if unpreserved. In the absence of supporting data, a limited shelf-life should be assigned based on the risk assessment of the individual drug. For drugs with narrow therapeutic indices, this may be as little as 2 or 3 days.

8.4.2 Dosage form

The choice of dosage form will depend on the clinical situation and also the patient (i.e. age, ability to swallow, etc.). The various practical options available – outlined in Chapter 2 – each have their own advantages and disadvantages in terms of clinical efficacy and practicality.

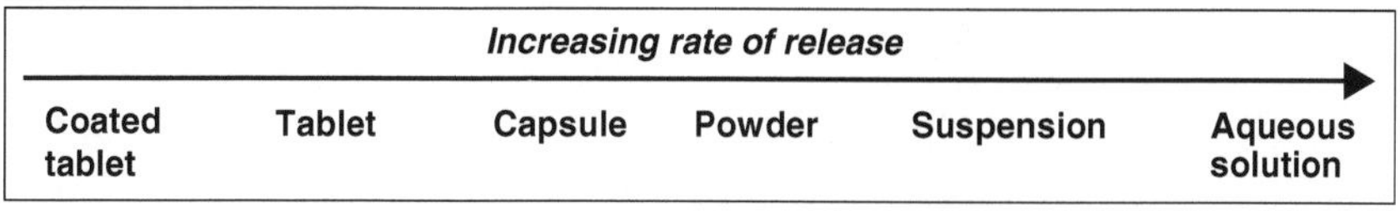

Increasing rate of release →					
Coated tablet	Tablet	Capsule	Powder	Suspension	Aqueous solution

Figure 8.1 Effect of formulation type on release rate and subsequent absorption of the medicine.

With regards to oral administration, since the medicine is required to be in solution for effective absorption, the type of dosage form employed will affect onset of action and bioavailability (Figure 8.1). This aspect should be considered if there is a need to alter the form (e.g. switching from solid to liquid dosage form if swallowing ability is impaired). In addition, the particle size in the formulation will also have a significant effect on bioavailability (e.g. powders and suspensions).

8.4.2.1 Solid dosage form (tablet/capsule/powder)

As stated in Chapter 2, tablets may often be dispersed, although part dosing through aliquots can lead to inaccuracies. Additionally, splitting tablets or making powders, although an option, may lack accuracy, and this practice should not be performed when functional coatings are present (e.g. modified release, pH sensitive).

8.4.2.2 Suspension

When producing suspensions, care should be taken to ensure uniform particle size and dispersion, where possible. This can either be achieved by purchasing the powder from a pharmaceutical manufacturer against a raw material specification, or more commonly by ensuring that tablets are ground to a fine uniform powder prior to suspending. This should be identified as a critical stage in the preparation process and suitable quality checks at this point should be made to ensure consistency.

The use of commercially and universally available suspending bases is encouraged, particularly where information on the stability of such bases is available in the literature. Suspending agents are also useful to suspend any insoluble tablet or capsule excipients, even where the drug itself is thought to be soluble at the chosen concentration.

Consideration must also be given to potential sedimentation – of drug and/ or excipients – over time, with labelling reflecting any specific requirements (i.e. shake well before use).

Commercially available suspending bases containing xanthan gum (e.g. Ora-Plus, 'Keltrol') tend to provide good suspending properties and are an excellent choice for the inexperienced formulator making one-off preparations. They can be used to suspend a variety of medicines, provide the formulation with a low sedimentation rate and easily re-disperse the medicine on shaking due to the thixotropic property of xanthan gum. However, some caution should be exercised when using xantham gum to ensure it is compatible with the medicine. It is an anionic material and may form a gelatinous precipitate if the formulation is highly alkaline or contains polyvalent metal ions (e.g. calcium).

Where xanthan gum cannot be used as the suspending base due to compatibility issues, there are other commercially available bases that will be compatible, but may not always have as good suspending properties. Bases containing methylcellulose as the suspending agent are generally compatible with most medicines, providing a good back-up option to xanthan gum.

Most commercial suppliers of suspending bases will have extensive compatibility data with their formulation and will also be able to provide alternative options in case of any incompatibility.

8.4.2.3 Solution

Although advantageous in terms of ease of administration and rapid and predictable absorption, drugs in solution are more prone to degradation than solids or suspensions and more susceptible to microbial contamination. Containers (usually bottles) are more bulky and less convenient for patients to carry. This may hinder compliance.

8.4.3 Organoleptic properties

Visual appearance, smell, taste and mouth feel all have a big influence on patient acceptance and compliance, especially in the paediatric population. For example, tablets/powders may have a chalky mouth feel, whereas drugs in suspension and solution in particular will often have a bitter taste if not flavoured or taste-masked.

Patients should be advised to look out for any changes that may indicate formulation failure (e.g. colour change) and also if any changes in organoleptic properties are likely to occur which do not have an adverse effect on therapy or formulation stability and integrity.

8.4.4 Excipients

McRorie (1996) suggests that, in general, one-time doses of medications with preservatives pose no risk of toxicity to even the tiniest premature neonate. However, multiple doses of the medicine (and preservative) can lead to life-threatening toxicity. In 1981 and 1982, the US Food and Drug Administration (FDA) received reports of 16 deaths in neonates attributed to benzyl alcohol found in sodium chloride flush solution (McRorie, 1996). Benzyl alcohol syndrome has since become a recognised event in premature infants (Brown *et al.*, 1982). The normal metabolic pathways for benzyl alcohol in adults are immature in premature infants, leading to the accumulation of benzyl alcohol and benzoic acid.

Chloroform is known to be hepatotoxic, renally toxic and potentially carcinogenic; its use as a preservative has been prohibited in the USA since 1976 (Hanson, 2003). However, many of the newer alternative preservatives are ineffective if the pH is above 7; therefore chloroform is still in use in some preparations in the UK (Hanson, 2003).

One of the more common preservative systems used is a combination of hydroxybenzoates (also known as 'parabens'). This preservative system can worsen asthma and has also been shown to promote hypersensitivity reactions (Golightly *et al.*, 1988).

Although propylene glycol is generally acknowledged to be safe in small quantities (Ruddick, 1972), larger amounts may cause side-effects such as hyperosmolarity, seizures, lactic acidosis and cardiac toxicity (in both children and adults) (Martin and Finberg, 1970; Arulanatham and Genel, 1978; Cate and Hedrick, 1980; Glasgow *et al.*, 1983). Propylene glycol may also act as an osmotic laxative. Preparations that contain large quantities may not therefore be suitable for neonates and young children.

Ethanol is known to have a CNS depressant effect, and its use is avoided wherever possible in extemporaneous preparations. However, limited quantities of ethanol may be justifiable under some circumstances, in order to prevent the use of other co-solvent systems associated with known toxicities. Limits for the inclusion of ethanol in paediatric formulations have been proposed in the USA (American Academy of Pediatrics Committee on Drugs, 1984; Anon, 1993).

Use of preparations that contain sugars, especially sucrose, are associated with the formation of dental cavities, particularly with long-term use.

These examples highlight the need for careful consideration of preservative systems and other excipients in extemporaneous formulations, particularly where young children are concerned. The *Handbook of Pharmaceutical Excipients* contains useful information for accountable pharmacists, and a list of excipients of known effect is available from the European Medicines Agency (EMEA).

8.4.5 Concentration

A change in concentration can either enhance or reduce stability. Oxidation and photodegradation, typically following zero-order kinetics, tend to be more significant at lower concentrations. Therefore, stability data may not be transferable from one concentration to another.

8.4.6 Dilution

Although dilution of more concentrated liquids may be possible in many instances, this can lead to potential calculation errors and may also reduce stability (see above) and impair preservative efficacy.

8.4.7 Ionic strength and pH

The degradation rate is often influenced by the ionic strength of the medium. For example, the rate of either acid- or base-catalysed hydrolysis of ampicillin in aqueous solution at constant pH is proportional to the concentration of the ion used as a buffer.

The rate of hydrolysis of many drugs is dependent on pH, with the reaction rate commonly increasing at either extreme of the scale.

8.4.8 Vehicle

Certain vehicles may be incompatible with some drugs, causing precipitation or degradation (e.g. polar solvents). Highly viscous liquids may pose issues pertaining to patient acceptance and difficulties in measuring doses accurately.

8.4.9 Storage

Consideration ought to be given to the nature of the container used to store the formulation. For instance, some substances may adsorb to plastic containers, leading to reduced homogeneity and potential degradation and, hence, inaccurate doses. Some drugs are sensitive to light, and as such may undergo significant photodegradation if not protected from it.

The amount of headspace in the final container should be limited, particularly for drugs prone to oxidation.

Conditions of storage will be determined by the nature of the drug and the formulation. Unpreserved formulations are generally stored in the fridge in order to minimise the risk of microbiological contamination, unless this is known to have an adverse effect on the formulation (e.g. inappropriate increase in viscosity or precipitation).

8.5 References

American Academy of Pediatrics Committee on Drugs (1984). Ethanol in liquid preparations intended for children. *Pediatrics* 73(3): 405–407.

Anon (1993). Panel recommends limits on alcohol content of non-prescription products. *Am J Hosp Pharm* 50: 400.

Arulanatham K, Genel M (1978). Central nervous system toxicity associated with ingestion of propylene glycol. *Pediatrics* 93: 515–516.
Brown WJ, Buist NRM, Gepson HT (1982). Fatal benzyl alcohol poisoning in a neonatal intensive care unit. *Lancet* 1: 1250.
Cate J, Hedrick R (1980). Propylene glycol intoxication and lactic acidosis. *N Engl J Med* 303: 1237.
Glasgow A, Boeckx R, Miller M, MacDonald M, August G, Goodman S (1983). Hyperosmolarity in small infants due to propylene glycol. *Pediatrics* 72: 353–355.
Golightly LK, Smolinske SS, Bennett ML, Sutherland EW, Rumack BH (1988). Pharmaceutical excipients. Adverse effects associated with 'inactive' ingredients in drug products. *Med Toxicol* 3: 128–165 (Part I) and 3: 209–240. (Part II).
Hanson G (2003). Bespoke pharmacy: tailoring medicines to the needs of patients – the pharmacy production unit's role. *Hosp Pharm* 10: 155–159.
Martin G, Finberg L (1970). Propylene glycol: a potentially toxic vehicle in liquid dosage form. *J Pediatr* 77: 877–878.
McRorie T (1996). Quality drug therapy in children: formulations and delivery. *Drug Inf J* 30:1173–1177.
Rowe RC, Sheskey PJ, Quinn ME, eds. (2009). *Handbook of Pharmaceutical Excipients*, 6th edn. London: Pharmaceutical Press.
Ruddick J (1972). Toxicology, metabolism, and biochemistry of 1,2-propanediol. *Toxicol Appl Pharmacol* 21:102–111.

8.6 Further reading

8.6.1 Databases and online resources

- Medline, EMBase and Pharmline databases
- Woods DJ (2001). *Formulation in Pharmacy Practice. eMixt, Pharminfotech*, 2nd edn. Dunedin, New Zealand. www.pharminfotech.co.nz/emixt
- Paddock Laboratories website. www.paddocklaboratories.com.

8.6.2 Data from NHS sources

- Dispensary in-house data
- NHS production units
- Quality control departments.

8.6.3 Textbooks

In addition to the British Pharmacopoeia, United States Pharmacopoeia and European Pharmacopoiea, the following textbooks are useful sources of information:

Connors KA, Amidon GL, Stella ZJ (1986). *Chemical Stability of Pharmaceuticals: A Handbook for Pharmacists*, 2nd edn. New York: Wiley Interscience. ISBN 0 471 87955 X.
Florey K, ed. (1972–1990). *Analytical Profiles of Drug Substances*, Vols 1–19. London: Academic Press.
Lund W, ed. (1994). *The Pharmaceutical Codex: Principles and Practice of Pharmaceutics*, 12th edn. London: Pharmaceutical Press. ISBN 0 85369 290 4.

Moffat AC, Osselton MD, Widdop B (2003). *Clarke's Analysis of Drugs and Poison*, 3rd edn. London: Pharmaceutical Press. ISBN 978 0 85369 473 1.

Nahata MC, Pai VB, Hipple TF (2004). *Pediatric Drug Formulations*, 5th edn. Cincinnati: Harvey Whitney Books Co. ISBN 0 929375 25 4.

Sweetman SC, ed. (2009). *Martindale: The Complete Drug Reference*, 36th edn. London: Pharmaceutical Press. ISBN 978 0 85369 840 1.

Trissel LA (2005). *Handbook on Injectable Drugs*, 13th edn. Bethesda, MD: American Society of Health System Pharmacists. ISBN 1 58528 107 7.

Trissel LA, ed. (2000). *Stability of Compounded Formulations*, 2nd edn. Washington DC: American Pharmaceutical Association, USA. ISBN 1 58212 007 2.

8.6.4 Other sources

- *International Journal of Pharmaceutical Compounding*
- *American Journal of Health System Pharmacists*
- Pharmaceutical companies
- Royal Pharmaceutical Society of Great Britain
- National Pharmaceutical Association.

9 Quality control

9.1 General

9.1.1 Quality control (QC) is that part of good preparation practice concerned with sampling, specifications and testing. It is also associated with the organisation, documentation and release procedures which ensure that appropriate tests are carried out and that products are not released for use until their quality has been judged to be satisfactory.

9.1.2 For extemporaneous products prepared for immediate dispensing to a single patient, quality control is necessarily restricted to testing of starting materials and documentation checks prior to release rather than testing of the finished product.

9.1.3 An assessment of the formulation and shelf-life should be carried out. This should include a review of the available stability information and an evaluation of the physico-chemical properties of each component making up the product. These factors should be considered as part of the risk assessment process (see Chapter 2).

9.2 Starting materials

9.2.1 All starting materials used in extemporaneous preparation should be from reputable sources and stored in their original containers. (Regional QA pharmacists can advise on the suitability of sources.) They should comply with their specification (see Chapter 6) and preferably be of pharmacopoeial grade or bear a marketing authorisation. Where a starting material is neither licensed nor to pharmacopoeial standard, a risk assessment should be undertaken.

9.2.2 There are two major areas of concern with starting materials. The first is identity of the contents of a container and the second is the quality of the substance.

9.2.3 The pharmacist must be assured that the product is what it is purported to be. Whenever possible all starting materials, with the exception of those with a marketing authorisation, should undergo a simple test to confirm their identity, as mislabelling of the containers constitutes the greatest risk. Such a test could be physical (e.g. visual, odour), provided that it is differential.

9.2.4 The assurance of quality may be achieved through the receipt of a suitable Certificate of Analysis. Certificates of Analysis declare every test carried out, the results obtained and, preferably, the limits applied. Certificates of Conformity are a declaration by the supplier that the product complies with the specification without elaborating on that specification nor the results obtained. Both are batch specific. This should be obtained for the batch in question and should be carefully evaluated before the starting material is used. (Chemicals for non-pharmaceutical purposes may contain toxic biocides or toxic trace elements.) A Certificate of Conformity may be an acceptable alternative if the specification with which the product is deemed to conform is clear and unambiguous. Certificates should be available through the supplier of the starting material.

9.2.5 It is permissible for a starting material to comply with a previous edition of an official pharmacopoeial monograph, but it must be clearly labelled in this way. The reason for the non-compliance with (or non-testing to) the current monograph should be investigated and a decision made on its suitability as an ingredient for extemporaneous dispensing.

9.2.6 The date of first opening should be marked on all starting materials. Where starting materials do not bear a manufacturer's expiry date, a maximum date of 2 years after first opening should be allocated. If considered cost-effective, materials may be re-tested if a suitable validated stability-indicating assay is available, up to a maximum expiry period of 5 years.

9.2.7 Starting materials must be appropriately stored in accordance with any specific storage conditions (e.g. refrigeration, avoiding light). The container should always be securely sealed after use.

9.2.8 Starting materials from natural sources or with a high level of hydration are particularly prone to microbiological contamination. In general, such materials are not suitable for storage for future use. Their suitability for further use should be carefully assessed prior to each use.

9.2.9 Water used for extemporaneous preparation should be potable or a sterile product, conforming to irrigation standard. In the latter case an expiry period of one working day after opening should be allocated to the container of water. Due consideration should be given to the nature of the product and the patient when considering the quality of the water used.

9.2.10 If the strength of a drug is normally stated in terms of the base and the starting material is a salt, or if it has water of crystallisation, this should be clear on the label, and in the formulae on the worksheet, to avoid potential medication errors during formulation at a later stage.

9.2.11 A Control of Substances Hazardous to Health (COSHH) assessment should be carried out on all materials to identify any hazards and necessary risk reduction measures.

9.2.12 All materials must be free from transmissible spongiform encephalopathy (TSE). Starting materials with a marketing authorisation will have this assurance; however, this cannot be assumed for other starting materials. TSE certification should be obtained from the supplier for each starting material (not each batch) to show that there is no risk of TSE contamination. This certification should be updated every 2 years for materials of animal origin (e.g. gelatin), and every 5 years for inorganic material (e.g. sodium chloride).

9.2.13 Expired or obsolete starting and packaging material should be destroyed in accordance with the Waste Regulations and the NHS Pharmaceutical Quality Assurance Committee's advisory document (NHS Pharmaceutical Quality Assurance Committee, 2009). The disposal should be recorded.

9.3 Finished products

9.3.1 Normally no analytical testing is performed on extemporaneously prepared products, and hence good preparation practice is of paramount importance. Any omissions or deviations must be carefully evaluated (see Chapter 3).

9.3.2 It is important that the formulation and its stability have been validated physically, chemically and microbiologically in relation to the clinical needs of the patient (see Chapter 8). It is recommended that formulations are from pharmacopoeial monographs or other reliable sources.

9.4 Release

9.4.1 Under current UK legislation, products prepared extemporaneously under a Section 10 exemption to the

Medicines Act 1968 are required to be released by a pharmacist who takes responsibility for the quality of the product.

9.4.2 The release of the product should be independent of the preparation activity and must be appropriately documented.

9.4.3 Product release should include the following:

- a visual inspection of the product
- a reconciliation of the starting materials and products (confirmation of identity)
- a label reconciliation.

9.4.4 Product release should also include verification that the product complies with specification (for example by checking that the worksheet is correctly completed) and that it is in accordance with the prescription. The release process should be documented in a procedure which should also identify actions to be taken in the case of product failure. The pharmacist should also be aware of any exceptions or deviations from validated procedures and should carefully assess their impact on product quality before they make the decision to release or reject the product.

Reference

NHS Pharmaceutical Quality Assurance Committee (2009). Guidelines on the handling and disposal of hospital pharmacy waste. Unpublished document available from regional quality assurance specialists in the UK or direct from the National QAWebsite: www.qainfozone.nhs.uk.

10

Complaints, product recalls and adverse events

10.1 Principles

10.1.1 Records should be kept to allow easy identification of the final recipient of all products and to allow rapid communication about recalls or quality problems.

10.1.2 All errors, defects, complaints and other indications of quality problems should be reviewed carefully by one or more designated people and reported to senior management.

10.1.3 There should be a written policy for timely assessment of the clinical significance and risk-potential of all defects identified.

10.1.4 A chronological log of all complaints, defects and recalls should be maintained to facilitate audit and detection of trends.

10.1.5 There should be a written policy for management and secure storage of all returned products.

10.1.6 All procedures and records should be subject to regular review and audit.

10.2 Quality problems

10.2.1 Measures should be in place to ensure that appropriate corrective and preventative action is taken. The source and nature of deficiencies, tests performed and remedial action taken should be documented with the preparation records and in the distribution record.

10.2.2 If a major product defect is suspected or identified, consideration should be given to checking if other products could be affected and to ceasing supply until the problem is fully investigated.

10.3 Complaints

10.3.1 All complaints should be investigated by a designated person.

10.3.2 Details of complainant and complaint should be recorded.

10.4 Recalls

10.4.1 If product deficiencies considered potentially harmful to health are identified, a product recall should be initiated immediately.

10.4.2 A written procedure for a recall should be in place. The procedure should identify designated senior members of staff authorised to initiate a recall.

10.5 Rejected and returned products

10.5.1 Defective products returned in response to a recall should be marked as such and stored in a secure, segregated area.

10.5.2 Safeguards must be in place to ensure that they cannot be supplied to another customer or patient in error.

10.5.3 The progress of the recall should be recorded. When the problem is satisfactorily resolved, a final report should be issued, including reconciliation between the delivered and recovered quantities of the products. In the UK, reports shall be retained for 5 years.

10.5.4 Products that have been returned after having been dispensed or issued for use (i.e. any that had left the control of the pharmacy in which they were prepared) should be destroyed.

11

Procurement and quality assessment of extemporaneously prepared medicines

11.1 Procurement

Extemporaneously prepared medicines are unlicensed medicines and are not subject to the same regulatory safeguards as those medicines which bear a marketing authorisation. However, any extemporaneously prepared medicines that have been made under a manufacturing 'Specials' licence will have been made in accordance with the principles of good manufacturing practice (GMP). Nevertheless, evidence of compliance with GMP should still be sought and evaluated by the purchasing pharmacist prior to use. It should be noted that for any unlicensed medicine, the purchaser retains the responsibility for the quality of the product.

The procurement of extemporaneously prepared medicines that have not been prepared under a current and appropriate manufacturing 'Specials' licence (MS) is not recommended. Care must be taken when procuring extemporaneously prepared medicines from 'Specials' manufacturers to ensure that the medicine has been made under the manufacturing licence (indicated by the presence of the 'MS' number clearly on the label of the finished product). Note: Caution – Some 'Specials' manufacturers may provide an extemporaneous dispensing service outside their licence, whereby products are 'dispensed' under Section 10. Always clarify the status of the medicine prior to purchase.

11.2 Quality assessment of extemporaneously prepared medicines from 'Specials' manufacturers

The purchasing pharmacist must take reasonable steps to ensure the quality and formulation of any procured medicines. Pharmacists should assure themselves that the product is of acceptable quality and has been made according to GMP by gathering and evaluating the appropriate evidence.

Evidence of compliance with GMP is usually readily available from reputable manufacturers and should be requested by the purchasing pharmacist in order to carry out a quality assessment prior to use. Evidence can be obtained either first hand by auditing the supplier, or by requesting appropriate documentation (e.g. licence details, information from the site master file, product specification, Certificate of Analysis or Conformity, audit reports, validation reports).

11.2.1 Quality assessment of the supplier

The 'Specials' manufacturer must have a current and appropriate manufacturing licence. This can be requested from the supplier or checked via the MHRA website, www.mhra.gov.uk. The 'Specials' manufacturer must also have evidence of a recent, satisfactory external audit to demonstrate GMP compliance. This may be from a variety of sources, including MHRA inspections, NHS regional quality assurance specialist inspections or a supplier audit, initiated by the purchaser.

Audit may be performed by the purchaser or on behalf of the purchaser by a suitably competent third party. In the NHS, this function is commonly carried out by regional quality assurance specialists on request. Colleagues in the NHS hospital trusts can access these reports by contacting their local lead quality assurance specialist. Reputable 'Specials' manufacturers may also make these audit reports available to customers on request.

11.2.2 Quality assessment of the product

The 'Specials' manufacturer should provide satisfactory stability data to support both the formula and shelf-life for any given product (see Chapter 8 for advice). Where the evidence base of the preparation is not supported by reasonable documentary confirmation of formulation stability, the purchasing pharmacist must carry out a risk assessment before purchasing the product.

The 'Specials' manufacturer should be able to supply a Certificate of Analysis or Conformity to demonstrate that the product has been prepared in accordance with its finished product specification (see Chapter 6).

The 'Specials' manufacturer should provide assurance that the starting materials are of pharmaceutical grade, and are accompanied by a transmissible spongiform encephalopathies (TSE) statement. This is in order to demonstrate that all reasonable measures have been taken to prevent the transmission of harmful contaminants from starting materials of animal origin.

Note: Such assurance should take into account the cross-contamination risk for those products which have been compounded on premises which handle starting materials of animal origin for other product lines.

Packaging and labelling of the medicine should be assessed by the purchaser to ensure that critical information is present, clear and unambiguous to minimise the risk of inadvertent medication error (see Chapter 6).

11.3 Contracting out the extemporaneous preparation service

Any arrangements for the provision of an extemporaneous preparation service from a third party should be supported by a written technical service level agreement (see Appendix 5 for an exemplar technical service level agreement). Such service level agreements must clearly state which party is responsible for ensuring that the preparation is clinically correct for the patient.

Note: It is important to note that products must only be supplied in response to prescriptions for individual patients.

11.4 Contracting out other services

Any work that could directly or indirectly affect the quality of the products, and which is contracted out, should be the subject of a written technical service level agreement. Examples of services include:

- maintenance of air handling systems, water systems or other utility systems
- maintenance of key equipment such as local exhaust ventilation systems, mixers and balances
- cleaning of components and consumables (e.g. clothing)
- environmental monitoring services (if applicable)
- supply of microbiological consumables (e.g. settle plates, swabs)
- waste management
- pest control
- quality control.

12

Audit and monitoring

12.1 Audit

12.1.1 Self audit should be completed on a regular basis and should cover all aspects of the quality assurance system including personnel, premises, equipment, documentation and record-keeping, production, quality control and environmental monitoring; also all aspects of the quality management system including deviations, complaints and work contracted out.

12.1.2 The frequency of self audit will be defined by the level of activity, but should be annually as a minimum.

12.1.3 There should be a detailed external independent review/audit carried out every 12–18 months by a designated competent auditor.

12.1.4 A suitable form for auditing the facility is outlined in Appendix 6.

12.1.5 It is incumbent on the accountable pharmacist to ensure that the self and independent audits are carried out by competent persons and in an impartial way.

12.1.6 Audit non-conformances should be considered and an action plan produced including timescales for corrective and preventative actions.

12.2 Supplier audit

12.2.1 If the service for extemporaneous preparation is outsourced to a third party then this party must be audited in line with this guidance document.

12.3 Monitoring

12.3.1 The level of environmental and final product monitoring required should be relevant to the type of products prepared and the scale of the preparation activity. Suitable standards should be incorporated in operating procedures.

12.3.2 Records of all monitoring activity, both environmental and product, should be kept and used to identify trends and in setting in-house standards and action limits.

12.3.3 All medicine storage areas should be temperature controlled and continuously monitored to ensure compliance with the appropriate temperature range.

12.3.4 The extent and frequencies of monitoring should be specified in written operating procedures.

12.3.5 Areas that should be considered for monitoring include:

- sinks and drains (microbiological)
- water supplies (microbiological/chemical)
- surfaces used for preparation (microbiological/chemical)
- air quality and devices used to clean air including extractor fans (microbiological/physical)
- cleaning and disinfection process (microbiological/chemical).

12.3.6 Action limits should be established. Action taken on exceeding these should be recorded. Corrective actions can include increased frequency of cleaning or use of an appropriate disinfectant.

12.3.7 All starting materials should be assessed for quality and suitability. This may include testing of starting materials and critical examination of Certificates of Analysis/Conformity.

12.3.8 Final products should only be tested where considered appropriate following a risk assessment using either unused products or extra samples specifically prepared.

12.3.9 Any testing laboratory used for analytical testing, stability assessment, etc. must be fully conversant with the technical background and requirements in preparation and should use validated methodology for the analysis of products.

PART B

Extemporaneous preparation formulary

13

Introduction

Part B contains information on some of the extemporaneous items most commonly prepared in UK NHS hospitals. It sets out some of the important information gathered during a UK research project entitled 'Improving the Quality and Formulation of Unlicensed, Non-Sterile, Oral Medicines Prepared in the NHS', described in Part A. The project was sponsored by the UK National Advisory Board for the Modernisation of NHS Hospital Medicines Manufacturing and Preparation Services.

This chapter attempts to begin to address the lack of information that has been documented for many years across the world. For example, Stewart and Tucker (1982) found a lack of physical and chemical data to support a range of extemporaneous preparations in Australia. Their survey found 116 different drugs in 270 different formulations. Furthermore, an unpublished survey of UK NHS trusts by the NHS Pharmaceutical Quality Assurance (QA) Committee (2003) found the extemporaneous dispensing of 209 different drugs, many of which are known to be formulated in several different ways.

On some occasions, although a suitable formula has been described, it is not used widely due to a problem obtaining raw materials. Therefore, there is a definite need for simple formulae with universally available components to allow preparation of reproducible formulations across the world (Woods, 2001). The quality of information relating to formulations in published paediatric trials also requires improvement (Standing *et al.*, 2005).

Data collection

From December 2005 to February 2006, a questionnaire was sent to all chief pharmacists at NHS trusts in Yorkshire, the North-East and London (all UK). The questionnaire requested information on extemporaneous dispensing activity in the preceding year, together with the origins of the formulae and shelf-lives assigned to the products.

The results of the survey showed that a total of 117 different drugs were prepared as oral medicines, some with several different formulations. The average number of different extemporaneous preparations made by each hospital trust was 10 (range 1–27). Table 13.1 shows the most common 50 extemporaneous oral liquid products reported.

Table 13.1 Extemporaneous dispensing activity (by number of units dispensed) in Yorkshire, the North-East and London (12 month activity, surveyed 2005–2006)

Drug	No. of units dispensed
Levothyroxine	880
Clobazam	743
Clozapine	693
Sodium chloride	621
Morphine sulfate	475
Ethambutol	371
Lorazepam	353
Pyrazinamide	330
Vancomycin	320
Knox mouthwash	281
Amiodarone	277
Azathioprine	277
Hydrocortisone	274
Clonidine	266
Sodium bicarbonate	261
Captopril	246
Acetazolamide	198
Midazolam	185
Tranexamic acid mouthwash	182
Magnesium glycerophosphate	173
Tacrolimus mouthwash	173
Allopurinol	149
Clonazepam	146
Quinine sulfate	131
Warfarin	131
Pyridoxine	128
Metformin	120
Dinoprostone	113

Table 13.1 *(continued)*

Sodium phenylbutyrate	112
Ergocalciferol	110
Omeprazole	108
Dexamethasone	107
Phenoxybenzamine	107
Diazoxide	99
Menadiol	95
Ubiquinone (co-enzyme Q10)	90
Thiamine	84
Potassium phosphate	77
Indometacin	76
Joulie's solution	69
Gliclazide	69
Primidone	69
Phenobarbitone	64
Isosorbide mononitrate	64
Gabapentin	64
Arginine	60
St Mark's solution	60
Sildenafil	57
Co-careldopa	53
Bendroflumethiazide	53

Table 13.2 demonstrates the varying quality of the data used to support the shelf-life assigned to the products.

Risk assessment

A risk assessment process was designed to prioritise the 50 most commonly dispensed drugs in terms of their potential to cause harm to patients. A new risk assessment process was needed as there is no published error rate for extemporaneous dispensing (Rennison and Portlock, 2003), and the system for pharmacovigilance for licensed medicines in the UK cannot easily be

Table 13.2 Origins of formulae and shelf-lives for extemporaneously prepared products in NHS trusts in Yorkshire, the North-East and London

Category	Definition	Proportion of total products
A	Arbitrary data – no chemical or microbiological data available	22.8%
B	Formulation and expiry date established by QC – data not available on file	11.5%
C	Formulation and expiry date established by QC – data available and copy report attached	3.0%
D	Formulation and expiry date included in formulary or database	8.1%
E	Formulation and expiry date derived from published paper	7.8%
F	Other information on stability (e.g. manufacturer's information) – data not on file	24.2%
G	Other information on stability (e.g. manufacturer's information) – data available	19.4%
Unknown	Correspondent reports that the source of the formula is unknown	3.2%

applied to unlicensed medicines. Therefore, there was also no known way of directly classifying or calculating the amount of risk associated with an extemporaneous preparation.

Some NHS trusts in the UK have qualitative decision trees or risk matrices that attempt to carry out a generic risk assessment. Experience and anecdotal reports would suggest that the majority of these tools are largely unproven and have not been properly validated. Such tools are usually designed to help the user at the point of use, comparing the risk of using an extemporaneous preparation with other clinical alternatives (or none). They do not attempt to consider the wider estimation of relative risk between different extemporaneous preparations.

The approach of using a scoring system devised by an expert panel is based on the original work done by Leventhal *et al.* (1979) and Naranjo *et al.* (1981) to develop methods of estimating the probability of adverse drug reactions, and similar attempts by other authors to refine the process (Lanctot and Naranjo, 1995).

It is equally important when estimating the risks associated with the clinical use of medicines to consider those risks associated with preparation in a manufacturing or dispensing environment (Daly and Marshall, 1996). In this environment, the possibility and the likely magnitude of harm from an unsafe medicine are important considerations in managing risk and arriving at sensible decisions (Daly and Marshall, 1996).

The risk associated with the extemporaneous preparation of each drug was therefore assessed by first splitting the overall risk into two equal components – technical risk and clinical risk – as follows:

CLINICAL RISK + TECHNICAL RISK = OVERALL RISK

The premise devised was that a drug has to be potentially clinically dangerous and technically difficult to formulate as an oral medicine, to be considered of the highest risk. The technical risk is the chance of an overdose or underdose occurring, while the clinical risk is the consequence(s) of receiving an underdose or overdose.

The risk assessment tool was devised in collaboration with a core team of senior pharmacists at Leeds Teaching Hospitals NHS Trust and the University of Bradford, UK. It was not practicable to consider the risks of every formulation, as some drugs had several formulations. Therefore, a method was devised to gauge the risks associated with a given drug (when formulated as an oral medicine). The risk assessment tool chosen comprised two matrices: one to evaluate clinical risk and one to evaluate technical risk.

Design of the technical risk assessment process

The technical risk of an error occurring was based on the complexity of the formulation. The premise established was that the risk of an error during preparation is a function of the number and complexity of the manipulations and calculations involved in the process. The ultimate outcomes of such an error occurring, in their simplest forms, are underdose and overdose Figure 13.1.

Where more than one worksheet was received during the data collection period for a given drug, the worksheet with the most complex calculations and manipulations was assessed. Each of the drugs was scored for technical risk by the core project team only, as the assessment was considered to be much more objective than the assessment of clinical risk. The risk scores associated with each box of the matrix were allocated by the core project team of senior pharmacists, acting as an expert group.

Design of the clinical risk assessment process

The clinical risk was regarded as the likely consequence of an underdose or overdose occurring. This is a function of the therapeutic index of the drug and its relative toxicity – minor, significant or death. Where the risks of underdose and overdose led to very different risks, the reviewer was directed to choose

Complexity of calculations	Number/complexity of manipulation(s)		
	Low	Moderate	High
Simple	Box A (score 2)	Box B (score 4)	Box C (score 6)
Moderate	Box D (score 4)	Box E (score 6)	Box F (score 8)
Highly complex	Box G (score 6)	Box H (score 8)	Box I (score 10)

Figure 13.1 Technical risk matrix for extemporaneous dispensing. **Manipulations:** Low complexity: comprises up to three simple physical steps in process (e.g. counting, crushing, diluting, suspending, dissolving). Moderate complexity: 4 or 5 simple steps in process. High complexity: preparations with multiple ingredients that require six or more steps in the process, or operations involving specialist equipment (e.g. electric blenders, heating devices). **Calculations:** Simple: calculations such as addition and subtraction (e.g. grams of powder, no. of millilitres required. Units do not change). Moderately difficult: calculations such as those that require 1,000-fold conversions but within the same expression of measurement (e.g. milligrams and micrograms). Highly complex: calculations with high error potential (e.g. dilution/use of concentrates (1 in *x* dilutions) and conversion between different expressions of strength such as millimoles and percentages or millimoles and micrograms.

the highest risk option. For example, overdose might be likely to yield minor toxicity, but underdose may lead to therapeutic failure for the typical indication. Figure 13.2 shows the risk matrix used.

In order to gather a consensus opinion, clinical risk was assessed by a reference group of 80 clinical pharmacists, using a clinical risk matrix questionnaire. The matrix was devised to consider first the therapeutic index and, second, the potential toxicity of each drug if an inaccurate dose was given (e.g. minor morbidity, significant morbidity, death). Each respondent was asked to choose the most appropriate box for each drug, based on the typical indication for use in their sphere of practice.

The risk score for each box was the same as for the corresponding boxes on the technical risk matrix. However, these scores were removed before the matrix was sent to correspondents, to prevent this leading to any bias in the results. When scoring the drug, the respondents were asked to consider the risks associated with each drug when used in its typical indication (as an extemporaneously dispensed oral medicine).

Dose-related outcomes	Therapeutic index (TI)		
	(a) **Wide TI** (Toxic dose much higher than therapeutic dose/ineffective dose much lower than therapeutic dose)	(b) **Medium TI**	(c) **Narrow TI** (Toxic dose close to therapeutic dose/ineffective close to therapeutic dose)
(A) Minor side-effects possible but significant morbidity unlikely	Box A (score 2)	Box B (score 4)	Box C (score 6)
(B) Significant morbidity likely from underdosing or overdose	Box D (score 4)	Box E (score 6)	Box F (score 8)
(C) Mortality likely from underdosing or overdose	Box G (score 6)	Box H (score 8)	Box I (score 10)

Figure 13.2 Clinical risk resulting from extemporaneous (oral dose) formulation failure.

Validation of the clinical risk matrix

In order to validate the clinical risk matrix, the boxes on the matrix were grouped as low, medium and high risk as in Table 13.3.

When the risk boxes are grouped together in this way, the mean percentage of respondents choosing the most common category is 59.6%. A majority of pharmacists (>50%) choose the same category in 35 of the 50 drugs (70%). One-way ANOVA statistical analysis with Bonferroni correction for the

Table 13.3 Pharmacists' choices of risk groups for the top 50 extemporaneously dispensed oral medicines in Yorkshire, the North-East and London

Drug	No. of respondents	Low risk A, B or D (%)	Medium risk C, E or G (%)	High risk F, H or I (%)
Levothyroxine	74	44.6	42.9	13.5
Clobazam	72	36.1	53.8	11.1
Clozapine	71	8.5	32.4	59.2
Sodium chloride	71	54.9	32.4	12.7
Morphine sulfate	73	12.3	31.5	56.2
Ethambutol	63	36.5	47.6	15.9
Lorazepam	69	44.9	42	13
Pyrazinamide	60	45	43.3	11.7
Vancomycin	71	71.8	19.7	8.5
Knox mouthwash	37	100	0	0
Amiodarone	71	14.1	29.6	56.3
Azathioprine	72	9.7	33.3	54.2
Hydrocortisone	70	52.8	44.3	2.9
Clonidine	62	46.8	46.8	6.5
Sodium bicarbonate	63	68.3	17.5	14.3
Captopril	71	53.5	36.6	10
Acetazolamide	48	62.5	35.4	2.1
Midazolam	56	33.9	41.1	25
Tranexamic acid mouthwash	63	92.1	7.9	0
Magnesium glycerophosphate	61	72.1	23	4.9
Tacrolimus mouthwash	45	37.8	20	42.2
Allopurinol	69	78.3	20.3	1.5
Clonazepam	68	36.8	50	13.2
Quinine sulfate	64	54.7	28.1	17.2
Warfarin	74	1.4	1.4	97.3

Table 13.3 *(continued)*

Pyridoxine	68	94.2	7.4	0
Metformin	67	49.3	34.3	16.4
Dinoprostone	30	26.7	33.3	40
Sodium phenylbutyrate	36	36.1	44.4	19.4
Ergocalciferol	59	83.1	15.3	1.7
Omeprazole	73	93.2	6.8	0
Dexamethasone	69	43.5	50.7	5.8
Phenoxybenzamine	36	11.1	63.9	25
Diazoxide	36	16.7	58.3	25
Menadiol	61	60.7	26.2	13.1
Ubiquinone	23	87	13	0
Thiamine	66	98.5	1.5	0
Potassium phosphate	58	32.8	37.9	29.3
Indometacin	65	41.5	38.5	20
Joulie's solution	40	50	37.5	15
Gliclazide	65	33.9	50.8	15.4
Primidone	54	20.4	40.7	38.9
Phenobarbitone	72	9.7	31.9	58.3
Isosorbide mononitrate	64	67.2	28.1	4.9
Gabapentin	60	36.7	50	13.3
Arginine	35	57.1	28.6	14.3
St Mark's solution	28	39.3	35.7	3.6
Sildenafil	64	34.4	31.3	34.4
Co-careldopa	64	30	50	20.3
Bendroflumethiazide	70	80	18.6	1.4

A, B and D = low clinical risk; C, E or G = medium clinical risk; F, H or I = High clinical risk.

number of data sets suggests that there is a statistically significant difference between the number of pharmacists choosing each of the three the different risk groups ($P < 0.001$). This would appear to suggest that the tool possesses consensual validity.

Risk assessment results

Technical risk results

A summary of the technical risk results is shown in Table 13.4.

Table 13.4 Technical risk scores assigned by the project team to the 50 most common extemporaneously dispensed oral liquids in Yorkshire, the North-East and London

Drug	Technical risk score
Azathioprine Ethambutol Phenobarbitone Pyrazinamide	10
Clobazam Joulie's solution Metformin Morphine Sodium chloride	8
Amiodarone Dexamethasone Dinoprostone Knox mouthwash Potassium phosphate Quinine sulfate Sodium bicarbonate	6
Acetazolamide Allopurinol Bendroflumethiazide Captopril Clonazepam Clozapine Diazoxide Gliclazide Hydrocortisone Levothyroxine Lorazepam Metformin Phenoxybenzamine Primidone Pyridoxine Sildenafil St Mark's solution Thiamine Warfarin	4

Table 13.4 *(continued)*

Arginine Clonidine Co-careldopa Ergocalciferol Gabapentin Indometacin Isosorbide mononitrate Magnesium glycerophosphate Menadiol Omeprazole Sodium phenylbutyrate Tacrolimus mouthwash Tranexamic acid mouthwash Ubiquinone Vancomycin	2

Clinical risk results

A summary of the clinical risk results is shown in Table 13.5.

The highest risk drug from a clinical perspective was found to be warfarin. This drug was taken as the reference against which the other 49 drugs could be compared. The STATA 9.2 statistical analysis computer program was used to carry a linear regression model analysis, using the clinical risk score as the outcome, and the drug as the single predictive factor. The difference between the clinical risk score for warfarin and each of the other drug's clinical risk score was found to be highly significant ($P < 0.001$), regardless of the number of replies.

Seventy-nine of the respondents supplied demographic data. These respondents had a mean 13 years of post-registration experience (range 1–30 years) and spent 53% (range 0–100%) of their time on 'clinical duties'. The pharmacists were asked to assess their own knowledge of extemporaneous dispensing. The results are shown below:

- Very good 1.3% ($n = 1$)
- Good 16.5% ($n = 13$)
- Satisfactory 39.2% ($n = 31$)
- Poor 35.4% ($n = 28$)
- Very poor 7.6% ($n = 6$).

A further linear regression analysis was carried out to investigate whether these three factors – knowledge, experience and percentage of time spent on clinical duties – had an impact on the way the respondents tended to assess the risk. The knowledge component did not have a statistically significant impact

Table 13.5 Mean clinical risk matrix scores for the 50 most common extemporaneously dispensed oral liquids

Drug	Number of replies (max 80)	Mean clinical risk score	Lower 95% confidence interval limit	Upper 95% confidence interval limit
Warfarin	75	**9.36**	8.95	9.77
Clozapine	71	**7.38**	6.78	7.97
Azathioprine	77	**7.08**	6.49	7.68
Phenobarbitone	71	**7.06**	6.46	7.65
Amiodarone	73	**7.04**	6.45	7.63
Morphine	75	**7.04**	6.45	7.63
Primidone	55	**6.36**	5.73	7.00
Phenoxybenzamine	37	**6.32**	5.60	7.05
Diazoxide	37	**6.11**	5.38	6.84
Dinoprostone	30	**6.07**	5.29	6.84
Tacrolimus mouthwash	45	**5.87**	5.19	6.54
Sildenafil	65	**5.85**	5.24	6.45
Potassium phosphate	59	**5.83**	5.21	6.46
Co-careldopa	64	**5.72**	5.11	6.33
Midazolam	64	**5.72**	5.11	6.33
Gliclazide	66	**5.58**	4.97	6.18
Indometacin	64	**5.56**	4.97	6.17
Sodium phenylbutyrate	36	**5.44**	4.71	6.17
Clobazam	72	**5.39**	4.80	5.98
Clonazepam	67	**5.37**	4.77	5.98
Ethambutol	64	**5.25**	4.64	5.86
Metformin	67	**5.16**	4.56	5.77
Gabapentin	73	**5.08**	4.49	5.68
Levothyroxine	75	**5.07**	4.48	5.65
Lorazepam	69	**5.04**	4.45	5.64
Pyrazinamide	60	**4.80**	4.18	5.42
Quinine sulfate	64	**4.78**	4.17	5.39

Table 13.5 *(continued)*

Clonidine	62	**4.74**	4.13	5.36
Hydrocortisone	71	**4.68**	4.08	5.27
Joulie's solution	40	**4.67**	3.96	5.38
Captopril	72	**4.64**	4.05	5.23
Arginine	35	**4.63**	3.89	5.36
Sodium chloride	71	**4.48**	3.89	5.07
Menadiol	59	**4.36**	3.74	4.98
St Mark's solution	28	**4.29**	3.49	5.10
Sodium bicarbonate	63	**4.19**	3.58	4.80
Isosorbide mononitrate	64	**4.09**	3.48	4.70
Acetazolamide	48	**4.08**	3.42	4.75
Magnesium glycerophosphate	62	**3.74**	3.29	4.52
Vancomycin	70	**3.74**	3.15	4.34
Bendroflumethiazide	72	**3.61**	3.02	4.20
Allopurinol	70	**3.49**	2.89	4.08
Ergocalciferol	60	**3.4**	2.78	4.02
Pyridoxine	69	**2.90**	2.30	3.50
Tranexamic acid mouthwash	64	**2.73**	2.12	3.34
Dexamethasone	69	**2.66**	2.07	3.25
Omeprazole	73	**2.66**	2.07	3.25
Ubiquinone	25	**2.64**	1.81	3.47
Thiamine	66	**2.30**	1.70	2.91
Knox mouthwash	37	**2.27**	1.55	2.99

on the overall risk score ($P = 0.780$). However, both of the other factors had a small but statistically significant predictive impact on the clinical risk score. For each extra percent of time spent on clinical duties, the clinical risk score is increased by 0.004 ($P = 0.003$). For every extra year of post-registration experience, the clinical risk score is increased by 0.017 ($P = 0.016$).

Overall risk results

Arbitrary risk categories were set as follows:

- Less than or equal to 5 = LOW
- More than 5 but less than or equal to 10 = MEDIUM
- More than 10 but less than or equal to 15 = HIGH
- More than 15 = VERY HIGH.

Table 13.6 combines and summarises the clinical and technical risk scores to show an overall risk score, and the corresponding overall risk classification for each drug.

Table 13.6 Table to show ranking of the 50 drugs by overall risk scores

Drug	Mean clinical risk score	Technical risk score	Overall risk score	Risk category
Azathioprine	7.08	10	**17.08**	Very high
Phenobarbitone	7.06	10	**17.06**	Very high
Ethambutol	5.25	10	**15.25**	Very high
Morphine	7.04	8	**15.04**	Very high
Pyrazinamide	4.80	10	**14.80**	High
Clobazam	5.39	8	**13.39**	High
Warfarin	9.36	4	**13.36**	High
Metformin	5.16	8	**13.16**	High
Amiodarone	7.04	6	**13.04**	High
Joulie's solution	4.67	8	**12.67**	High
Sodium chloride	4.48	8	**12.48**	High
Dinoprostone	6.07	6	**12.07**	High
Potassium phosphate	5.83	6	**11.83**	High
Clozapine	7.38	4	**11.38**	High
Primidone	6.36	4	**10.36**	High
Phenoxybenzamine	6.32	4	**10.32**	High
Sodium bicarbonate	4.19	6	**10.19**	High
Diazoxide	6.11	4	**10.11**	High
Sildenafil	5.85	4	**9.85**	Medium
Midazolam	5.72	4	**9.72**	Medium
Gliclazide	5.58	4	**9.58**	Medium

Table 13.6 *(continued)*

Clonazepam	5.37	4	**9.37**	Medium
Levothyroxine	5.07	4	**9.07**	Medium
Lorazepam	5.04	4	**9.04**	Medium
Hydrocortisone	4.68	4	**8.68**	Medium
Dexamethasone	2.66	6	**8.66**	Medium
Captopril	4.64	4	**8.64**	Medium
St Mark's solution	4.29	4	**8.29**	Medium
Knox mouthwash	2.27	6	**8.27**	Medium
Acetazolamide	4.08	4	**8.08**	Medium
Tacrolimus mouthwash	5.87	2	**7.87**	Medium
Co-careldopa	5.72	2	**7.72**	Medium
Bendroflumethiazide	3.61	4	**7.61**	Medium
Indometacin	5.56	2	**7.56**	Medium
Allopurinol	3.49	4	**7.49**	Medium
Sodium phenylbutyrate	5.44	2	**7.44**	Medium
Gabapentin	5.08	2	**7.08**	Medium
Pyridoxine	2.90	4	**6.90**	Medium
Clonidine	4.74	2	**6.74**	Medium
Arginine	4.63	2	**6.63**	Medium
Menadiol	4.36	2	**6.36**	Medium
Thiamine	2.30	4	**6.30**	Medium
Isosorbide mononitrate	4.09	2	**6.09**	Medium
Magnesium Glycerophosphate	3.74	2	**5.74**	Medium
Vancomycin	3.74	2	**5.74**	Medium
Ergocalciferol	3.40	2	**5.40**	Medium
Tranexamic acid mouthwash	2.73	2	**4.73**	Low
Omeprazole	2.66	2	**4.66**	Low
Ubiquinone	2.64	2	**4.64**	Low

Discussion

The use of risk assessment matrices for clinical and technical risk allowed the ranking of the products according to a perceived risk to patient safety. The tools had limitations and were not designed as a decision-making tool to help in an individual clinical scenario. They may, however, prove useful for helping to decide which products could be safely moved back to ward preparation.

The sensitivity of the tools, especially the clinical matrix, is limited by the need for a generic tool to apply to any drug in any situation. Furthermore, the tools had to be applied to the drug rather than the formulation, as each drug could have several formulations. While Table 13.6 gives an approximate ranking of the product based on the limited number of criteria in each matrix, a full literature review would be needed to highlight more complex or subtle issues which are specific to a particular drug (e.g. stereochemistry issues).

As with all risk assessments, the final scores are subjective in nature. However, the relatively large size of the sample population helps to demonstrate a consensual approach. Warfarin, as the drug with the highest clinical risk, was used as the comparator drug for the statistics, as there are no acknowledged standards against which these scores can be compared. Although the statistics show that the difference between the clinical risk of warfarin and any other drug is statistically significant, the difference between the other drugs is more debatable. Therefore, the results can only be taken as an approximate ranking of the products.

The results also suggested that the background and experience of the respondent will have an impact on the way in which a pharmacist assesses risk. For example, some of the drugs assessed have more than one acknowledged indication. The way in which a pharmacist assesses the risk will depend heavily on the way that they have seen it used in their own field of practice. Some drugs may, therefore, show a bimodal distribution for the risk assessment scores. An example of this phenomenon may be sildenafil, which has two different indications. Although licensed for the treatment of male impotence, it is used widely as an extemporaneous preparation in paediatric cardiology. Therefore, the way in which an extemporaneous preparation of such a drug is scored depends greatly on the background and knowledge of the particular pharmacist involved.

Some of the drugs had relatively few respondents due to the highly specialist nature of their use (e.g. dinoprostone, used in tertiary paediatric cardiothoracic surgery for maintaining ductal patency). However, the agreement seen for dinoprostone still appears to be significant. For such specialist drugs, there may be less doubt as to the intended indication or the associated risk. The pharmacists who feel confident and competent enough to rate such a drug are likely to have a highly specialist knowledge of pharmacotherapy in this area, potentially leading to a more accurate response when compared with more common drugs that may have several potential indications as an extemporaneous medicine.

Forty-three per cent of the respondents rated their knowledge of extemporaneous dispensing as poor or very poor. These data seem to confirm that the limited knowledge of this area of pharmacy practice may constitute a risk to patients' safety. This lack of knowledge has been described by several other authors (Crawford and Dombrowski, 1991; Pappas *et al.*, 2002; Cruikshank, 2003; Newton, 2003; Rennison and Portlock, 2003; Chowdury *et al.*, 2003; Kirsch, 2005). This poor knowledge may lead to an elevated risk to patients in daily practice.

With regard to technical risk, those drugs which require simple manipulations and few calculations were perceived as low risk in terms of potential operator error. However, when a drug is relatively unstable, the formulation required is often more complex as preservatives and stabilising agents are added, leading to an increased risk of errors.

The lack of sensitivity of the tool is acknowledged. It is designed as a starting point to target resources for further work only. For example, the risk associated with extemporaneous preparation of omeprazole may have been underestimated. Although the formulation for this drug tends to be simple, there are serious concerns with regard to its bioavailability when formulated as an oral liquid (Balaban *et al.*, 1997; Sharma *et al.*, 2000).

Conclusion

The 50 most commonly prepared extemporaneous preparations have been ranked in terms of their approximate risk to patient safety. Discussions with key stakeholders (e.g. UK National Paediatric Pharmacists' Group (NPPG), NHS Pharmaceutical Quality Assurance Committee) suggest that the risk assessment tool has successfully ranked those drugs which are perceived to be 'risky' in clinical practice as 'very high' or 'high' risk (e.g. warfarin, phenobarbital, azathioprine). Resources can now be targeted in a structured way to investigate those products which are thought to pose the highest risks.

References

Balaban DH, Duckworth CW, Peura DA (1997). Nasogastric omeprazole: effects on gastric pH in critically ill patients. *Am J Gastroenterol* 92: 79–93.

Chowdury T, Taylor KMG, Harding G (2003). Teaching of extemporaneous preparation in UK schools of pharmacy. *Pharm Educ* 3: 229–236.

Crawford SY, Dombrowski SR (1991). Extemporaneous compounding activities and the associated needs of pharmacists. *Am J Hosp Pharm* 48: 1205–1210.

Daly MJ, Marshall IW (1996). Risk management in hospital pharmacy. *Pharm Man* 12: 44–47.

Kirsch L (2005). Extemporaneous quality. *J Pharm Sci Technol* 59: 1–2.

Lanctot KL, Naranjo CA (1995). Pharmacoepidemiology and drug utilization: comparison of the bayesian approach and a simple algorithm for assessment of adverse drug events. *Clin Pharmacol Ther* 58: 692–698.

Leventhal JM, Hutchison TA, Kramer MS, Feinstein AR (1979). An algorithm for the operational assessment of adverse drug reactions. III Results of tests among clinicians. *JAMA* 242: 1991–1994.

Naranjo CA, Busto U, Sellers EM, Sandor P, Ruiz I, Roberts EA et al. (1981). A method for estimating the probability of adverse drug reactions. *Clin Pharmacol Ther* 30: 239–245.

Pappas A, McPherson R, Stewart K (2002). Extemporaneous dispensing: whatever happened to it? A survey of Australian general practitioners *J Pharm Res* 32: 310–314.

Rennison S.M., Portlock J.C. (Sept 2003). Is it time to stop dispensing extemporaneously in community pharmacy? *Int J Pharm Pract* R68(suppl).

Sharma VK, Peyton B, Spears T, Raufman J-P, Howden CW (2000). Oral pharmacokinetics of omeprazole and lansoprazole after single and repeated doses as intact capsules or as suspensions in sodium bicarbonate. *Aliment Pharmacol Ther* 14: 887–892.

Standing JF, Khaki ZF, Wong ICK (2005). Poor formulation in published pediatric drug trials. *Pediatrics* 116: 1013.

Stewart PJ, Tucker IG (1982). A survey of current extemporaneously prepared paediatric formulations. *Aust J Hosp Pharm* 12: 64–68.

Woods DJ (2001). *Formulation in Pharmacy Practice. eMixt, Pharminfotech*, 2nd edn. Dunedin, New Zealand. www.pharminfotech.co.nz/emixt (accessed 18 April 2006).

14

Formulary of extemporaneous preparations

Standardisation and rationalisation are critical processes in achieving an improvement in the quality and formulation of extemporaneous preparations. The stability summaries in this chapter are set out in a standardised way, beginning with an executive summary of the information, underpinned by a more thorough discussion of all the available supporting data.

The standard approach of the editorial board has been to minimise the technical risks by using a simple formula with the minimum number of raw materials and calculations to produce a high-quality product (including acceptable organoleptic properties). This approach also ensures that the product can be prepared in any dispensary equipped to carry out extemporaneous preparation. Typically this will involve the grinding of tablets (as these are a universally available source of the active ingredient) in a pestle and mortar and subsequently mixing and diluting the resulting powder with a suspending agent.

As the formulations include tablets as the starting material, our approach is to utilise suspending agents as the vehicle whether or not the active ingredient is soluble. This is to ensure uniformity of dose should there be any adsorption of the drug to insoluble excipients in the tablets. Appendix 7 provides additional information relating to the most common suspending agents used in these formulations.

Where any controversy exists with regard to the quality of the data, the approach has been to limit shelf-lives as a safety precaution (e.g. to 7 days or less). Unpreserved formulations are allocated a maximum in-use shelf-life of 7 days (refrigerated) on microbiological grounds.

It should also be noted that the standard headings in the stability summaries include a section for bioavailability. Often, there is no information available with regard to absorption, kinetics or a proven clinical effect. Pharmacists should make prescribers aware of this lack of data, and include extra monitoring for clinical and adverse effects in pharmaceutical care plans (where applicable).

Although these summaries attempt to recommend formulae of satisfactory pharmaceutical quality, the decision as to whether to use these formulae or other options remains with the accountable pharmacist. Whether the formulation is fit-for-purpose for a given patient depends on the individual clinical circumstances of that patient (e.g. the nature of the excipients in an adult preparation may not be suitable for use in neonates).

As stated earlier in the section on risk management (Chapter 2), the responsibility for ensuring that the formulation used for any extemporaneously prepared medicine is effective rests with the pharmacist, who should ensure that the patient's treatment is closely monitored and any associated adverse events are appropriately reported.

Acetazolamide oral liquid

Risk assessment of parent compound

A risk assessment survey completed by clinical pharmacists suggested that acetazolamide has a medium therapeutic index. In general, inaccurate doses were thought to be associated with minor morbidity (Lowey and Jackson, 2008). However, acetazolamide is also used in situations where an inaccurate dose could lead to severe morbidity. Current formulations used at NHS trusts were generally of a low to moderate technical risk, based on the number and/or complexity of manipulations and calculations. Acetazolamide is therefore generally regarded as a moderate-risk extemporaneous preparation (Box 14.1)

Summary

There is a wide body of evidence to suggest that acetazolamide can be extemporaneously prepared in a range of simple and complex formulations. Appropriate, simple suspending agents may be used to give a typical

Box 14.1 *Preferred formula*

Using a typical strength of acetazolamide 25 mg/mL as an example:

- acetazolamide tablets 250 mg × 10
- Ora-Plus 50 mL
- Ora-Sweet SF to 100 mL.

Method guidance

Tablets can be ground to a fine, uniform powder in a pestle and mortar. A small amount of Ora-Plus may be added to form a paste, before adding further portions of Ora-Plus up to 50% of the final volume. Transfer to a measuring cylinder. The Ora-Sweet can be used to wash out the pestle and mortar before making the suspension up to 100% volume. Transfer to an amber medicine bottle.

Shelf-life

28 days at room temperature in amber glass. **Shake the bottle.**

four-week expiry for an extemporaneous product, providing the formulation is effectively preserved. Xanthan gum-based vehicles are known to be physically compatible with acetazolamide (Nova Laboratories Ltd, 2003), and have been used widely in the NHS (Lowey and Jackson, 2008).

The preparation based on Ora-Plus and Ora-Sweet Sugar Free (SF) is recommended for simplicity. The acidic pH of Ora-Sweet SF/Ora-Plus (approximate pH 4.2; Paddock Laboratories, 2006) is also likely to limit hydrolysis and solubility. The cherry flavour of Ora-Sweet SF may help to mask the drug's bitter taste. Note that the pH and specification of 'Keltrol' (xanthan gum suspending agent) preparations may vary with supplier. If 'Keltrol' is chosen as the suspending agent, the formula and shelf-life should be assessed by a suitably competent pharmacist before use.

Clinical pharmaceutics

Points to consider

- At typical paediatric strengths (10–50 mg/mL), most or all of the drug should be in solution.
- Ora-Plus and Ora-Sweet SF are free from chloroform and ethanol.
- Ora-Plus and Ora-Sweet SF contain hydroxybenzoate preservative systems.
- Ora-Sweet SF has a cherry flavour which may help to mask the bitter taste of the drug.
- There is no bioavailability information available.

Other notes

- Acetazolamide is reported to have a bitter taste (Parasrampuria and Das Gupta, 1989).
- Actezolamide injection is reported to have been given orally but must be diluted first (Guy's and St. Thomas' and Lewisham Hospitals, 2005).
- Acetazolamide base is used for tablets and the sodium salt for injectables (Anon, 1997).
- Benzoic acid, chloroform, orange syrup and tragacanth have been used in suspensions of acetazolamide prepared from crushed tablets (Lund, 1994).
- Acetazolamide is reported to be physically compatible with 'Keltrol' (Nova Laboratories Ltd, 2003).
- Scored tablets can be used for whole-tablet or half-tablet doses. Tablets can be crushed and/or dispersed under water.

Technical information

Structure

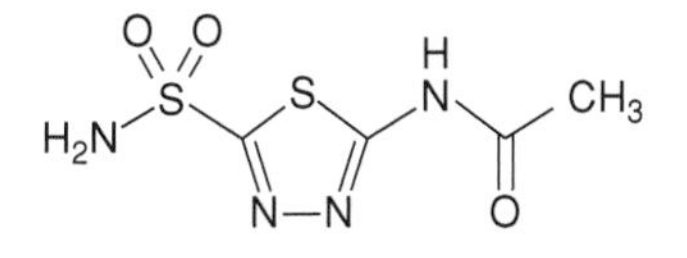

Empirical formula

$C_4H_6N_4O_3S_2$ (British Pharmacopoeia, 2007).

Molecular weight

222.2 (British Pharmacopoeia, 2007).

Solubility

1 in 1400 of water (Lund, 1994); 72 mg/mL (Parasrampuria and Das Gupta, 1990).

pK_a

7.2 and 9.0 at 25°C (Lund, 1994).

Optimum pH stability

pH 4.0–5.0 (Lund, 1994).

Stability profile

Physical and chemical stability

Acetazolamide is a weak acid, and degrades by hydrolysis of the amide side-chain. Given its solubility in water of approximately 72 mg/mL, the strength of the formulation will have a major impact on the proportion of the drug that is held in suspension. Typical paediatric preparations of 10–50 mg/mL will contain most or all of the drug in solution, and are thus vulnerable to hydrolysis. Solubility increases in solutions of alkali hydroxides (Woods, 2001).

Hydrolysis in solution has been reported to occur more rapidly at pH 1–3 or 8–12 (Parasrampuria and Das Gupta, 1990). It has been suggested that the aqueous stability of acetazolamide sodium decreases at pH values over 9 (Trissel, 2005).

Degradation products

Breakdown products are acetic acid and 5-amino-1,3,4-thiadiazole-2 sulfonamide (Parasrampuria and Das Gupta, 1989; Lund, 1994). No information is known to be available with regard to toxicity or activity of these breakdown products.

Stability in practice

There is a significant body of evidence to support the stability of acetazolamide when formulated as an oral liquid, including two key published articles.

Allen and Erickson (1996) examined the stability of 25 mg/mL acetazolamide base (derived from tablets) in the following formulations:

- 1:1 mixture of Ora-Sweet and Ora-Plus
- 1:1 mixture of Ora-Sweet Sugar Free (SF) and Ora-Plus
- cherry syrup.

A simple manual method of preparation was used with stepwise additions of vehicle, and details of the stability-indicating high-performance liquid chromatography (HPLC) methodology are supplied. The liquids were stored in amber clear plastic bottles (polyethylene terephthalate) with a low-density polyethylene foam cap lining. Three bottles of each suspension were stored at 5°C and three at 25°C. The initial pH values for the 'Ora' preps were 4.3–4.4 and 3.5 for the cherry syrup formulation. None of the suspensions showed a pH change of more than 0.5 pH units over the test period.

Any changes in appearance, pH, odour and drug content were monitored over a 60-day period. All suspensions showed at least 94% of the starting concentration remained after 60 days, regardless of storage temperature. The suspensions were protected from light. Details of the sampling procedure were not provided.

Alexander *et al.* (1991) investigated the stability of a 25 mg/mL suspension derived from powdered acetazolamide tablets. Seventy per cent sorbitol solution was used to wet the powder, which was then mixed with a suspension vehicle of aluminium magnesium trisilicate, carboxymethylcellulose, water and syrup. Other excipients were glycerol, methyl and propyl hydroxybenzoates, propylene glycol, Allura Red (E129) and strawberry flavour. The pH was adjusted to 5 with hydrochloric acid. Three bottles were stored at each of the following temperatures: 5, 22, 30, 40 and 50°C. Samples were checked for appearance, ease of pouring and redispersion, microbial growth, pH and drug content over a period of 79 days.

Each bottle to be assayed was shaken for 10 minutes. A 1 mL portion was removed from the suspension and diluted to 50 mL; a stability-indicating HPLC method was used.

Less than 6% of the initial concentration of acetazolamide was lost after storage in amber bottles at 5°C, 22°C and 30°C for 79 days. The suspensions stored at 40°C and 50°C degraded to 90% of the starting concentration by day 32 and 79, respectively. The authors suggest an activation energy of 14.88 kcal/mol for the hydrolysis reaction, yielding a calculated shelf-life of 371 days at 22°C. The corresponding figure for a suspension stored in a fridge is estimated at 4 years (Lund, 1994). The suspensions also reportedly remained homogeneous and showed good redispersibility.

While this work provides further evidence as to the relative stability of acetazolamide, the method of preparation suggested by Alexander *et al.* (1991) involved automated equipment and a complex formula which may not be readily accessible to the majority of pharmacists in the UK. The depth from which the sample was taken was not stated, and a 1 mL sample dilution may be open to some experimental error.

Parasrampuria and Das Gupta (1990) have reported stability data on two oral liquid dosage forms of acetazolamide. The formulae are based on using polyethylene glycol 400 (PEG 400; 7% v/v) as the solubilizing agent and propylene glycol (53% v/v) as the co-solvent to keep acetazolamide in solution. Sweetening agents (syrup, sorbitol, saccharin and aspartame), colours and flavours (raspberry, sweet flavour and menthol) were used to improve the organoleptic properties of the formulations. Ethanol was used as a solubilizing agent for the menthol, and the solutions were preserved with sodium benzoate. The two final formulations differed only in their choice of phosphate or citrate buffer (formulations adjusted to pH to minimise hydrolysis). Further formulations were also produced without some of the inactive ingredients, to establish if the presence of the excipients affected the stability in any way.

The solubility of acetazolamide was very high (87.8 mg/mL) in PEG 400 versus only 7.4 mg/mL in propylene glycol. The solubility was greater above pH 8.2, but hydrolysis is known to be accelerated at this pH (Parasrampuria and Das Gupta, 1989).

Solutions were stored at 37°C and 25°C, and assayed over time using a previously published HPLC method (Das Gupta and Parasrampuria, 1987). The formulations showed less than 3% degradation occurred in any of the formulations at 25°C over 179 days. At 37°C, the formulations showed approximately 10% degradation over the same time period. The authors suggest a tentative extrapolated shelf-life of 2 years for the two final formulations when stored at 25°C (Parasrampuria and Das Gupta, 1990; Lund, 1994).

No details were given as to the sampling procedures. The complexity of the formulae prohibit their use at a dispensary level, but give further information on the stability of acetazolamide in solution. This work built on previously published quantitative analysis data (Das Gupta and Parasrampuria, 1987).

The toxicities of some of the excipients used by the authors should be noted. For example, ethanol is acknowledged to be a CNS depressant. Although propylene glycol is generally acknowledged to be safe in small quantities (Ruddick, 1972), larger amounts may cause side-effects such as hyperosmolarity, seizures, lactic acidosis and cardiac toxicity (in both children and adults) (Martin and Finberg, 1970; Arulananatham and Genel, 1978; Cate and Hedrick, 1980; Glasgow *et al.*, 1983). Propylene glycol and polyethylene glycol may also act as osmotic laxatives (Price, 2002). Preparations that contain large quantities may not therefore be suitable for neonates and young children. Propylene glycol is estimated to be one third as intoxicating as alcohol, with administration of large volumes being associated with adverse effects most commonly on the CNS, especially in neonates and children (Martin and Finberg, 1970; Arulanantham and Genel, 1978; MacDonald *et al.*, 1987).

Preformulation studies by the same authors investigated the effect of pH, phosphate and citrate buffer concentrations, ionic strength and temperature on the stability of acetazolamide using a stability-indicating HPLC method (Parasrumpuria and Das Gupta, 1989). An internal standard of sulfamerazine was used. The decomposition of acetazolamide solution (derived from pharmaceutical grade powder) followed a first-order equation and the solutions showed maximum stability at pH ~4.

Khamis *et al.* (1993) report a stability-indicating first-derivative spectrophotometric assay for use in dissolution and kinetic studies. The author suggests that acetazolamide solution shows maximum stability at pH 4. Acetazolamide powder in 0.01 M sodium hydroxide showed a degradation rate constant of 3.51×10^{-3} per day and a degradation half-life of 8.23 days. The acetazolamide hydrolysed quickly over the pH range 7.8–12. Degradation rate constants at elevated temperatures and for sonicated solutions are also stated.

Bioavailability data

No information.

References

Alexander KS, Haribhakti RP, Parker G (1991). Stability of acetazolamide in suspension compounded from tablets. *Am J Health Syst Pharm* 48: 1241–1244.

Allen LV, Erickson MA (1996). Stability of acetazolamide, allopurinol, azathioprine, clonazepam and flucytosine in extemporaneously compounded oral liquids. *Am J Health Syst Pharm* 53: 1944–1949.

Anon (1997). Acetazolamide 25 mg/ml oral liquid. *Int J Pharm Compound* 1: 101.

Arulananatham K, Genel M (1978). Central nervous system toxicity associated with the ingestion of propylene glycol. *J Pediatr* 93: 515–516.

British Pharmacopoeia (2007). *British Pharmacopoeia*, Vol. 1. London: The Stationery Office, pp. 45–46.

Cate J, Hedrick R (1980). Propylene glycol intoxication and lactic acidosis. *N Engl J Med* 303: 1237.

Das Gupta V, Parasrampuria J (1987). Quantitation of acetazolamide in pharmaceutical dosage forms using high-performance liquid chromatography. *Drug Dev Ind Pharm* 13: 147–157.

Glasgow A, Boeckx R, Miller M, MacDonald M, August G, Goodman S (1983). Hyperosmolarity in small infants due to propylene glycol. *Pediatrics* 72: 353–355.

Guy's and St. Thomas' and Lewisham Hospitals (2005). *Paediatric Formulary*. London: Guy's and St Thomas' and Lewisham NHS Trust.

Khamis EF, Abdel-Hamid M, Hassan EM, Eshra A, Elsayed MA (1993). A stability-indicating first-derivative spectrophotometric assay of acetazolamide and its use in dissolution and kinetic studies. *J Clin Pharm Ther* 18: 97–101.

Lowey AR, Jackson MN (2008). A survey of extemporaneous dispensing activity in NHS trusts in Yorkshire, the North-East and London. *Hosp Pharmacist* 15: 217–219.

Lund W, ed. (1994). *The Pharmaceutical Codex: Principles and Practice of Pharmaceutics*, 12th edn. London: Pharmaceutical Press, pp. 708–711.

MacDonald MG, Getson PR, Glasgow AM *et al.* (1987). Propylene gycol: increased incidence of seizures in low birth weight infants. *Pediatrics* 79: 622–625.

Martin G, Finberg L (1970). Propylene glycol: a potentially toxic vehicle in liquid dosage form. *J Pediatr* 77: 877–878.

Nova Laboratories Ltd (2003). Information Bulletin – Suspension Diluent Used at Nova Laboratories.

Paddock Laboratories (2006). Ora-Plus and Ora-Sweet SF Information Sheets. www.paddock-laboratories.com (accessed 12 April 2006).

Parasrampuria J, Das Gupta V (1989). Preformulation studies of acetazolamide. Effect of pH, two buffer species, ionic strength and temperature of its stability. *J Pharm Sci* 78: 855–857.

Parasrampuria J, Das Gupta V (1990). Development of oral liquid dosage forms of acetazolamide. *J Pharm Sci* 79: 835–836.

Price JC (2002). Polyethylene glycol. In: Rowe RC, Sheskey PJ, Weller PJ, eds. *Handbook of Pharmaceutical Excipients*, 4th edn. London: Pharmaceutical Press and Washington DC: American Pharmaceutical Association.

Ruddick J (1972). Toxicology, metabolism, and biochemistry of 1,2-propanediol. *Toxicol Appl Pharmacol* 21: 102–111.

Trissel LA (2005). *Handbook on Injectable Drugs*, 13th edn. Bethesda, MD: American Society of Health System Pharmacists.

Woods DJ (2001). *Formulation in Pharmacy Practice. eMixt, Pharminfotech*, 2nd edn. Dunedin, New Zealand. www.pharminfotech.co.nz/emixt (accessed 18 April 2006).

Allopurinol oral liquid

Risk assessment of parent compound

A survey of clinical pharmacists has suggested that allopurinol generally has a medium therapeutic index and the potential to cause minor side-effects if inaccurate doses are administered (Lowey and Jackson, 2008). However, there are clinical situations, such as in the prevention of tumour lysis syndrome, where the dosage and administration accuracy may become more critical. Current formulations used at NHS trusts were generally of a moderate technical risk, based on the number and/or complexity of manipulations and calculations. Allopurinol is therefore regarded as a moderate risk extemporaneous preparation (Box 14.2).

Summary

There is a wide body of evidence to suggest that allopurinol can be formulated as a stable oral liquid. This product is now known to be available from reputable 'Specials' manufacturers, which should be considered for purchase in preference to the preparation of an extemporaneous item. The quality and formulation of the 'Special' product should be checked by an appropriate person before purchase.

Clinical pharmaceutics

Points to consider

- Due to poor aqueous solubility of allopurinol, the vast majority of drug is likely to be suspension rather than solution at a typical strength of 20 mg/mL. **Shake the bottle before use to improve dosage uniformity.**
- Ora-Plus, Ora-Sweet and Ora-Sweet SF are free from ethanol and chloroform. Ora-Plus and Ora-Sweet SF are sucrose-free.
- Ora-Sweet and Ora-Sweet SF are cherry-flavoured.
- 'Keltrol' is made to slightly different specifications by several manufacturers, but typically contains ethanol and chloroform. The exact specification should be assessed before use by a suitably competent pharmacist. Ethanol is a CNS depressant. Chloroform has been classified as a class 3 carcinogen (i.e. substances that cause concern owing to possible carcinogenic effects but for which available information is not adequate to make satisfactory assessments) (CHIP3 Regulations, 2002).

Box 14.2 Preferred formulae

Note: 'Specials' are now available from reputable manufacturers. Extemporaneous preparation should only occur where there is a demonstrable clinical need.

Using a typical strength of 20 mg/mL as an example:

- allopurinol tablets 300 mg × 6
- allopurinol tablets 100 mg × 2
- Ora-Plus to 50 mL
- Ora-Sweet Sugar Free (SF) or Ora-Sweet to 100 mL.

Method guidance

Tablets can be ground to a fine, uniform powder in a pestle and mortar. A small amount of Ora-Plus may be added to form a paste, before adding further portions of Ora-Plus up to 50% of the final volume. Transfer to a measuring cylinder. The Ora-Sweet SF can be used to wash out the pestle and mortar before making the suspension up to 100% volume. Transfer to an amber medicine bottle.

Shelf-life

28 days at room temperature in amber glass. **Shake the bottle.**

Alternative

Again, a typical strength of 20 mg/mL:

- allopurinol tablets 300 mg × 6
- allopurinol tablets 100 mg × 2
- 'Keltrol' 0.5% suspending agent (or similar) to 100 mL.

Shelf-life

28 days at room temperature in amber glass. **Shake the bottle.**

Note: Xanthan gum formulations vary slightly between manufacturers, and may contain chloroform and ethanol. The individual specification should be checked by the appropriate pharmacist. Some 'Keltrol' formulations may contain flavouring agents.

- Tablets (100 mg and 300 mg) may be crushed/dispersed under water for whole-tablet doses.
- Bioavailability data are not available for extemporaneously prepared allopurinol liquids.

Technical information

Structure

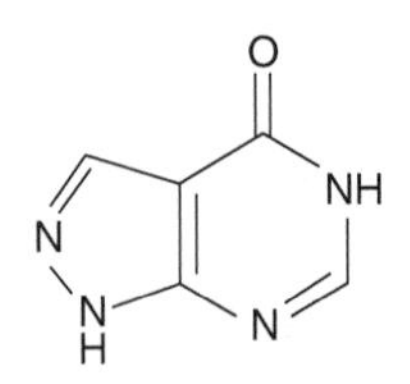

Empirical formula

$C_5H_4N_4O$ (British Pharmacopoeia, 2007)

Molecular weight

136.1 (British Pharmacopoeia, 2007)

Solubility

0.48 mg/mL in water at 25°C (Benezra and Bennett, 1978).

pK_a

Reported as 9.4 (McEvoy, 1999) and 10.2 (Benezra and Bennett, 1978; Budavari, 1996).

Optimum pH stability

3.1–3.4 at 105°C (Beaumont *et al.*, 1978).

Stability profile

Physical and chemical stability

Allopurinol is a white or almost white powder with a slight odour (Benezra and Bennett, 1978). It is very slightly soluble in water (British Pharmacopoeia, 2007). Its aqueous solubility is reported to be increased at alkaline pH (Woods, 2001).

The primary degradant 3-aminopyrazole-4-carboxamide decomposes to 3-aminopyrazole-4-carboxylic acid in basic solutions. This does not occur in acidic solution (Benezra and Bennett, 1978).

Beaumont *et al.* (1978) have reportedly predicted the shelf-life of a 2% suspension (20 mg/mL) at pH 9.5 and 25°C to be 5.4 years. However, no data or details are available to support this claim.

At 105°C, it has been reported that the maximum stability of allopurinol in solution occurs at pH 3.1–3.4 (Beaumont *et al.*, 1978). Rapid breakdown is reported to occur at high pH, but no details are available.

Gressel and Gallelli (1968) investigated the alkaline stability of allopurinol solutions, at concentrations of 1 mg/mL and 5 mg/mL. Kinetic studies showed good adherence to first-order kinetics. The authors suggest a first-order dependency on hydroxyl ion concentration, and that for allopurinol injections, the lowest pH commensurate with the desired solubility should be used. At pH 10.8, the unbuffered solution of allopurinol has a t_{90} of approximately 150 days.

Allopurinol tablets should be stored in well-closed, light-resistant containers at controlled room temperature (McEvoy, 1999).

Degradation products

Primary decomposition product in acidic and basic solutions is 3-aminopyrazole-4-carboxamide (Gressel and Gallelli, 1968; Benezra and Bennett, 1978). See above for further information.

Stability in practice

There is a significant body of evidence available to support the stability of allopurinol, when formulated as an oral liquid. This evidence has allowed the manufacture of allopurinol oral liquid by 'Specials' manufacturers in the UK.

Allen and Erickson (1996) have investigated the stability of allopurinol at a concentration of 20 mg/mL in three different vehicles:

- 1:1 mixture of Ora-Plus and Ora-Sweet
- 1:1 mixture of Ora-Plus and Ora-Sweet SF
- cherry syrup (cherry syrup concentrate diluted 1:4 with simple syrup).

Standard 300 mg tablets were crushed in a mortar and ground to a fine powder. Aliquots of the appropriate vehicle(s) were added to form a paste and then to bring to volume before transfer to a calibrated bottle. Enough suspension was made to yield six 120 mL amber clear plastic (polyethylene terephthalate) prescription bottles with a low-density polyethylene foam cap lining.

The bottles were evenly split between between storage temperatures of 5°C and 25°C (all protected from light). After agitation of the bottles, 5 mL samples were withdrawn on days 1, 2, 7, 10, 28, 35 and 60. Baseline pH values were taken and compared against results on days 30 and 60. Physical characteristics were also noted at each sample time.

A stability-indicating HPLC assay was used; details are available (Allen and Erickson, 1997). The initial pH values were 4.3 and 3.2 for the Ora-based and cherry syrup-based suspensions, respectively. Retested pH values at day 30 and day 60 did not change by more than 0.5 pH units.

Results showed that all of the suspensions maintained more than 95% of the initial concentration after 60 days, regardless of storage temperature or suspending agent(s). Even after taking into account the lower confidence intervals for standard deviations, well over 90% of the initial amount was remaining. Microbial growth was not studied.

Dressman and Poust (1983) have also investigated the stability of allopurinol (20 mg/mL) in suspension. The vehicle consisted of a Cologel (methylcellulose) suspending agent flavoured with a 2:1 mixture of simple syrup and wild cherry syrup.

Tablets were crushed in a pestle and mortar before transfer to a calibrated amber glass bottle and then addition of the suspending agent. Thorough mixing was achieved by around 30 seconds of vigorous manual shaking, followed by ultrasonication. Only two suspensions were made (of only 10 mL each), and divided between storage at 5°C and 'room temperature' (details not stated).

A stability-indicating HPLC methodology was followed and some details are stated in the paper. The suspension excipients were found not to interfere with the assay. Less than 5% of the initial allopurinol concentration was lost at either temperature after 56 days.

The number of samples run on each test day, the standard deviations and the initial actual concentrations are not stated. Therefore, the possible loss of drug due to the initial transfer of dry powder from pestle to bottle is not quantifiable. Only four data points are available for the room temperature sample, and only three data points are available for the 5°C sample. Microbiological stability was not assessed.

Although the paper lacks some detail and statistical accuracy, the general findings appear to be in agreement with the work carried out by Allen and Erickson (1996), and the theory stated above.

Woods (2001) suggests a similar formula based on methylcellulose suspending agent, with the addition of citric acid and syrup. However, there are no data to support the stated shelf-life.

Alexander *et al.* (1997) evaluated the stability of a complex 20 mg/mL allopurinol oral suspension. The formula is as follows:

- allopurinol 300 mg tablets × 200
- veegum 22.5 g
- sodium carboxymethylcellulose 22.5 g
- lycasin syrup 1,500 g
- sodium bisulfite 1.5 g
- saccharin sodium 1.5 g
- parabens stock solution 60 mL (methyl paraben 10% and propyl paraben 2% in neat propylene glycol)

- wild cherry flavour 2 mL
- vanilla flavour 2 mL
- distilled water to 3,000 mL.

After packaging in 120 mL portions in amber glass bottles with child-resistant caps, the suspension was stored at temperatures ranging from 50°C to 80°C. The authors reported no problems with physical stability, except for a colour change from off-white to yellow. Interpretation of accelerated stability study data suggested a shelf-life (time to achieve 10% degradation) of 8.3 years at 25°C.

While the formula may be too complex to be applicable on a wider scale, the results of a stability-indicating HPLC analysis provided further evidence for the relative stability of allopurinol as an oral liquid.

Reports in the pharmaceutical literature have claimed that allopurinol is physically and chemically stable in 'Keltrol' (xanthan gum suspending agent) for 12 weeks (Anon, 1986). Allopurinol suspensions in 'Keltrol' are widely used in the NHS (Lowey and Jackson, 2008). However, no data have been published to support the stability of 'Keltrol'-based products.

Bioavailability data

No information available.

References

Alexander KS, Davar N, Parker GA (1997). Stability of allopurinol suspension compounded from tablets. *Int J Pharm Compound* 1: 128–131.

Allen LV, Erickson MA (1996). Stability of acetazolamide, allopurinol, azathioprine, clonazepam, and flucytosine in extemporaneously compounded oral liquids. *Am J Health Syst Pharm* 53: 1944–1949.

Anon (1986). 'Extremely useful' new suspending agent. *Pharm J* 237: 665.

Beaumont TG, Coomber TS, Conroy AP, Gandhi RP, Hill GT, Burroughs Wellcome (1978). Personal communication cited in Benezra SA, Bennett TR (1978). Allopurinol. In: Florey K, ed. *Analytical Profiles of Drug Substances*, Vol. 7. London: Academic Press, pp. 1–17.

Benezra SA, Bennett TR (1978). Allopurinol. In: Florey K, ed. *Analytical Profiles of Drug Substances*, Vol. 7. London: Academic Press, pp. 1–17.

British Pharmacopoeia (2007). *British Pharmacopoeia*, Vol. 1. London: The Stationery Office, pp. 86–88.

Budavari S, ed. (1996). *The Merck Index*, 12th edn. Rahway, NJ: Merck and Co.

CHIP3 Regulations (2002). Chemicals (Hazard Information and Packaging for Supply) Regulations. SI 2002/1689.

Dressman JB, Poust RI (1983). Stability of allopurinol and of five antineoplastics in suspension. *Am J Hosp Pharm* 40: 616–618.

Gressel PD, Gallelli JF (1968). Quantitative analysis and alkaline stability studies of allopurinol. *J Pharm Sci* 57: 335–338.

Lowey AR, Jackson MN (2008). A survey of extemporaneous dispensing activity in NHS trusts in Yorkshire, the North-East and London. *Hosp Pharmacist* 15: 217–219.

McEvoy J, ed. (1999). *AHFS Drug Information.* Bethesda, MD: American Society of Health System Pharmacists.

Woods DJ (2001). *Formulation in Pharmacy Practice. eMixt, Pharminfotech*, 2nd edn. Dunedin, New Zealand. www.pharminfotech.co.nz/emixt (accessed 18 April 2006).

Amiodarone hydrochloride oral liquid

Risk assessment of parent compound

A risk assessment survey of clinical pharmacists suggested that amiodarone hydrochloride had a medium-to-narrow therapeutic index, with the potential to cause significant morbidity or death if inaccurate doses were to be administered (Lowey and Jackson, 2008). Current formulations used at NHS trusts show a low-to-moderate technical risk, based on the number and/or complexity of the required manipulations and calculations. Amiodarone is therefore regarded as a high-risk extemporaneous product (Box 14.3).

Summary

There appears to be a wide body of evidence to support the extemporaneous formulation of amiodarone hydrochloride as an oral liquid. Due to its high clinical risk and relative stability, amiodarone may be a possible candidate for purchase from a 'Specials' manufacturer. Longer shelf-lives may be possible if prepared in suitable premises.

Clinical pharmaceutics

Points to consider

- At typical paediatric strengths (10–40 mg/mL), almost all of the drug will be in suspension. **Shake the bottle to ensure uniformity of drug throughout the suspension.**
- Ora-Plus and Ora-Sweet SF are free from chloroform and ethanol.
- Ora-Plus and Ora-Sweet SF contain hydroxybenzoate preservative systems.
- Ora-Sweet SF has a cherry flavour.
- No bioavailability data are available for the extemporaneously prepared suspensions.

***Box 14.3** Preferred formulae*

Using a typical strength of 40 mg/mL as an example:

- amiodarone hydrochloride 200 mg tablets × 20
- Ora-Plus 50 mL
- Ora-Sweet SF to 100 mL.

Method guidance

Tablets can be ground to a fine, uniform powder in a pestle and mortar. A small amount of Ora-Plus may be added to form a paste, before adding further portions of Ora-Plus up to 50% of the final volume. Transfer to a measuring cylinder. The Ora-Sweet can be used to wash out the pestle and mortar before making the suspension up to 100% volume. Transfer to an amber medicine bottle.

Shelf-life

28 days refrigerated or at room temperature in amber glass. **Shake the bottle.**

Alternative

An alternative formula would be:

- amiodarone hydrochloride 200 mg tablets × 20
- methylcellulose 1% or similar to 50 mL
- Syrup BP to 100 mL.

Shelf-life

7 days at room temperature. **Shake the bottle.**

Note: This shelf-life is restricted due to concerns regarding microbial contamination. Dilution of the syrup will result in the loss of its inherent anti-microbial effect, as well as dilution of the preservatives therein.

Other notes

Suspensions containing carob bean gum have also been widely used in the NHS, based on historical advice from Sanofi-Aventis (now Sanofi-Synthelabo; Medicines Information Department, personal communication 2006) and information from Nova Laboratories Ltd (2003). The data from Sanofi-Synthelabo are no longer available.

Technical information

Structure

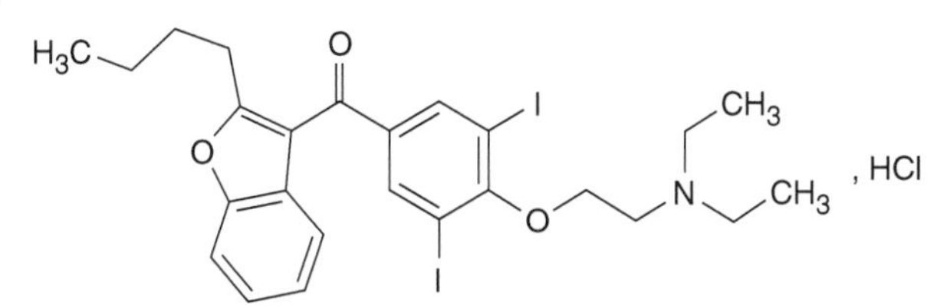

Empirical formula

$C_{25}H_{30}ClI_2NO_3$ (British Pharmacopoeia, 2007)

Molecular weight

682 (British Pharmacopoeia, 2007)

Solubility

Very slightly soluble in water (approximately 0.72 mg/mL at 25°C) (Martindale, 2005; British Pharmacopoeia, 2007).

pK_a

Reported as 5.6 (Moffat, 1986). Reported elsewhere as 6.6 (Xu and Trissel, 1999).

Optimum pH stability

Not known.

Stability profile

Physical and chemical stability

Amiodarone is a white or almost white, fine crystalline powder. It is slightly soluble in water and sparingly soluble in alcohol (British Pharmacopoeia, 2007). This limited solubility should help to minimise degradation in an extemporaneous formulation.

There is little detail available regarding potential degradation mechanisms. Alexander and Thyagarajapuram (2003) suggest that amiodarone hydrochloride is cleaved at both the ester and ketone groups during accelerated stability testing.

Amiodarone hydrochloride powder should be protected from light, and stored at a temperature not exceeding 30°C (British Pharmacopoeia, 2007). The pH of the injection formulation is reported to be 4.08 (Trissel, 2005).

Degradation products

No information on pharmacology or toxicity. Alexander and Thyagarajapuram (2003) suggest possible structures for four potential degradation products. Christopherson *et al.* (2004) suggest that (2-butyl-1-benzofuran-3-yl)(4-hydroxy-3,5-diiodophenyl)methanone (diiodo BBFA) and (2-butyl-1-benzofuran-3-yl)(4-hydroxyphenyl)methanone (BBFA) may be possible degradants.

Stability in practice

There are a range of published papers to support the stability of amiodarone when formulated as an oral liquid.

Nahata *et al.* (1997) investigated the stability of an amiodarone oral suspension derived from commercially available tablets. The tablets were crushed in a mortar and added to a vehicle comprising a 1:1 mixture of methylcellulose 1% and syrup (final strength 5 mg/mL). The suspensions were stored in 10 plastic bottles and 10 glass bottles, equally divided into storage at 4°C in a fridge and 25°C in a water bath. Samples were collected on days 0, 7, 14, 28, 42, 56, 70 and 91. Each day, three samples from each bottle were analysed under each study condition, in duplicate. A stability-indicating HPLC methodology was followed for the analysis (Christopherson *et al.*, 2004).

The amiodarone concentration remained above 90% for 91 days in both glass and plastic containers at 4°C. At 25°C, the concentration remained above 90% for at least 42 days in both glass and plastic bottles, with approximately 15% of potency lost by day 91. pH values remained within the range 4.2–4.4. Microbial growth was not assessed. Intra-day and inter-day variations appear acceptable.

Nahata *et al.* (1999) then published a further stability study using a 1:1 mixture of Ora-Plus and either Ora-Sweet or Ora-Sweet Sugar Free (SF). The authors comment on the previous study, acknowledging the difficulty of obtaining 1% methycellulose commercially in some countries. Licensed Cordarone tablets were manually crushed and added to the suspending agents to yield a 5 mg/mL suspension. The pH of the suspensions was adjusted to between 6 and 7 using sodium bicarbonate.

The suspensions were stored in plastic prescription bottles for three months. Five bottles were stored at 4°C and a further five at 25°C in a water bath. Three samples were collected from each bottle on days 0, 7, 14, 28, 56, 70 and 91 and analysed using a stability-indicating HPLC method.

Mean concentrations of the bottles stored at 25°C remained above 90% for at least 56 days, and for at least 91 days for the bottles stored in the fridge. No changes in physical appearance or odour were noted, and the pH ranged from 6.3 to 6.4 throughout the study. Taking into account the standard deviations, the authors suggest that the 25°C bottles could be allocated an expiry of 42 days rather than 56 days. There were no noticeable differences between the type of Ora-Sweet vehicle used (Ora-Sweet or Ora-Sweet SF). Microbiology studies were not carried out.

Although the authors report anecdotal successful use of the formulations, pharmacokinetic and pharmacodynamic studies have not been carried out.

Alexander and Thyagarajapuram (2003) have performed accelerated stability studies for a 20 mg/mL extemporaneous suspension of amiodarone hydrochloride, following guidelines set by the International Conference of Harmonisation. Two hundred and fifty Procarone tablets were crushed and triturated using a pestle and mortar. The suspending agent was made up of:

- 18.75 g of carboxymethylcellulose sodium (CMC) in warm water
- 18.75 g of Veegum High Viscosity (colloidal magnesium aluminium silicate) weighed and hydrated in a separate portion of distilled water overnight
- 1 L of syrup 70% w/v.

Tween 80 (1.25 mL) was added to the crushed tablets before the addition of the CMC. The Veegum was used to wash out any remaining CMC from the stock container and the suspension was mixed using an electric mixer. A 25 mL portion of paraben concentrate (methylparaben 10%, propylparaben 2%, propylene glycol to 1,000 mL) was added as a preservative.

The suspensions were transferred to amber medical glass bottles and then stored in the fridge at 4–6°C. Three bottles were maintained at each of the following temperatures and analysed weekly over 90 days: 4, 30, 40, 50 and 60°C.

Analysis and characterisation of the degradation products was undertaken using stability-indicating high-performance liquid chromatographic-mass spectrometry (HPLC-MS). The suspensions were reported not to settle out, even after four weeks storage at room temperature. A slight darkening of the suspension stored at 60°C was the only change in physical appearance.

The initial concentration was found to be 19.92 mg/mL, taken as 100%. The percentage of amiodarone hydrochloride remaining after 91 days was above 90% for the suspensions stored at 4, 30 and 40°C, though the authors do not state standard deviations. The sampling technique and number are also not stated. After 91 days, the percentage remaining in the 50 and 60°C suspensions were 80.41% and 75.52% respectively.

The shelf-life for a suspension stored at 25°C was found to be 193 days. The corresponding figure for storage at 4°C was 677 days.

Although these times are considerably longer than those calculated in the previous work by Nahata (1997), the complexity of the formulation and its preparation precludes its use in dispensaries in the UK. However, the work by Alexander and Thyagarajapuram (2003) does add further evidence for the relative stability of amiodarone when formulated as an oral suspension.

Moreover, Alexander and Thyagarajapuram (2003) investigate the basic properties of four possible degradation products, indicating possible cleavage of the amiodarone molecule at both its ester and ketone groups. The need for further work to investigate degradation mechanisms is acknowledged.

Suspensions containing carob bean gum have also been widely used in the NHS, based on advice from Sanofi-Aventis (now Sanofi-Synthelabo; Medicines Information Department, personal communication, 2006) and Nova Laboratories Ltd. The source of this advice is no longer available from Sanofi-Synthelabo.

Kopelent-Frank and Schimper (1999) investigated the stability of amiodarone hydrochloride in 5% dextrose and 0.9% sodium chloride. In the process of developing a new HPTLC method, 1 mg/mL formulations were exposed to elevated temperatures of 50°C. After 6 hours, less than 3% loss had occurred in the dextrose, with no appreciable loss in sodium chloride. Artificial sunlight (Suntest) provoked significantly more degradation. The authors suggest that amiodarone preparations should be protected from sunlight when in use. The use of amber glass bottles for the extemporaneous formulation would appear a sensible practical step to limit such degradation.

Bioavailability data

No data exist for the extemporaneous preparations mentioned above. Pourbaix *et al.* (1985) present some data for a 400 mg dose administered to healthy adults as 5 mL of a 16% glycolic solution. When compared with two different tablet formulations, the authors found the oral solution to be bioequivalent, suggesting that the incomplete bioavailability of the parent drug is a function of first-pass metabolism rather than the dissolution characteristics of the tablets. However, no stability data for the solution are presented.

References

Alexander KS, Thyagarajapuram N (2003). Formulation and accelerated stability studies for an extemporaneous suspension of amiodarone hydrochloride. *Int J Pharm Compound* 7:389–393.

British Pharmacopoeia (2007). *British Pharmacopoeia*, Vol. 1. London: The Stationery Office, pp. 127–128

Christopherson MJ, Yoder KJ, Miller RB (2004). Validation of a stability-indicating HPLC method for the determination of amiodarone HCl and its related substances in amiodarone HCl injection. *J Liq Chromatogr Relat Technol* 27:95–111.

Kopelent-Frank H, Schimper A (1999). HPTLC-based stability assay for the determination of amiodarone in intravenous admixtures. *Pharmazie* 54:542–544.

Lowey AR, Jackson MN (2008). A survey of extemporaneous dispensing activity in NHS trusts in Yorkshire, the North-East and London. *Hosp Pharmacist* 15:217–219.

Moffat AC, Osselton MD, Widdop B (2003). *Clarke's Analysis of Drugs and Poisons*, 3rd edn. London: Pharmaceutical Press. ISBN 978 0 85369 473 1.

Nahata MC (1997). Stability of amiodarone in an oral suspension stored under refrigeration and at room temperature. *Pediatrics* 31:851–852.

Nahata MC, Morosco RS, Hipple TF (1999). Stability of amiodarone in extemporaneous oral suspensions prepared from commercially available vehicles. *J Pediatr Pharm Pract* 4: 186–189.

Nova Laboratories Ltd (2003). Information Bulletin – Suspension Diluent Used at Nova Laboratories.

Pourbaix S, Berger Y, Desager J-P, Pacco M, Harvengt C (1985). Absolute bioavailability of amiodarone in normal subject. *Clin Pharmacol Ther* 37:118–123.

Sweetman SC (2009). *Martindale: The Complete Drug Reference*, 36th edn. London: Pharmaceutical Press.

Trissel LA (2005). *Handbook on Injectable Drugs*, 13th edn. Bethesda, MD: American Society of Health System Pharmacists.

Xu QA, Trissel LA (1999). *Stability-Indicating HPLC Methods for Drug Analysis*. Washington DC: American Pharmaceutical Association and London: Pharmaceutical Press.

L-Arginine hydrochloride oral liquid

Risk assessment of parent compound

A risk assessment survey completed by clinical pharmacists suggested that L-arginine hydrochloride has a wide therapeutic index, with the potential to cause significant side-effects if inaccurate doses were to be administered (Lowey and Jackson, 2008). Current formulations used at NHS trusts for the mouthwash show a low technical risk, based on the number and complexity of manipulations and calculations. L-Arginine hydrochloride is therefore regarded as a low-risk extemporaneous product (Box 14.4).

Summary

No published stability data could be found, although there are several reports of successful clinical use. Arginine hydrochloride is available as a preserved oral solution and injection solution from several 'Specials' manufacturers. These manufacturers should be able to provide adequate justification of the shelf-life they have assigned to their product. Higher strengths of the oral solution are also known to be manufactured (D. Barber, Stockport Pharmaceuticals, Stepping Hill Hospital, personal communication, 2007).

A 7-day expiry is assigned to the extemporaneous preparation due to the absence of an effective preservative system.

Clinical pharmaceutics

Points to consider

- The arginine will be in solution at usual concentrations (12.5–40%). There should be no dosage uniformity concerns.

Box 14.4 *Preferred formula*

Using a typical strength of 100 mg/mL as an example:

- L-arginine hydrochloride powder 10 g
- distilled water to 100 mL.

Shelf-life

7 days refrigerated in amber glass (unpreserved solution).

- There are no bioavailability data available for the extemporaneous preparation.

Other notes

- There are reports of the oral use of unlicensed injections produced by 'Specials' manufacturers (Guy's and St. Thomas' and Lewisham Hospitals, 2005; L. Lawrie, Bradford Hospitals NHS Trust, personal communication, 2006).

Technical information

Structure

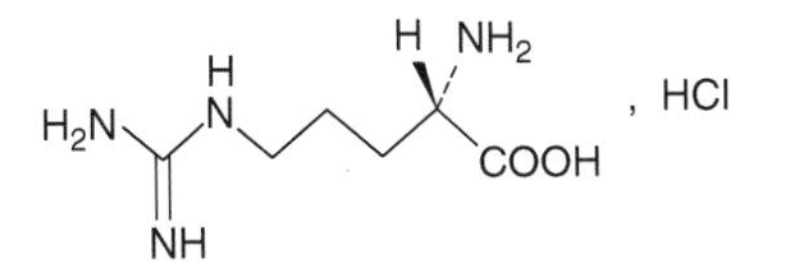

Empirical formula

$C_6H_{14}N_4O_2$,HCl (British Pharmacopoeia, 2007).

Molecular weight

210.7 (British Pharmacopoeia, 2007).

Solubility

Freely soluble in water (British Pharmacopoeia, 2007).

pK_a

Not known.

Optimum pH stability

Not known.

Stability profile

Physical and chemical stability

Arginine hydrochloride is a white or almost white, crystalline powder or colourless crystals, freely soluble in water, very slightly soluble in alcohol (British Pharmacopoeia, 2007).

There are no published reports of the route of degradation.

Arginine hydrochloride should be stored protected from light (British Pharmacopoeia, 2007).

Degradation products

No data.

Stability in practice

No published stability data could be found, although there are several reports of successful clinical use. Arginine hydrochloride is available as a preserved oral solution from several 'Specials' manufacturers. These manufacturers should be able to provide adequate justification of the shelf-life they have assigned to their product.

Bioavailability data

No data.

References

British Pharmacopoeia (2007). *British Pharmacopoeia*, Vol. 1. London: The Stationery Office, pp. 773–774.

Guy's and St. Thomas' and Lewisham Hospitals (2005). *Paediatric Formulary*. London: Guy's and St Thomas' and Lewisham NHS Trust.

Lowey AR, Jackson MN (2008). A survey of extemporaneous dispensing activity in NHS trusts in Yorkshire, the North-East and London. *Hosp Pharmacist* 15: 217–219.

Azathioprine oral liquid

Risk assessment of parent compound

Azathioprine has a very high clinical risk, with a narrow therapeutic window and the potential to cause death if inaccurate dosages are administered. Formulations vary considerably across the country from the simple to complex. Current formulations used at NHS trusts include those with a high technical risk, based on the number and complexity of calculations and manipulations. Azathioprine is therefore regarded as a very high-risk extemporaneous product (Box 14.5).

Summary

There is a wide body of evidence to support the stability and formulation of azathioprine as an oral suspension. Given the stability of the drug, its cytotoxic nature, and the associated high clinical risk, the commissioning of azathioprine suspensions from a 'Specials' manufacturer should be considered. This may help improve the reproducibility of the production process.

Given the narrow therapeutic index, it may be prudent to monitor the patient more closely than usual.

The crushing of cytotoxic tablets carries significant health and safety implications and should only be undertaken after a local risk assessment, and where the appropriate equipment is available.

Azathioprine is known to be available as a suspension from 'Specials' manufacturers. The formulation of these products should be checked by a suitably competent pharmacist before purchase. Extemporaneous preparation should only be considered where the 'Special' is inappropriate or unavailable.

Clinical pharmaceutics

Points to consider

- Azathioprine is insoluble and will therefore be held in suspension. Dosage uniformity will therefore be an issue. **Shake the bottle before use to improve dosage uniformity.**
- Ora-Plus and Ora-Sweet SF are free from chloroform and ethanol.
- Ora-Plus and Ora-Sweet SF contain hydroxybenzoate preservative systems.

***Box 14.5** Preferred formulae*

Note: A 'Special' is now available from a reputable manufacturer in the UK. Extemporaneous preparation should only occur, therefore, where there is a demonstrable clinical need.

Using a typical strength of 5 mg/mL as an example:

- azathioprine 50 mg tablets × 10
- Ora-Plus to 50 mL
- Ora-Sweet SF to 100 mL.

Method guidance

Tablets can be ground to a fine, uniform powder in a pestle and mortar. A small amount of Ora-Plus may be added to form a paste, before adding further portions of Ora-Plus up to 50% of the final volume. Transfer to a measuring cylinder. The Ora-Sweet can be used to wash out the pestle and mortar before making the suspension up to 100% volume. Transfer to an amber medicine bottle.

Shelf-life

28 days at room temperature in amber glass. **Shake the bottle.**

Alternative

An alternative formula would be:

- azathioprine 50 mg tablets × 10
- xanthan gum 0.5% (or similar) with preservative to 100 mL.

Note: Caution cytotoxic – handling precautions apply.

Shelf-life

28 days at room temperature in amber glass. **Shake the bottle.**

- 'Keltrol' is made to slightly different specifications by several manufacturers, but typically contains ethanol and chloroform. The exact specification should be assessed before use by a suitably competent pharmacist. Ethanol is a CNS depressant. Chloroform has been classified as a class 3 carcinogen (i.e. substances that cause concern owing to possible carcinogenic effects but for which available information is not adequate to make satisfactory assessments) (CHIP3 Regulations, 2002).

- There are no bioavailability data available for extemporaneous preparations.

Technical information

Structure

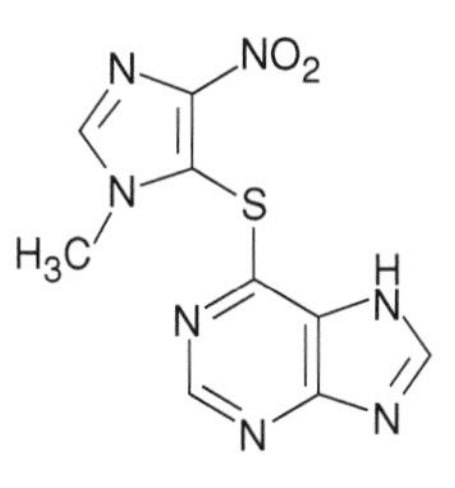

Empirical formula

$C_9H_7N_7O_2S$ (British Pharmacopoeia, 2007).

Molecular weight

277.3 (British Pharmacopoeia, 2007).

Solubility

Practically insoluble in water (Lund, 1994).

pK_a

8.2 at 25°C (Lund, 1994). Also cited as 7.87 (Connors *et al.*, 1986).

Optimum pH stability

5.5–6.5 (Mitra and Narurkar, 1986a; Lund, 1994).

Stability profile

Physical and chemical stability

Azathioprine is a pale-yellow powder, practically insoluble in water and in alcohol (British Pharmacopoeia, 2007). The limited solubility of azathioprine should help to minimise the drug's susceptibility to degradation mechanisms when formulated as an oral liquid.

The primary route of degradation is cleavage of the sulfide bond which links the imidazole and purine rings, yielding either 6-mercaptopurine and 1-methyl-4-nitro-5-hydroxyimidazole or hypoxanthine and 1-methyl-4-nitro-5-thioimidazole. Further degradation steps then occur. Stability may be maximised by avoiding conditions which are known to cleave this bond: alkaline pH, intense ultraviolet light and high temperatures (Connors *et al.*, 1986).

Singh and Gupta (1988) found that kinetic studies at pH 7 or above produced colourless samples, while those below pH 7 were light yellow to yellow in colour. The authors suggest that the colourless samples contained 6-mercaptopurine and the associated degradant 1-methyl-4-nitro-5-hydroxyimidazole, while the yellow samples contained hypoxanthine and 1-methyl-4-nitro-5-thioimidazole, the latter being responsible for the yellow colour. The imidazole appears to be relatively unstable and may degrade further to non-chromophoric products (Gupta *et al.*, 1987).

Gupta *et al.* (1987) investigated the breakdown of azathioprine dissolved in buffer solutions of known pH and stored at 80°C final drug strength 10^{-2} M). Comparing solutions at pH 4.35 and pH 11.2, the authors surmise that the dissimilarity in the spectral transformations suggests that azathioprine degrades in the two pH regions through different reaction pathways. Thin-layer chromatography (TLC) studies were reported to confirm the spectral observations.

Azathioprine is soluble in dilute solutions of alkali hydroxides, allowing the drug to undergo slow decomposition to mercaptopurine (Connors *et al.*, 1986). Elion *et al.* (1961) suggested that the rate of decomposition increases as the alkali concentration increases. However, when formulated in acid or neutral conditions, the poor solubility of azathioprine will help to limit degradation rates.

Data produced by Fell and Plag (1979) suggest that exposure to light may increase degradation rates. Moreover, it is recommended that azathioprine BP in the solid state is stored in light-resistant containers (Lund, 1994).

Mitra and Narurkar (1986a) have investigated the kinetics of the breakdown of azathioprine to 6-mercaptopurine and 1-methyl-4-nitro-5-hydroxyimidazole in buffered aqueous solution at 73°C over the pH range 1 to 13 at an ionic strength of 0.5. Maximum stability was achieved at pH 5.5–6.5, with degradation rates increasing significantly at pH <4 and >8. When the degradation rate was determined at 55, 65, 73 and 85°C at pH 1.2 and ionic strength 0.5, the activation energy was calculated as 80.48 kJ/mol.

The presence of sulfhydryl compounds such as cysteine, glutathione and hydrogen sulfide may also result in hydrolysis (Bresnick, 1959; Mitra and Narurkar, 1986b; Lund, 1994; Trissel, 2000). Metabolic studies have revealed that the breakdown of azathioprine to 6-mercaptopurine occurs mainly in the red blood cells and in the liver cells which are rich in glutathione (de Miranda *et al.*, 1973).

Azathioprine base is used for the tablets, but the sodium salt is used for injections. The typical pH of azathioprine sodium injection is 9.8–11 (Lund, 1994). Azathioprine should be stored protected from light (British Pharmacopoeia, 2007).

Degradation products

The main degradation product found is the cytotoxic agent 6-mercaptopurine. Other degradants include 1-methyl-4-nitro-5-hydroxyimidazole, hypoxanthine and 1-methyl-4-nitro-5-thioimidazole.

Stability in practice

Several studies have demonstrated the relative stability of azathioprine suspension. Shelf-lives of in excess of one month at room temperature have been documented.

Allen and Erickson (1996) have investigated the stability of azathioprine 50 mg/mL in three different suspending vehicles: a 1:1 mixture of Ora-Plus and Ora-Sweet, a 1:1 mixture of Ora-Plus and Ora-Sweet SF, and a simple cherry syrup. The source of the drug was commercially available 50 mg tablets, crushed and ground to a fine powder in a mortar.

Sufficient product was prepared for six 120 mL amber clear plastic (polyethylene terephthalate) prescription bottles with a low-density polyethylene lining. Half the products were stored at 5°C and half at 25°C, protected from light. After agitation on a rotating mixer for 30 minutes, 5 mL samples were removed on days 0, 1, 2, 7, 10, 14, 28, 35 and 60. The authors do not describe where the sample was removed from. In addition to stability-indicating HPLC analysis, subjective checks were also made on appearance and odour. The validity of the HPLC method was demonstrated by degrading drug samples using heat, acid, base, oxidising agent and light; the chromatograms showed that the degradants did not interfere with the detection of the parent drug. Two injections were made for each of the three samples ($n = 6$) and the samples were diluted in mobile phase before injection.

No changes were noted in the physical appearance of the suspensions over the 60-day period. The initial pH values were 4.3 for the Ora-based vehicles and 3.5 for the cherry syrup suspension. These values changed by less than 0.5 pH units throughout the study.

All of the formulations retained at least 96% of the initial concentration, regardless of storage temperature. The intra-day and inter-day coefficients of variation were 1.3% and 1.4%, respectively. Microbial growth was not studied.

Dressman and Poust (1983) studied the stability of 50 mg/mL azathioprine in cologel (methylcellulose), brought to final volume with a flavouring agent consisting of a 2:1 mixture of simple syrup and wild cherry syrup. The suspensions were prepared extemporaneously using azathioprine tablets. Thorough mixing was achieved by manual shaking for about 30 seconds until the powder was wetted, followed by ultrasonication until the powder was well dispersed. Only one suspension was prepared for storage at 5°C and one for storage at ambient room temperature. Samples were removed on days 0, 7, 28 and 56 for the suspension stored at ambient temperature and days 0, 56 and 84 for the suspension stored in the fridge.

A corresponding suspension without the active drug was prepared and subjected to the test method to ensure that the non-active ingredients did not interfere with the assay. However, the assay used for azathioprine was a semi-quantitative ultraviolet scan, based on the premise that studies of degradation in azathioprine solutions at Burroughs Wellcome showed a dramatic decrease in the UV absorbance at 280 nm as azathioprine decomposes, with a new peak appearing at 240 or 320 nm depending on the pH of the medium.

No new peaks occurred at either 240 or 320 nm and no change was found in the UV maximum wavelength. The authors suggest that the suspensions retained 97% and 99.6% of the labelled strength for 56 days at room temperature and for 84 days at 5°C, respectively. Although the semi-quantitative nature of the assay must be regarded as a limitation for this piece of research, the results correlate with the data published by Allen and Erickson (1996).

UV spectroscopy was also used in a study conducted by Escobar-Rodriguez *et al.* (1995). A 10 mg/mL strength was prepared using methylcellulose, simple syrup and lemon essence. The suspensions were reported to be stable for 50 days when stored at room temperature (25°C) or in the fridge (4–8°C). However, the suspension stored at room temperature did show considerable inter-day variation.

Stability-indicating HPLC analysis demonstrated that a 1 mg/mL solution of azathioprine in 0.02 M sodium hydroxide was stable for 8 days when stored in the dark at ambient temperatures (Fell and Plag, 1979). The stability was reported to be reduced to 4 days when the solution was exposed to diffuse daylight.

The following excipients are reported to have been used in suspensions of azathioprine: carmellose sodium, cologel, glycerol, methylcellulose, methyl hydroxybenzoate, syrup, vanillin and wild cherry syrup (Lund, 1994).

Bioavailability data

No information available.

References

Allen LV Jr, Erickson MA III (1996). Stability of acetazolamide, allopurinol, azathioprine, clonazepam, and flucytosine in extemporaneously compounded oral liquids. *Am J Health Syst Pharm* 53: 1944–1949.

Bresnick E (1959). The metabolism in-vitro of antitumour imidazolyl derivatives of mercaptopurine. *Fed Proc* 18: 371.

British Pharmacopoeia (2007). *British Pharmacopoeia*, Vol. 1. London: The Stationery Office, p. 198.

CHIP3 Regulations (2002). Chemicals (Hazard Information and Packaging for Supply) Regulations. SI 2002/1689.

Connors KA, Amidon GL, Stella ZJ (1986). *Chemical Stability of Pharmaceuticals: A Handbook for Pharmacists*, 2nd edn. New York: Wiley Interscience.

de Miranda P, Beacham LM III, Creagh TH, Elion GB (1973). The metabolic fate of the methyl nitroimidazole moiety of azathioprine in the rat. *J Pharmacol Exp Ther* 187: 588–601.

Dressman JB, Poust RI (1983). Stability of allopurinol and of five antineoplastics in suspension. *Am J Hosp Pharm* 40: 616–618.

Elion GB, Callahn S, Bieber S, Hitchings GH, Rundles RW (1961). A summary of investigations with 6-[(1-methyl-4-nitro-5-imidazolyl)thio]purine (B.W. 57–322). *Cancer Chemother Rep* 14: 93.

Escobar-Rodriguez I, Ferrari-Piquero JM, Ordonez-Soto A, Herreros de Tejada A, Garcia-Benayas C (1995). Oral azathioprine suspension for paediatric use: preparation and stability (Spanish). *Farm Hosp* 19: 219–221.

Fell AF, Plag SM (1979). Stability-indicating assay for azathioprine and 6-mercaptopurine by reversed-phase high-performance liquid chromatography. *J Chromatogr* 186: 691.

Gupta RL, Kumar M, Singla RK, Singh S (1987). Degradative behaviour of azathioprine in aqueous solutions. *Indian J Pharm Sci* 49: 169–171.

Lowey AR, Jackson MN (2008). A survey of extemporaneous dispensing activity in NHS trusts in Yorkshire, the North-East and London. *Hosp Pharmacist* 15: 217–219.

Lund W, ed. (1994). *The Pharmaceutical Codex: Principles and Practice of Pharmaceutics*, 12th edn. London: Pharmaceutical Press.

Mitra AK, Narurkar MM (1986). Kinetics of azathioprine degradation in aqueous solution. *Int J Pharm* 35: 165–171.

Mitra AK, Narurkar MM (1986). Effect of mercaptan nucleophiles on the degradation of azathioprine in aqueous solution. *In J Pharm* 28: 119–124.

Singh S, Gupta RL (1988). A critical study on degradation of azathioprine in aqueous solutions. *Int J Pharm* 42: 263–266.

Trissel LA, ed. (2000). *Stability of Compounded Formulations*, 2nd edn. Washington DC: American Pharmaceutical Association.

Bendroflumethiazide oral liquid

Risk assessment of parent compound

A risk assessment survey completed by clinical pharmacists suggested that bendroflumethiazide has a wide to medium therapeutic index, with the potential to cause minor side-effects if inaccurate doses were to be administered (Lowey and Jackson, 2008). Current formulations used at NHS trusts generally show a low to medium technical risk, based on the number and/or complexity of manipulations and calculations. Bendroflumethiazide is therefore regarded as a medium-risk extemporaneous preparation (Box 14.6).

Summary

NHS centres have a reasonable amount of experience to support the successful clinical use of bendroflumethiazide oral liquid. The drug is likely to be insoluble in these preparations and significant degradation is unlikely. However, due to the paucity of data, a conservative shelf-life of 7 days is recommended.

***Box 14.6** Preferred formula*

Using a typical strength of 500 micrograms/mL as an example:

- bendroflumethiazide tablets 2.5 mg × 20 (or 5 mg × 10)
- xanthan gum 0.5% (with preservative) or similar to 100 mL.

Method guidance

Tablets can be ground to a fine, uniform powder in a pestle and mortar. A small amount of xanthan gum may be added to form a paste, before adding further portions up to 50% of the final volume. Transfer to a measuring cylinder. The remaining xanthan gum can be used to wash out the pestle and mortar before making the suspension up to 100% volume. Transfer to an amber medicine bottle.

Shelf-life

7 days at room temperature in amber glass. **Shake well before use.**

Patients should be monitored carefully and consideration should be given to alternative treatment strategies until further data are available.

Clinical pharmaceutics

Points to consider

- No published stability data are available to support formulation stability – monitor patient carefully.
- Bendroflumethiazide is insoluble and will therefore be held in suspension at a typical strength of 500 micrograms/mL. Dosage uniformity may therefore be an issue. **Shake the bottle before use to improve dosage uniformity.**
- 'Keltrol' (xanthan gum suspending agent) is made to slightly different specifications by several manufacturers, but typically contains ethanol and chloroform. The exact specification should be assessed before use by a suitably competent pharmacist. Ethanol is a CNS depressant. Chloroform has been classified as a class 3 carcinogen (i.e. substances that cause concern owing to possible carcinogenic effects but for which available information is not adequate to make satisfactory assessments) (CHIP3 Regulations, 2002).
- Bendroflumethiazide has a slight floral odour (Florey and Russo-Alesi, 1976).
- No information regarding toxicity or activity of breakdown products is available.
- No bioavailability information is available.
- Alternative diuretics are also available, depending on the indication for use. Some may be available as 'Specials' or imports. Alternatives also include the halving of tablets and dispersion in water if needed (A Fox, Principal Pharmacist, Southampton University Hospitals NHS Trust, personal communication, 2007).

Technical information

Structure

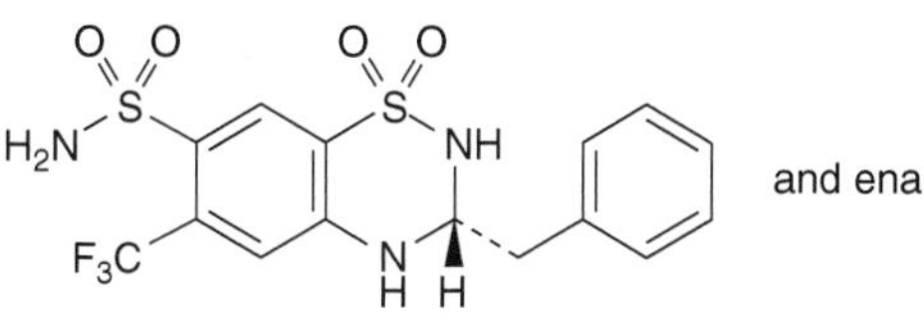

Empirical formula

$C_{15}H_{14}F_3N_3O_4S_2$ (British Pharmacopoeia, 2007).

Molecular weight

424.1 (British Pharmacopoeia, 2007).

Solubility

Practically insoluble in water; soluble in ethanol (British Pharmacopoeia, 2007).

pK_a

8.5 (25°C) (Moffat, 1986).

Optimum pH stability

Not known.

Stability profile

Physical and chemical stability

Bendroflumethiazide is a white, or almost white, crystalline powder. It is practically insoluble in water but freely soluble in acetone or alcohol (British Pharmacopoeia, 2007).

Benzothiadiazines are known to degrade via hydrolysis (Mollica, 1971). Perlman *et al.* (1984) also suggest that bendroflumethiazide is sensitive to light.

Degradation products

These may include hydroflumethiazide (3,4-dihydro-6-trifluoromethyl-2H-1,2,4-benzothiadiazine-7-sulfonamide 1,1-dioxide) and 5-trifluoromethyl-2,4-disulfamoylaniline (TFSA) (Perlman *et al.*, 1984; Frontini and Mielck, 1992). Hassan (1983) suggests that the latter is not pharmacologically active.

TFSA has been described elsewhere in alternative nomenclature as 4-amino-6-trifluoromethyl-1,3-benzenedisulfonamide (Florey and Russo-Alesi, 1976; Hennig *et al.*, 1981; Barnes and Nash, 1993, 1994).

Stability in practice

There are no known published or unpublished studies to provide evidence for the stability of bendroflumethiazide when formulated as an oral liquid.

As the solid, bendroflumethiazide has been exposed to 60°C for two weeks with no decomposition measured via IR and modified Bratton-Marshal reaction (Florey and Russo-Alesi, 1976). Florey and Russo-Alesi (1976) also report that bendroflumethiazide in ethanol (1 mg/mL) showed 25% decomposition at the end of two weeks at 60°C. In aqueous suspension, bendroflumethiazide is reported to undergo almost complete breakdown to the disulfonamide by the end of one week at 60°C.

In solution at pH 12, bendroflumethiazide undergoes complete hydrolysis to the disulfonamide in 1 hour at 35°C. However, Turner *et al.* (1970) also report bendroflumethiazide to be unstable in certain acid conditions, such as when treated with nitrous acid and/or heat.

Frontini and Mielck (1992) have described a method for the determination and quantitation of bendroflumethiazide and its degradation products using HPLC. They suggest that a low pH (pH 2), combined with the use of dark flasks, are important in maintaining the stability of samples during automatic injection. The solvent for samples was prepared by dissolving 750 mg of KCl in 10 mL of 1 N HCl and about 400 mL of distilled water in a 1,000 mL volumetric flask, adding 400 mL of methanol, mixing under cooling and diluting with water to volume (MeOH/buffer 40% v/v).

Perlman *et al.* (1984) have commented on the problems of lack of stability of liquid samples for HPLC, stating that conditions must be chosen to minimise or eliminate *in situ* formation of disulfonamide. The stability of bendroflumethiazide in methanol (in the dark) was confirmed by observing no change in the disulfonamide content after 60 hours. Barnes and Nash (1993) found that methanolic standards and samples were generally stable for up to 5 days at ambient temperature in the dark followed by 1 day in diffuse light in full clear glass autosampler vials. However, stability was variable with part-filled vials, with corresponding production of 4-amino-6-trifluoromethyl-1,3-benzenedisulfonamide.

Related preparations

Barnes and Nash (1994) have reported the stability of bendroflumethiazide in low-dose extemporaneously prepared 1.25 mg capsules. The capsules were formulated with the idea of administering the drug by opening the capsule shell and dispersing the contents in an infant's feed. Bendroflumethiazide tablets were crushed and ground and then diluted with lactose before manual addition to hard gelatine capsule shells. The capsules were then stored at ambient temperature exposed to light or 75% relative humidity, and at 45 or 60°C. Analysis was via a stability-indicating reverse-phase HPLC method (Barnes and Nash, 1993).

The content of one capsule was dispersed with stirring into 25 mL of water and a qualitative assessment of each dispersion was made. The capsule contents remained dispersed in water for capsules stored up to 1 year for all conditions except at 45 and 60°C, where the contents of the capsules aggregated and were difficult to re-disperse in water after three months, resulting in the liquid having a yellow discoloration. The extraction of powder from each capsule was also assessed, and the values were greater than 95% of the theoretical weight.

Insufficient degradation was observed to allow the kinetic order to be determined. Average estimates of the decreases in bendroflumethiazide assayed over the length of the study (1 year) ranged from 4% to 7%, depending on the storage conditions. However, the amount of bendroflumethiazide lost was not accounted for by a corresponding increase in the expected degradation product, 4-amino-6-trifluoromethyl-1,3-benzenedisulfonamide. The authors therefore suggest an unknown route of degradation.

Data analysis by the authors suggests a shelf-life for the capsules of more than seven months for both the light-exposed and 75% relative humidity conditions. This high level of stability as a solid is expected, as the solid has been exposed to 60°C for two weeks with no decomposition measured via IR and modified Bratton-Marshal reaction (Florey and Russo-Alesi, 1976).

Bioavailability data

No data available.

References

Barnes AR, Nash S (1993). HPLC determination of bendrofluazide in capsules. *Int J Pharm* 94: 231–234.

Barnes AR, Nash S (1994). Stability of bendrofluazide in a low-dose extemporaneously prepared capsule. *J Clin Pharm Ther* 19: 89–93.

British Pharmacopoeia (2007). *British Pharmacopoeia*, Vol. 1. London: The Stationery Office, pp. 224–225.

Florey K, Russo-Alesi FM (1976). In: Florey K, ed. *Analytical Profiles of Drug Substances*, Vol. 5. London: Academic Press, pp. 1–19 (erratum 1977, Vol. 6, p. 598).

Frontini R, Mielck JB (1992). Determination and quantitation of bendroflumethiazide and its degradation products using HPLC. *J Liq Chromatogr* 15: 2519–2528.

Hassan SM (1983). A stability-indicating assay for bendrofluazide using high-performance liquid chromatography. *Chromatography* 17: 101–103.

Hennig UG, Moskalyk RE, Chatten LG, Chan SF (1981). Semiaqueous potentiometric determination of apparent pK_a values for benzothiadiazines and detection of decomposition during stability during solubility variation with pH studies. *J Pharm Sci* 70: 317–319.

Lowey AR, Jackson MN (2008). A survey of extemporaneous dispensing activity in NHS trusts in Yorkshire, the North-East and London. *Hosp Pharmacist* 15: 217–219.

Moffat AC, Osselton MD, Widdop B (2003). *Clarke's Analysis of Drugs and Poisons*, 3rd edn. London: Pharmaceutical Press. ISBN 978 0 85369 473 1

Mollica JA, Rehm CR, Smith JB, Govan HK (1971). Hydrolysis of benzothiadiazines. *J Pharm Sci* 60: 1380.

Perlman S, Szyper M, Kirschbaum JJ (1984). High-performance liquid chromatographic analysis of nadolol and bendroflumethiazide combination tablet formulations. *J Pharm Sci* 73: 259–261.

Turner JC, Nichols AW, Sloman JE (1970). Free primary aromatic amine or bendrofluazide? *Pharm J* June 6: 622.

Captopril oral liquid

Risk assessment of parent compound

A risk assessment survey completed by clinical pharmacists suggested that captopril has a wide therapeutic index, but with the potential to cause significant side-effects if inaccurate doses were to be administered (Lowey and Jackson, 2008). Current formulations used at NHS trusts generally show a low technical risk, based on the number and/or complexity of manipulations and calculations. Captopril is therefore generally regarded as a medium-risk extemporaneous preparation (Box 14.7).

Box 14.7 *Preferred formula*

Using a typical strength of 1 mg/mL as an example:

- captopril tablets 25 mg × 4
- Ora-Plus to 50 mL
- Ora-Sweet to 100 mL.

Method guidance

Tablets can be ground to a fine, uniform powder in a pestle and mortar. A small amount of Ora-Plus may be added to form a paste, before adding further portions of Ora-Plus up to 50% of the final volume. Transfer to a measuring cylinder. The Ora-Sweet can be used to wash out the pestle and mortar before making the suspension up to 100% volume. Transfer to an amber medicine bottle.

Shelf-life

7 days refrigerated in amber glass. **Shake the bottle**

Alternative

An alternative formula would be:

- captopril tablets 25 mg × 4
- xanthan gum suspending agent 0.5% ('Keltrol') to 100 mL.

Shelf-life

7 days refrigerated in amber glass. **Shake the bottle.**

Summary

The evidence for aqueous formulations of captopril is wide ranging and often conflicting. The stability of the drug is affected by a number of variables, which makes accurate comparison of the published studies problematic. The variation in captopril formulations used to treat children with heart failure has been acknowledged in the medical and general press (Mulla *et al.*, 2007; BBC News Online, 2007).

Licensed formulations of captopril liquid are reported to be available in some countries, including Australia (Woods, 2001; Mulla *et al.*, 2007; H. Pretorius, IDIS Importers, personal communication, 2007). Importation of these preparations should be considered as a priority. The appropriateness of the formulation for the target patient should be assessed by a suitably competent pharmacist before use.

Purchase of captopril as an oral liquid from a 'Specials' manufacturer may also be considered. The manufacturer should be able to provide evidence to support their formulation. Oily (non-aqueous) preparations, solutions, suspensions, low-strength tablets and powders for reconstitution are known to be available. The evidence to support formulations from 'Specials' manufacturers should be assessed by a suitably competent pharmacist before purchase.

As an interim measure only, a simple extemporaneous formulation may be considered. A limited shelf-life should be allocated in order to minimise risk. Alternative ACE inhibitors may also be considered for use in some clinical situations.

Clinical pharmaceutics

Points to consider

- Captopril should be in solution at all usual concentrations, given its solubility.
- Captopril possesses two *S,S* optically active centres (Kadin, 1982). It also exists in two polymorphic forms, with different physical properties and melting points (Kadin, 1982).
- It would appear that the excipients in a captopril formulation can have a significant impact on drug degradation rates. The exact details of the formulation must be approved before use. Any changes to the formulation will require re-validation.
- Captopril has a strong sulfur taste and odour (Lye *et al.*, 1997). Aqueous solutions of captopril in water retain this sulfur taste and may not be considered to be palatable (Woods, 2001). The sulfur taste and odour do not necessarily indicate captopril degradation.

- Captopril degradation is known to be inversely proportional to drug concentration (Connors *et al.*, 1986).
- There are no known data available for bioavailability.

Technical information

Structure

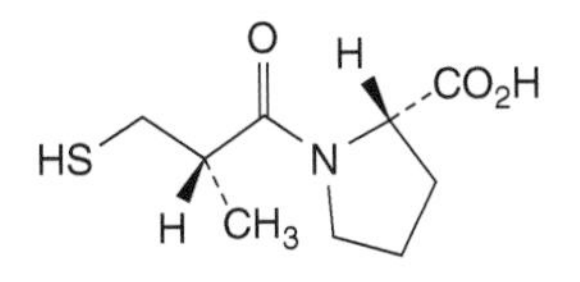

Empirical formula

$C_9H_{15}NO_3S$ (British Pharmacopoeia, 2007).

Molecular weight

217.3 (British Pharmacopoeia, 2007).

Solubility

160 mg/mL at 25°C (Kadin, 1982).

pK_a

3.7 (carboxyl group); 9.8 (sulfhydryl group) (Kadin, 1982).

Optimum pH stability

Less than 3.5 (Kadin, 1982).

Stability profile

Physical and chemical stability

Captopril is a white or almost white crystalline powder. It is freely soluble in water and in methanol, and may have a characteristic sulfide-like odour (Martindale, 2005; British Pharmacopoeia, 2007). It dissolves in solutions of alkali hydroxides (British Pharmacopoeia, 2007). Under ordinary conditions, captopril is not hygroscopic (Kadin, 1982). Above 40°C, captopril shows extraordinarily high water solubility (Kadin, 1982).

As a solid, captopril exhibits excellent stability, with no significant decomposition detected for up to six months in bulk samples stored at 5, 33 and 50°C, or when exposed to 900 foot-candles in a light box for 30 days (Shilkowski, 1982). Captopril powder should be stored in an airtight container (British Pharmacopoeia, 2007).

When in solution, captopril undergoes an oxygen-facilitated, first-order, free radical oxidation at its thiol, yielding captopril disulfide (Kadin, 1982). The reaction is complex (a combination of autooxidation and metal ion-catalysed oxidation) and the postulated reaction mechanism is based on the ionisation of the thiol function only (Connors *et al.*, 1986). Although hydrolysis at the amide link is possible, this only occurs when forced. Oxidation is reported to be delayed by lowering pH, adding chelating or antioxidant agents, increasing the captopril concentration, and the utilisation of nitrogen or low oxygen head spaces (Kadin, 1982). Oxidation is also reported to occur less readily in methanol, with no degradation evident for a 40 micrograms/mL solution stored for up to two weeks at 5°C (Roberts, 1982).

Kadin (1982) quotes oxidation rate constants at various pH measurements, suggesting that captopril is most stable in solution at a pH under 3.5. Above pH 4, rate constants rise rapidly.

The oxidation is catalysed by transition metal ions through a recycling of oxygen free radicals (Wang, 1982). Copper and iron are reportedly the most effective catalysts, and as little as 1 ppm of copper has been observed to catalyse oxidation of captopril in solution (Wang, 1982). Diethylenetriamine pentaacetic acid (DTPA), disodium edetate (EDTA), citric acid and oxalic acid have all been reported to have been used as stabilising agents (Kadin, 1982).

The effect of concentration on the stability of the solution is marked. No significant degradation was found in a 250 mg/mL solution at pH 12.5–14 after being left overnight in uncovered beakers. However, a 25–50 microgram/mL solution at pH 13.5 stored overnight at room temperature in open tubes lost 84% of its sulfhydryl activity (Kadin, 1982). This work has been confirmed since by other authors (Lee and Notari, 1987; Chen *et al.*, 1985; Matthew and Das Gupta, 1996). The stoichiometry of the oxidative breakdown of captopril shows that 2 moles of captopril are lost for every 0.5 mole of oxygen consumed. Therefore, increased stability is found with a higher captopril concentration (Matthew and Das Gupta, 1996).

Lee and Notari (1987) studied the kinetics and mechanism of captopril oxidation in aqueous solution under controlled oxygen partial pressure (90–760 mmHg). Captopril disulfide was found to be the only degradation product. As the drug concentration decreased, the reaction changed from first order to zero order. This threshold concentration was in turn dependent on pH, oxygen partial pressure and cupric ion concentration.

Chen *et al.* (1995) studied the initial degradation rates of low concentrations of captopril (0.1 mg/mL) in acetate, citric and phosphate buffer solutions with different buffer concentrations at 80°C and pH 6.0. The degradation appeared to be first order in nature, with rates of degradation increasing with increasing buffer concentrations. Degradation rates were lower with the citrate buffer, suggesting that the chelating effect of this agent may be useful for improving the stability of captopril in solution.

Degradation products

The major oxidative degradation product is the dimer, captopril disulfide.

Stability in practice

There is a significant amount of stability data available for captopril preparations. These data are complex and conflicting in nature.

Captopril in water

The data for captopril solutions is wide-ranging and contains conflicting opinions. Research has included investigations of the stability of captopril in tap water (Pereira and Tam, 1992; Pramar *et al.*, 1992; Anaizi and Swenson, 1993; Escribano-Garcia *et al.*, 2005). The usefulness of this work is severely limited by the geographical differences in the composition of tap water around the world. Some of the work is also limited by ad hoc sampling routines, small numbers of samples and a lack of detail.

Further research has been conducted on the stability of captopril in sterile and/or distilled water. However, despite the more reproducible standards of the water vehicles, stability results still show a marked variation.

A range of researchers have analysed the stability of captopril at a concentration of 1 mg/mL in sterile water for irrigation or distilled water (Andrews and Essex, 1986; Anaizi and Swenson, 1993; Nahata *et al.*, 1994). The shelf-lives recommended varied from 3 to 14 days when stored in the fridge (4 or 5°C). However, the 14-day recommendation by Nahata *et al.* (1994) may be open to some doubt due to large standard deviations.

Other researchers have recommended much longer shelf-lives for similar formulations. Using various pH buffers, Schlatter *et al.* (1997) investigated the stability of 1 mg/mL solutions made from triturated tablets in sterile water. At pH 3 and pH 5, the solution was found to be stable for at least 28 days at 4°C. Using pharmaceutical grade powder as the raw material, Escribano-Garcia *et al.* (2005) found a 1 mg/mL captopril solution to be stable for 30 days at 4°C, when stored in PVC containers and protected from light. Nahata *et al.* (1994) also showed a significant beneficial effect for the addition of sodium ascorbate.

However, many of the papers show inconsistencies and unexpected or conflicting results with regard to the presence or absence of excipients and concentration effects. For example, Chan *et al.* (1994) found degradation did not follow the hypothesis that degradation is inversely proportional to concentration. The authors suggest that the increased amount of tablet excipients at higher strengths may be partially responsible.

The use of water-based captopril solutions is not recommended. Pharmaceutically, there are many disadvantages to this approach, including problems with insoluble tablet excipients and an unacceptable taste. Near-to-patient manipulation of captopril tablets (e.g. dispersing a tablet in water at the bedside) may be an option for a limited number of patients. However, many patients require part-tablet doses, for which this approach is not acceptable.

Captopril in syrup

Various researchers have studied the stability of captopril in syrup and other sugar solutions, on occasion mixed with other agents such as methylcellulose (Nahata *et al.*, 1994; Lye *et al.*, 1997; Liu *et al.*, 1999). Although promising data are found in some of the papers, the formulations are often too complex to be considered for extemporaneous preparation. Controversial findings are also seen. For example, Lye *et al.* (1997) found that tablet-based formulations were more stable than those based on pharmaceutical grade powder.

The use of syrup as a vehicle for captopril is not recommended. It has poor suspending capabilities with regard to insoluble tablet excipients, some of which may bind active drug. Therefore, dose uniformity is a serious concern. Moreover, syrup formulations may be vulnerable to microbial contamination when diluted with other agents. Any dilution steps also carry the risk of the introduction of unwanted metal ions, thus catalysing drug degradation. The use of syrup-based formulations also carries long-term dental health concerns.

Captopril in a suspending agent

The use of a suspending agent has some distinct advantages. If tablets are used as the raw material source, as is common practice in the NHS, an appropriate agent will suspend insoluble excipients as well as any drug bound to these particles. Pre-prepared suspending agents can also be buffered to improve drug stability, and flavoured to mask the unpleasant taste of captopril. The thickening effect of suspending agents may also be useful for some patients.

Allen and Erickson (1996) investigated the stability of captopril 750 micrograms/mL in three different vehicles, using tablets as the raw material.

- 1:1 mixture of Ora-Sweet and Ora-Plus
- 1:1 mixture of Ora-Sweet Sugar Free (SF) and Ora-Plus
- cherry syrup (cherry syrup concentrate diluted 1:4 with simple syrup).

The initial pH measurements of the suspensions were in the range 4.1–4.2 and changed by less than 0.5 pH units during the study.

Significant degradation occurred in all of the bottles over 60 days at 5°C and 25°C. The degradation was most apparent in the cherry syrup formulation, which maintained stability for only 2 days. The Ora-Plus and Ora-Sweet formulation displayed the best stability, maintaining 93.2% (± 1.1%) drug concentration by day 14 when stored at 5°C, and 90.3% (± 0.6%) by day 10 at 25°C.

Given that the bottles were only sampled every few days, and that this would not have simulated true 'in-use' conditions and normal exposure to the air, it would seem prudent to recommend a conservative shelf-life (e.g. 7 days expiry for captopril 750 micrograms/mL in Ora-Plus 1:1 Ora-Sweet (or Ora-Sweet SF) stored at 5°C). The low concentration of the suspension may have contributed to the limited stability, given that captopril degradation is known to be inversely proportional to drug concentration (Connors *et al.*, 1986).

Two unpublished studies were carried out by a regional NHS Quality Control Laboratory in 1992 and 1995 (B. McBride, personal communication, 2006), using 'Keltrol' as a suspending agent. Concentrations of 2.5 mg and 5 mg/mL were investigated at 4°C and room temperature. The baseline pH of the 5 mg/mL suspensions was 2.962, and this did not change significantly after two weeks storage at either temperature.

The authors of the reports recommended shelf-lives of 7 days and 28 days for formulations stored at ambient and fridge temperatures, respectively, with reference to the 2.5 mg/mL formulation. As expected, the stability of the higher strength formulation was slightly improved, allowing a 14 day shelf-life to be recommended at room temperature and 28 days in the fridge.

However, 'Keltrol' is now a lapsed brand name. This xanthan gum-based suspending agent is now prepared to different specifications and pHs by a range of manufacturers. Typically, the formulation contains ethanol and chloroform. Ethanol is a CNS depressant and chloroform has been classified as a class 3 carcinogen (i.e. substances that cause concern owing to possible carcinogenic effects but for which available information is not adequate to make satisfactory assessments) (CHIP3 Regulations, 2002).

Given the variation in specification and the potential toxicity of the ingredients, 'Keltrol' should only be used as a suspending agent after consultation with a suitably competent pharmacist. As well as being free from chloroform and ethanol, the Ora-based products have a reproducible formulation, and are supported by a significant evidence base for a wide range of drugs. However, such products do need to be imported from the USA and are therefore relatively expensive. They also contain saccharin and sorbitol.

Other approaches

Other researchers have investigated more complex formulation approaches, including buffer systems, antioxidants and co-solvent systems (Nahata *et al.*, 1994; Matthew and Das Gupta, 1996; Liu *et al.*, 1999). The data from these studies may be useful for the development of 'Special' preparations of captopril, but the formulations are deemed too complex for application to extemporaneous practice in a typical NHS dispensary.

Bioavailability data

There is no information available for the unlicensed preparations. Pharmacokinetic information in adults may be available for imported products that hold marketing authorisations outside the UK (e.g. Capoten Oral Liquid, Bristol Myers Squibb, Australia).

References

Allen LV Jr Erickson MA (1996). Stability of baclofen, captopril, diltiazem hydrochloride, dipyridamole, and flecainide acetate in extemporaneously compounded oral liquids. *Am J Health Syst Pharm* 53: 2179–2184.

Anaizi NH, Swenson C (1993). Instability of aqueous captopril solutions. *Am J Hosp Pharm* 50: 486–488.

Andrews CD, Essex A (1986). Captopril suspension (letter). *Pharm J* 237: 734–735.

BBC News Online (2007). Warning on child heart drug doses. http://news.bbc.co.uk/1/hi/health/6446875.stm (accessed 24 May 2007).

British Pharmacopoeia (2007). *British Pharmacopoeia*, Vol. 1. London: The Stationery Office, pp. 366–367.

Chan DS, Sato AK, Claybaugh JR (1994). Degradation of captopril in solutions compounded from tablets and standard powder. *Am J Hosp Pharm* 51: 1205–1207.

Chen D, Chen H, Ku H (1995). Degradion rates of captopril in aqueous medium through buffer-catalysis oxidation. *Drug Dev Ind Pharm* 21: 781–792.

CHIP3 Regulations (2002). Chemicals (Hazard Information and Packaging for Supply) Regulations. SI 2002/1689.

Connors KA Amidon GL, Stella ZJ (1986). *Chemical Stability of Pharmaceuticals: A Handbook for Pharmacists*, 2nd edn. New York: Wiley Interscience.

Escribano-Garcia MJ, Torrado-Duran S, Torrado-Duran JJ (2005). Stability study of an aqueous formulation of captopril at 1 mg/ml. *Farm Hosp* 29: 30–36.

Kadin H (1982). Captopril. In: Florey K, ed. *Analytical Profiles of Drug Substances*, Vol. 11. London: Academic Press, pp. 79–137.

Lee T-K, Notari RE (1987). Kinetics and mechanism of captopril oxidation in aqueous solution under controlled oxygen partial pressure. *Pharm Res* 4: 98–103.

Liu J, Chan SY, Ho PC (1999). Effects of sucrose, citric buffer and glucose oxidase on the stability of captopril in liquid formulations. *J Clin Pharm Ther* 24: 145–150.

Lowey AR, Jackson MN (2008). A survey of extemporaneous dispensing activity in NHS trusts in Yorkshire, the North-East and London. *Hosp Pharmacist* 15: 217–219.

Lye MYF, Yow KL, Lim LY, Chan SY, Chan E, Ho PC (1997). Effects of ingredients on stability of captopril in extemporaneously prepared oral liquids. *Am J Health Syst Pharm* 54: 2483–2487.

Matthew M, Das Gupta V (1996). The stability of captopril in aqueous systems. *Drug Stability* 1: 161–165.

Mulla H, Tofeig M, Bu'Lock F, Samani N, Pandya HC (2007). Variations in captopril formulations used to treat children with heart failure: a survey in the United Kingdom. *Arch Dis Child* 92: 409–411.

Nahata MC, Morosco R, Hipple TF (1994). Stability of captopril in three liquid dosage forms. *Am J Hosp Pharm* 51: 95–96.

Pereira CM, Tam YK (1992). Stability of captopril in tap water. *Pharm J* 49: 612–615.

Pramar Y, Das Gupta V, Bethea C (1992). Stability of captopril in some aqueous systems. *J Clin Pharm Ther* 17: 185–189.

Roberts HR (1982). Personal communication, May 1977, cited in: Florey K, ed. *Analytical Profiles of Drug Substances*, Vol. 11. London: Academic Press, pp. 79–137.

Schlatter J, Sola A, Saulnier JL (1997). Stabilité d'une solution orale de captopril 1 mg/ml. *J Pharm Clin* 16: 125–128.

Shilkowski ER (1982). Personal communication, October 1980, cited in Kadin H (1982). Captopril. In: Florey K, ed. *Analytical Profiles of Drug Substances*, Vol.11. London: Academic Press, pp. 79–137.

Sweetman SC (2009). *Martindale: The Complete Drug Reference*, 36th edn. London: Pharmaceutical Press.

Wang YJ (1982). Personal communication, February 1980, cited in Kadin H (1982). Captopril. In: Florey K, ed. *Analytical Profiles of Drug Substances*, Vol. 11. London: Academic Press, pp. 79–137.

Woods DJ (2001). *Formulation in Pharmacy Practice. eMixt, Pharminfotech*, 2nd edn. Dunedin, New Zealand. www.pharminfotech.co.nz/emixt (accessed 18 April 2006).

Clobazam oral liquid

Risk assessment of parent compound

A risk assessment survey completed by clinical pharmacists suggested that clobazam has a moderate therapeutic index. Inaccurate doses were thought to be associated with significant morbidity (Lowey and Jackson, 2008). Current formulations used at NHS trusts included those with a high technical risk, based on the number and/or complexity of manipulations and calculations. Clobazam is therefore regarded as a high-risk extemporaneous preparation (Box 14.8).

Summary

There is a small amount of data available to support the use of clobazam as an extemporaneous suspension. Data from regional NHS Quality Control laboratories would appear to justify a cautious 14-day expiry at room temperature.

Box 14.8 *Preferred formula*

Using a typical strength of 1 mg/mL as an example:

- clobazam 10 mg tablets × 10
- xanthan gum 0.5% ('Keltrol' or similar) to 100 mL.

Method guidance

Tablets can be ground to a fine, uniform powder in a pestle and mortar. A small amount of suspending agent may be added to form a paste, before adding further portions up to approximately 75% of the final volume. Transfer to a measuring cylinder. The remaining suspending agent can be used to wash out the pestle and mortar before making the suspension up to 100% volume. Transfer to an amber medicine bottle.

Shelf-life

14 days at room temperature in amber glass. **Shake the bottle.**

Note: 'Keltrol' xanthan gum formulations vary slightly between manufacturers, and may contain chloroform and ethanol. The individual specification should be checked by the appropriate pharmacist.

'Specials' may also be available for clobazam oral liquid. The supplier should be able to justify their shelf-life. The formulation and specification should be reviewed by a suitably competent pharmacist before use.

Clinical pharmaceutics

Points to consider

- Due to the poor solubility of clobazam, most of the drug is likely to be in suspension rather solution. **Shake the bottle before use to improve dosage uniformity.**
- Clobazam displays polymorphism. In theory, changes in polymorphic form could alter particle size (Han *et al.*, 2006). Whether this occurs with clobazam is not known.
- 'Keltrol' is made to slightly different specifications by several manufacturers, but typically contains ethanol and chloroform. The exact specification should be assessed before use by a suitably competent pharmacist. Ethanol is a CNS depressant. Chloroform has been classified as a class 3 carcinogen (i.e. substances that cause concern owing to possible carcinogenic effects but for which available information is not adequate to make satisfactory assessments) (CHIP3 Regulations, 2002).
- There are no bioavailability data available for clobazam extemporaneous formulations.

Technical information

Structure

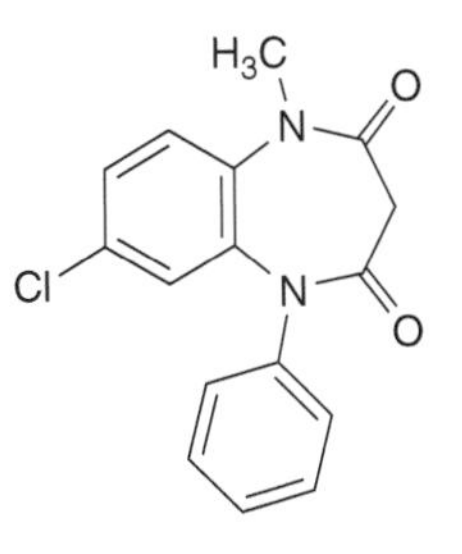

Empirical formula

$C_{16}H_{13}ClN_2O_2$ (British Pharmacopoeia, 2007).

Molecular weight

300.7 (British Pharmacopoeia, 2007).

Solubility

Slightly soluble (British Pharmacopoeia, 2007).

pK_a

Not known.

Optimum pH stability

Not known.

Stability profile

Physical and chemical stability

Clobazam is a white or almost white, crystalline powder (British Pharmacopoeia, 2007). At typical concentrations of 1 mg/mL, most of the drug is likely to be in suspension. Therefore, significant degradation is not expected. The structure suggests that there is potential for oxidation and hydrolysis of clobazam, but this is not expected to be problematic given that the majority of the drug is not likely to be in solution.

Clobazam is structurally similar to diazepam (Kuch, 1979). It is more soluble but would appear to be less susceptible to oxidation. Therefore, the stability profile could be expected to be comparable to that of diazepam. Diazepam is one of the more stable substituted 1,4-benzoadiazepines. It undergoes hydrolysis in aqueous solution, with the pH of maximum stability thought to be around pH 5 (Connors *et al.*, 1986).

Degradation products

No information available.

Stability in practice

There are no known published stability reports for suspensions of clobazam.

Two broad formulation approaches were found in a recent survey (Lowey and Jackson, 2006), based on xanthan gum-based preparations or on a historical formula from 1985 attributed to Hoescht Medical Information.

The majority of the xanthan gum preparations were based on unpublished advice from Nova Laboratories (Nova Laboratories Ltd, 2003; J. Cox, Nova Laboratories Ltd, personal communication, 2004). Clobazam is known to be physically compatible with 'Keltrol' (Nova Laboratories Ltd, 2003). Nova

Laboratories have previously recommended a shelf-life of 28 days in the fridge, or 14 days at room temperature (J. Cox, Nova Laboratories Ltd, personal communication, 2004).

An unpublished, informal quality control report from North-West QC Laboratories contains a limited amount of data over a test period of 5 days (D. Ireland, personal communication, 2006). The formula followed was as follows:

- clobazam tablets 10 mg × 10
- xanthan gel (Diluent A) 1% 50 mL
- purified water to 100 mL.

Comparator samples were made up using clobazam powder. Over the 5 days, minimal degradation was deemed to have occurred. Based on these results, a 14-day shelf-life was suggested by the author.

The Hoescht formula is reported as follows:

- clobazam (as Frisium 10 mg capsules – now discontinued) × 10
- 1% peppermint oil in ethanol BP 2.5 mL
- glycerol 6 mL
- syrup BP with preservative 35 mL
- distilled water to 100 mL.

It has been suggested that the clobazam should be triturated in the 1% peppermint oil in ethanol before dilution (St Mary's Stability Database, 2006). However, there are no data available to support this formula.

Bioavailability data

No data available.

References

British Pharmacopoeia (2007). *British Pharmacopoeia*, Vol. 1. London: The Stationery Office, pp. 534–535.

CHIP3 Regulations (2002). Chemicals (Hazard Information and Packaging for Supply) Regulations. SI 2002/1689.

Connors KA, Amidon GL, Stella ZJ (1986). *Chemical Stability of Pharmaceuticals: A Handbook for Pharmacists*, 2nd edn. New York: Wiley Interscience.

Han J, Beeton A, Long PF, Wong I, Tuleu C (2006). Physical and microbiological stability of an extemporaneous tacrolimus suspension for paediatric use. *J Clin Pharm Ther* 31: 167–172.

Kuch H (1979). Clobazam: chemical aspects of the 1,4- and 1,5-benzodiazepines. *Br J Clin Pharmacol* 7: 17S–21S.

Lowey AR, Jackson MN (2008). A survey of extemporaneous dispensing activity in NHS trusts in Yorkshire, the North-East and London. *Hosp Pharmacist* 15: 217–219.

Nova Laboratories Ltd (2003). Information Bulletin – Suspension Diluent Used at Nova Laboratories.

St Mary's Online Stability Database (2006). http://www.stmarys.demon.co.uk/(no longer available) (accessed 7 April 2006).

Clonazepam oral liquid

Risk assessment of parent compound

A risk assessment survey has suggested that clonazepam has a medium therapeutic index with the potential to cause moderate to severe morbidity if inaccurate doses are administered as an extemporaneous preparation (Lowey and Jackson, 2008). Current formulations used at NHS trusts were generally of a low technical risk, based on the number and/or complexity of manipulations and calculations. Clonazepam is therefore regarded as a medium-risk extemporaneous preparation (Box 14.9).

Summary

There would appear to be a reasonable body of evidence to support the extemporaneous formulation of clonazepam in a suitable vehicle. Formulations are now available for purchase from 'Specials' manufacturers. The suppliers should be able to justify the assigned shelf-life. The formulation should be reviewed by a suitably competent pharmacist before purchase.

Clinical pharmaceutics

Points to consider

- The amount of drug in suspension will depend on the strength of the formulation (typical strengths 100 micrograms/mL to 2 mg/mL). As a precaution, **shake the bottle to ensure uniformity of drug throughout the suspension.**
- Ora-Plus and Ora-Sweet SF are free from chloroform and ethanol.
- Ora-Plus and Ora-Sweet SF contain hydroxybenzoate preservative systems.
- Ora-Sweet SF has a cherry flavour which may help to mask the bitter taste of the drug.
- Clonazepam has a faint odour.
- There is no bioavailability information available for these preparations.
- 'Keltrol' is made to slightly different specifications by several manufacturers, but typically contains ethanol and chloroform. The exact specification should be assessed before use by a suitably competent pharmacist. Ethanol is a CNS depressant. Chloroform has been classified

Box 14.9 *Preferred formulae*

Note: 'Specials' are now available from reputable manufacturers. Extemporaneous preparation should only occur where there is a demonstrable clinical need.

Using a typical strength of 100 micrograms/mL as an example:

- clonazepam 2 mg tablets × 5
- Ora-Plus to 50 mL
- Ora-Sweet or Ora-Sweet SF to 100 mL.

Method guidance

Tablets can be ground to a fine, uniform powder in a pestle and mortar. A small amount of Ora-Plus may be added to form a paste, before adding further portions of Ora-Plus up to 50% of the final volume. Transfer to a measuring cylinder. The Ora-Sweet can be used to wash out the pestle and mortar before making the suspension up to 100% volume. Transfer to an amber medicine bottle.

Shelf-life

28 days at room temperature in amber glass. Avoid PVC containers. **Shake the bottle.**

Alternative

An alternative formula would be:

- clonazepam 2 mg tablets × 5
- xanthan gum suspending agent 0.5% ('Keltrol' or similar) to 100 mL.

Shelf-life

28 days at room temperature in amber glass. Avoid PVC containers. **Shake the bottle.**

Note: Xanthan gum formulations vary slightly between manufacturers, and may contain chloroform and ethanol. The individual specification should be checked by the appropriate pharmacist.

as a class 3 carcinogen (i.e. substances that cause concern owing to possible carcinogenic effects but for which available information is not adequate to make satisfactory assessments) (CHIP3 Regulations, 2002). Flavoured versions may also be available.

- Clonazepam injection has also been administered orally (Guy's and St Thomas' and Lewisham Hospitals, 2005; C. Norton, Lead Pharmacist for Clinical Trials and Adult Psychiatry and WMLRN Medicines for Children Pharmacist Adviser, Birmingham Children's Hospital, 2007).

Other notes

- Avoid the use of PVC containers.

Technical information

Structure

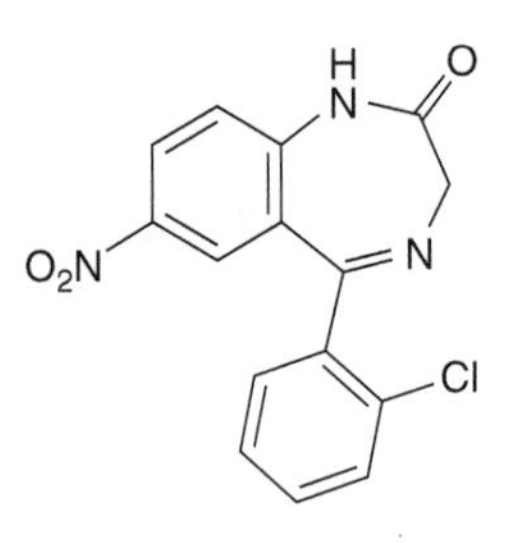

Empirical formula

$C_{15}H_{10}ClN_3O_3$ (British Pharmacopoeia, 2007)

Molecular weight

315.7 (British Pharmacopoeia, 2007)

Solubility

Practically insoluble in water (British Pharmacopoeia, 2007); <0.1 mg/mL in water at 25°C (Winslow, 1977).

pK_a

1.5 and 10.5 (Moffat, 1986).

Optimum pH stability

Not known.

Stability profile

Physical and chemical stability

Clonazepam is a slightly yellowish, crystalline powder, practically insoluble in water, slightly soluble in alcohol and in methanol, very slightly soluble in ether (British Pharmacopoeia, 2007). Clonazepam exhibits no optical rotation (Winslow, 1977). It is known to be physically compatible with some xanthan gum vehicles (Nova Laboratories Ltd, 2003).

Although clonazepam is practically insoluble in water, the amount of drug in solution and suspension will depend heavily on the strength of the preparation. At low strengths, such as 100 micrograms/mL, a portion of the drug may be in solution and therefore open to aqueous degradation kinetics. The exact solubility is unknown (stated as <0.1 mg/mL), which makes accurate predictions difficult.

At higher strengths, such as 2 mg/mL, the vast majority of the drug is likely to be in suspension and therefore overall levels of degradation are likely to be low.

Degradation occurs principally via hydrolysis (Winslow, 1977). The degradation products are listed below.

An ethanolic solution of clonazepam is reported to change from colourless to brown after a short time, with a corresponding decrease in concentration (Wad, 1986). Bares *et al.* (2004) have studied the stability of clonazepam 100 mg/mL in bidistilled water (pH 6.5) as part of the development of a rapid RP-HPLC method for the determination of clonazepam in human plasma. The authors found that clonazepam degraded rapidly when stored in daylight following a first-order kinetic rate with a half-life of 87 minutes. Corresponding samples in the dark, or in plasma showed no significant levels of degradation.

Clonazepam solutions exhibit loss due to sorption to PVC containers (Roy and Besner, 1997).

Degradation products

Major breakdown products are reported to be 2-amino-2′-chloro-5-nitrobenzophenone and 3-amino-4-(2-chlorophenyl)-6-nitrocarbostyril (Winslow, 1977).

Stability in practice

There are two published stability studies for clonazepam extemporaneous preparations. The data suggest that shelf-lives of in excess of 28 days are achievable.

Allen and Erickson (1996) have investigated the stability of clonazepam 100 micrograms/mL in three different suspending agents:

- 1:1 mixture of Ora-Sweet and Ora-Plus
- 1:1 mixture of Ora-Sweet SF and Ora-Plus
- cherry syrup (cherry syrup concentrate diluted 1:4 with simple syrup).

The tablets were crushed into a fine powder in a pestle and mortar before the vehicle(s) were added in a stepwise manner. Six 120 mL suspensions were prepared in amber clear plastic (polyethylene terphthalate) 'prescription ovals' with a low-density polyethylene foam cap lining. Three suspensions were stored at 5°C and three at 25°C (in the absence of light). After agitating on a rotating mixer for 30 minutes, 5 mL samples were removed on days 0, 1, 2, 7, 10, 14, 28, 35 and 60.

Samples were stored at –70°C until analysis by a stability-indicating HPLC method. There were no changes reported over 60 days with regard to physical characteristics (including visual and olfactory observations). The suspensions prepared in the Ora diluents had a starting pH of 4.1–4.2 and those prepared in cherry syrup had an initial pH of 2.9. These values changed by less than 0.5 pH units over 60 days.

On average, less than 5% drug degradation occurred in 60 days in any of the suspending agents, regardless of storage temperature.

Roy and Besner (1997) have studied an alternative formula of clonazepam 100 micrograms/mL in an HSC (Hospital for Sick Children) vehicle containing simple syrup and methylcellulose:

- clonazepam 500 micrograms tablet × 20
- sterile water for irrigation 3 mL
- vehicle HSC to 100 mL.

HSC vehicle is a 70:30 mixture of 1% methylcellulose and simple syrup, respectively.

The HPLC assay results showed a wide inter-day and intra-day variation. For example, 92.6% (± 3.2%) of drug was remaining at day 45. However, the corresponding figure for day 60 was 107.1% (± 10.4%). The authors claim that the suspension is stable for 60 days, but such an expiry may not be justifiable on these results alone.

Full details of the specificity of HPLC are not stated, and microbial stability was not assessed.

Bioavailability data

No information.

References

Allen LV, Erickson MA (1996). Stability of acetazolamide, allopurinol, azathioprine, clonazepam, and flucytosine in extemporaneously compounded oral liquids. *Am J Health Syst Pharm* 53: 1944–1949.

Bares IF, Pehourcq Jarry C (2004). Development of a rapid RP-HPLC method for the determination of clonazepam in human plasma. *J Pharm Biomed Anal* 36: 865–869.

British Pharmacopoeia (2007). *British Pharmacopoeia*, Vol. 1. London: The Stationery Office, pp. 546–547.

CHIP3 Regulations (2002). Chemicals (Hazard Information and Packaging for Supply) Regulations. SI 2002/1689.

Guy's and St Thomas' and Lewisham Hospitals (2005). *Paediatric Formulary*. London: Guy's and St Thomas' and Lewisham NHS Trust.

Lowey AR, Jackson MN (2008). A survey of extemporaneous dispensing activity in NHS trusts in Yorkshire, the North-East and London. *Hosp Pharmacist* 15: 217–219.

Moffat AC, Osselton MD, Widdop B (2003). *Clarke's Analysis of Drugs and Poisons*, 3rd edn. London: Pharmaceutical Press. ISBN 978 0 85369 473 1.

Nova Laboratories Ltd (2003). Information Bulletin – Suspension Diluent Used at Nova Laboratories.

Roy JJ, Besner J-G (1997). Stability of clonazepam suspension in HSC vehicle. *Int J Pharm Compound* 1: 440–441.

Wad N (1986). Degradation of clonazepam in serum by light confirmed by means of a high performance liquid chromatographic method. *Ther Drug Monit* 8: 358–360.

Winslow WC (1977). Clonazepam. In: Florey K, ed. *Analytical Profiles of Drug Substances*, Vol. 6. London: Academic Press, pp. 61–81.

Clonidine hydrochloride oral liquid

Risk assessment of parent compound

A risk assessment survey completed by clinical pharmacists suggested that clonidine hydrochloride has a medium therapeutic index. Inaccurate doses were thought to be associated with significant morbidity (Lowey and Jackson, 2008). Current formulations used at NHS trusts were generally of a low technical risk, based on the number and/or complexity of manipulations and calculations. Clonidine hydrochloride is therefore regarded as a moderate-risk extemporaneous preparation (Box 14.10).

Summary

Some data are available to support the stability of clondine hydrochloride in injectable and oral products. The typical approach in the NHS is to use a xanthan gum-based suspending agent. Although clonidine is freely soluble, the use of an appropriate suspending agent will suspend the insoluble tablet excipients. Extremes of pH should be avoided. The use of syrup as a vehicle is not recommended as syrup has questionable ability to suspend the insoluble tablet excipients (and any drug bound thereto). If the use of syrup is unavoidable, the substitution of clonidine injection in place of tablets may be considered.

The commissioning of this product as a 'Specials' may be an option, as stability would appear to be relatively good. 'Specials' manufacturers should be able to provide justification for the shelf-life attributed.

Clinical pharmaceutics

Points to consider

- Clonidine is freely soluble. Most, if not all, of the drug should be in solution rather than suspension at typical strengths (5–150 micrograms/mL). Dosage uniformity is not likely to be an issue.
- Ora-Plus and Ora-Sweet SF are free from chloroform and ethanol (Paddock Laboratories, 2006).
- Ora-Plus and Ora-Sweet SF contain hydroxybenzoate preservative systems.
- Ora-Sweet SF has a cherry flavour which may help to mask the bitter taste of the drug.

***Box 14.10** Preferred formulae*

Using a typical strength of 10 micrograms/mL as an example:

- clonidine hydrochloride 100 micrograms tablets × 10
- Ora-Plus to 50 mL
- Ora-Sweet to 100 mL.

Method guidance

Tablets can be ground to a fine, uniform powder in a pestle and mortar. A small amount of Ora-Plus may be added to form a paste, before adding further portions of Ora-Plus up to 50% of the final volume. Transfer to a measuring cylinder. The Ora-Sweet can be used to wash out the pestle and mortar of any remaining drug before making the suspension up to 100% volume. Transfer to an amber medicine bottle.

Shelf-life

28 days refrigerated in amber glass. **Shake the bottle.**

Alternative

An alternative formula would be:

- clonidine hydrochloride 100 micrograms tablets × 10
- xanthan gum 0.5% suspending agent (or similar) to 100 mL.

Shelf-life

28 days refrigerated in amber glass. **Shake the bottle.**

Note: 'Keltrol' xanthan gum formulations vary slightly between manufacturers, and may contain chloroform and ethanol. The individual specification should be checked by the appropriate pharmacist.

- 'Keltrol' is made to slightly different specifications by several manufacturers, but typically contains ethanol and chloroform. The exact specification should be assessed before use by a suitably competent pharmacist. Ethanol is a CNS depressant. Chloroform has been classified as a class 3 carcinogen (i.e. substances that cause concern owing to possible carcinogenic effects but for which available information is not adequate to make satisfactory assessments) (CHIP3 Regulations, 2002).
- There are no details available with regard to bioavailability.

Other notes

- Serious errors have been reported with the preparation and administration of clonidine extemporaneous preparations (Erickson and Duncan, 1998; Kappagoda *et al.*, 1998; Romano and Dinh, 2001; Suchard and Graeme, 2002).
- The 100 micrograms tablet may be crushed and dispersed in water (Guy's and St Thomas' and Lewisham Hospitals, 2005).
- The hydrochloride is used for both tablet and injectable formulations. It has a bitter taste (Martindale, 2005).

Technical information

Structure

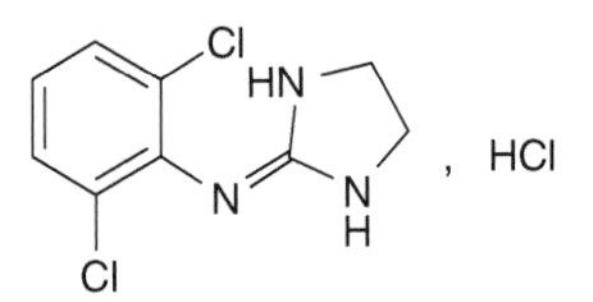

Empirical formula

$C_9H_9Cl_2N_3$,HCl (British Pharmacopoeia, 2007)

Molecular weight

266.66 (British Pharmacopoeia, 2007).

Solubility

1 in 13 in water (Trissel, 2000; Woods, 2001) (77 mg/mL).

pK_a

8.2 (Clarke, 1986).

Optimum pH stability

Not known.

Stability profile

Physical and chemical stability

Clonidine hydrochloride is freely soluble. Therefore, the active drug is typically in solution at standard strengths 5 to 25 micrograms/mL.

Although it is possible to force the cleavage of the CN bond by hydrolysis (Kostecka *et al.*, 1998), this does not occur under normal ambient conditions. Photolytic cleavage of the CN bond is also theoretically possible, but has not been shown to be a problem, either in laboratory tests or during the normal shelf-life of the injection product (Kostecka *et al.*, 1998).

Kostecka *et al.* (1998) suggest that, although clondine is stable, two routes of degradation are possible. It is theoretically possible to cleave the CN bond by hydrolysis; however, as clonidine is a Schiff base, this process would require catalysis by acid, base, alkaline hydrogen peroxide, ozone, or one of several other catalysts. In laboratory tests, the only degradation of the injection that could be achieved was via the presence of hydrogen peroxide at elevated temperatures.

The natural pH of solutions containing 0.15 to 1.5 mg/mL in sodium chloride 0.9% is pH 6.0 to 6.5 (Trissel *et al.*, 1997). The BP injection is reported to have a pH of 4–7 (Woods, 2001). There is less information about the stability of clonidine hydrochloride injection outside these pH values.

Degradation products

No information.

Stability in practice

The published evidence to support the stability of clonidine in suspension is limited to one paper.

Levinson and Johnson (1992) examined the investigation of a 100 micrograms/mL formulation of crushed clonidine hydrochloride tablets in simple syrup. The tablets were crushed and ground in a glass mortar, and a small volume of purified water was added to form a fine paste. Stepwise additions of syrup occurred to complete the volume.

Although clonidine is freely soluble, the final formulation contained insoluble tablet excipients. Three 60 mL samples were produced and transferred to amber, type III glass prescription bottles with child-resistant closures. The samples were then stored in the dark at 4°C. Three corresponding samples were prepared by dissolving analytical grade clonidine hydrochloride powder in water before the stepwise additions of syrup as previously.

The bottles were sampled by HPLC over a period of 28 days. Further details are given of the HPLC methodology. Before analysis, the samples were mixed thoroughly for 30 minutes on a rotating mixer. A 0.5 mL aliquot was withdrawn and diluted to an expected concentration of 5 micrograms/mL and passed through a 0.22 micrometre filter. The samples made from clonidine hydrochloride powder were inspected for any evidence of precipitate. All samples were assessed for changes in colour, odour and pH; no 'appreciable changes' were found.

Although all the samples were documented to have retained at least 92% of the parent drug after 28 days, the HPLC results show a considerable variation over the test period for the formulations from both the tablets and analytical grade powder. Only five data points are published, and a relatively large drop in concentration remaining occurs between day 21 and day 28. Sources of experimental error may include the use of a 0.5 mL sample for dilution for use in the HPLC assay, and the potential viscosity of the formulations. Microbiological testing was not carried out.

The authors conclude that a 100 micrograms/mL clonidine hydrochloride liquid made from crushed tablets is stable for at least 28 days when stored in amber glass bottles at 4°C.

Related preparations

Kostecka *et al.* (1998) describe several stability studies that investigate the stability of clonidine hydrochloride injection, using an HPLC assay. The injection used was formulated to a pH of approximately 6 and contained sodium chloride and water. Purging samples with oxygen did not demonstrate any noticeable degradation after two weeks. The presence of oxygen or the absence of oxygen in the headspace of the vials had no influence upon product stability. The authors conclude that clonidine hydrochloride injection is not sensitive to oxygen.

Vials placed in a light chamber for two weeks showed no difference from controls with regard to colour, clarity or potency. Slight differences in pH (up to 0.4 pH units) were not deemed to be significant. A small new peak on the chromatogram was found to account for 0.13% of the starting drug, and the overall potency did not show a decrease. Given that the light chamber exposed the drug to extraordinary amounts of light, the authors conclude that the drug does not require protection from light in practice.

Injections subject to higher or lower pHs using hydrochloric acid or sodium hydroxide (pH 4 to pH 8) showed a small pH shift over time towards neutral. The sample had a pH of 4.22 after three weeks and 4.56 after six weeks. There was no difference between any of the samples in terms of clarity, colour, or potency. Despite these accelerated conditions, there was no discernible degradation. No stopper effects were seen with inverted samples and freezing-thawing produced no effect on pH, colour, clarity or potency.

The authors conclude that clonidine hydrochloride injections are stable for at least 36 months.

Trissel *et al.* (1997) have evaluated the stability of intrathecal injections of clonidine hydrochloride prepared extemporaneously from powder

(150 micrograms/mL, 500 micrograms/mL and 1.5 mg/mL). The injection was brought to volume with sodium chloride and pH adjusted to 6.5 with sodium hydroxide if necessary. After filtration, the vials were sterilised by autoclaving at 121°C for 30 minutes before storage at 4, 23 and 37°C. Stability-indicating HPLC analysis revealed no loss of drug due to the autoclaving process. Moreover, no loss occurred after three months of storage at 37°C and up to two years at 4 and 23°C.

Bioavailability data

No information.

References

British Pharmacopoeia (2007). *British Pharmacopoeia*, Vol. 1. London: The Stationery Office, pp. 547–548.

CHIP3 Regulations (2002). Chemicals (Hazard Information and Packaging for Supply) Regulations. SI 2002/1689.

Erickson SJ, Duncan A (1998). Clonidine poisoning – an emerging problem: epidemiology, clinical features, management and preventative strategies. *J Paediatr Child Health* 34: 280–282.

Guy's and St Thomas' and Lewisham Hospitals (2005). *Paediatric Formulary*. London: Guy's and St Thomas' and Lewisham NHS Trust.

Kappagoda C, Schell DN, Hanson RM, Hutchins P (1998). Clonidine overdose in childhood: implications of increased prescribing. *J Paediatr Child Health* 34: 508–512.

Kostecka D, Duncan MR, Wagenknecht D (1998). Formulation of a stable parenteral product; clonidine hydrochloride injection. PDA. *J Pharm Sci Technol* 52: 320–325.

Levinson ML, Johnson CE (1992). Stability of an extemporaneously compounded clonidine hydrochloride oral liquid. *Am J Hosp Pharm* 49: 122–125.

Lowey AR, Jackson MN (2008). A survey of extemporaneous dispensing activity in NHS trusts in Yorkshire, the North-East and London. *Hosp Pharmacist* 15: 217–219.

Paddock Laboratories (2006). Ora-Plus and Ora-Sweet SF Information Sheets. www.paddock-laboratories.com. (accessed 12 April 2006).

Romano MJ, Dinh A (2001). A 1000-fold overdose of clonidine caused by a compounding error in a 5-year-old child with attention-deficit/hyperactivity disorder. *Pediatrics* 108: 471–472.

Suchard JR, Graeme KA (2002). Pediatric clonidine poisoning as a result of pharmacy compounding error. *Pediatr Emerg Care* 18: 295–296.

Sweetman SC (2009). *Martindale: The Complete Drug Reference*, 36th edn. London: Pharmaceutical Press.

Trissel LA, ed. (2000). *Stability of Compounded Formulations*, 2nd edn. Washington DC: American Pharmaceutical Association.

Trissel LA, Xu QA, Hassenbuch SJ (1997). Development of clonidine hydrochloride injections for epidural and intrathecal administration. *Int J Pharm Compound* 1: 274–277.

Woods DJ (2001). *Formulation in Pharmacy Practice. eMixt, Pharminfotech*, 2nd edn. Dunedin, New Zealand. www.pharminfotech.co.nz/emixt. (accessed 18 April 2006).

Clozapine oral liquid

Risk assessment of parent compound

A risk assessment survey completed by clinical pharmacists suggested that clozapine has a narrow therapeutic index. Inaccurate doses were thought likely to lead to significant morbidity (Lowey and Jackson, 2008). However, the correlation between dose and clinical effect is not good, with wide inter-patient variation in plasma levels (Whiskey and Taylor, 2004). Current formulations used at NHS trusts were generally of a moderate technical risk, based on the number and/or complexity of manipulations and calculations. Clozapine is therefore regarded as a high-risk extemporaneous preparation (Box 14.11).

Box 14.11 *Preferred formulae*

Using a typical strength of 20 mg/mL as an example:

- clozapine (Clozaril) tablets 100 mg × 20
- Ora-Plus 50 mL
- Ora-Sweet to 100 mL.

Method guidance

Tablets can be ground to a fine, uniform powder in a pestle and mortar. A small amount of Ora-Plus may be added to form a paste, before adding further portions of Ora-Plus up to 50% of the final volume. Transfer to a measuring cylinder. The Ora-Sweet can be used to wash out the pestle and mortar before making the suspension up to 100% volume. Transfer to an amber medicine bottle.

Shelf-life

28 days in amber glass, refrigerated or at room temperature. **Shake the bottle.**

- clozapine (Clozaril) tablets 100 mg × 20
- xanthan gum 0.5% ('Keltrol' or similar) to 100 mL.

Note: 'Keltrol' xanthan gum formulations vary slightly between manufacturers, and may contain chloroform and ethanol. The individual specification should be checked by the appropriate pharmacist.

Summary

There is a wide body of evidence to support the stability of clozapine in suspension. This may allow the purchase of an appropriate suspension from a 'Specials' manufacturer. However, the bioavailability (and hence clinical effectiveness) of these suspensions currently remains unproven. Commissioning a standardised product from a 'Specials' manufacturer would help to minimise any issues relating to the formulation and manufacturing process.

Current best practice for extemporaneous preparation would appear to be the use of a suspending agent based on xanthan gum (e.g. Ora-Plus, 'Keltrol').

Clinical pharmaceutics

Points to consider

- Clozapine is insoluble and will therefore be held in suspension. Dosage uniformity will therefore be an issue if the drug particles settle. **Shake the bottle before use to improve dosage uniformity.**
- Ora-Plus and Ora-Sweet SF are free from chloroform and ethanol.
- Ora-Plus and Ora-Sweet SF contain hydroxybenzoate preservative systems.
- Ora-Sweet SF has a cherry flavour.
- 'Keltrol' is made to slightly different specifications by several manufacturers, but typically contains ethanol and chloroform. The exact specification should be assessed before use by a suitably competent pharmacist. Ethanol is a CNS depressant. Chloroform has been classified as a class 3 carcinogen (i.e. substances that cause concern owing to possible carcinogenic effects but for which available information is not adequate to make satisfactory assessments) (CHIP3 Regulations, 2002).
- Bioavailability issues exist for patients swapping from tablets to liquid (or vice versa) as well as for patients swapping brands of clozapine tablets. Close monitoring of the patient is recommended.

Technical information

Structure

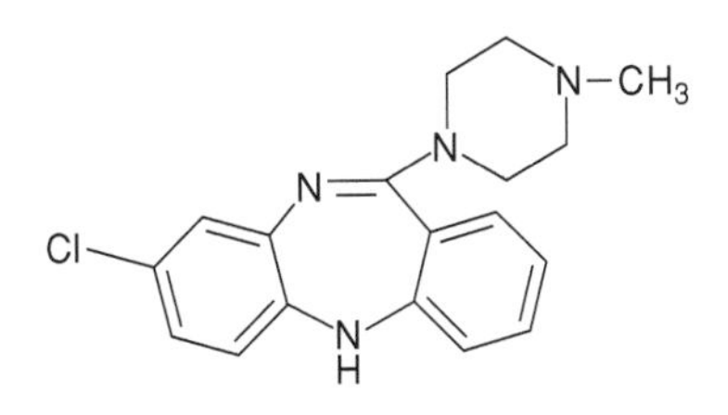

Empirical formula

$C_{18}H_{19}ClN_4$ (British Pharmacopoeia, 2007)

Molecular weight

326.8 (British Pharmacopoeia, 2007)

Solubility

Practically insoluble in water (0.0118 mg/mL) (British Pharmacopoeia, 2007).

pK_a

Not known.

Optimum pH stability

Not known.

Stability profile

Physical and chemical stability

Clozapine is a yellow crystalline powder with poor solubility (British Pharmacopoeia, 2007), The drug's poor solubility may help to minimise degradation in extemporaneous formulations. However, in common with other dibenzodiazepines, clozapine may degrade via ring opening at the R2=N bond with extensive secondary products (Anon, 2001).

Nagiba *et al.* (2002) have demonstrated five methods for assaying clozapine in the presence of its main acid-induced degradation product.

Degradation products

No information available.

Stability in practice

There is a significant amount of evidence to support the relative stability of clozapine when formulated as a suspension. The quality of the data does, however, differ between authors.

Walker *et al.* (2005) have published a stability report for clozapine 20 mg/mL suspensions, using six different vehicles:

- Ora-Sweet
- Ora-Plus
- 1:1 mixture of Ora-Sweet and Ora-Plus
- suspending agent as used by the Hospital for Sick Children (HSC Suspending Agent), Toronto, Canada; a mixture of simple syrup and methylcellulose; not commercially available
- simple syrup
- Guy's paediatric mixture (a preserved syrup-based vehicle).

A stability-indicating HPLC method was used to assay the suspensions. One hundred millilitres was prepared of each of the preparations, split into three equal portions and stored in 60 mL low-density polyethylene amber containers (total 18 containers). All the containers were stored at room temperature (taken as 23°C), with no protection from ambient light.

Undisclosed samples were removed for analysis on days 0, 3, 6, 14, 28 and 63. Physical changes such as caking and consistency were noted as well as changes in colour. Standard curves were constructed via dilution of a clozapine standard. Samples were prepared by dissolving 1 mL of suspension in 100 mL methanol to prepare a solution with a nominal concentration of 0.2 mg/mL. All samples were analysed in duplicate. Linear regression statistics were used to compare mean results from different days for each container, and the 95% confidence interval lower limit had to be over 90% of the starting concentration for the concentration to be considered acceptable. Analysis of variance was used to test differences in degradation rate between different vehicles, and multiple linear regression detected any effect of time and suspension vehicle on concentration during the study period.

During the study, intra-day coefficients of variation averaged 1.7% for samples and 1.4% for standards. However, the degradation levels observed were very low, meaning experimental error made it difficult to determine a definite trend of degradation in some of the suspensions. Multilpe linear regression detected a small but statistically significant average decrease of 2.56%.

The lower limit of the 95% confidence interval remained above 90% for all the suspensions tested up to and including day 63, and the mean concentration remained above 95%. Although the final concentration appeared to be slightly lower for the suspensions containing Ora-Plus alone, Ora-Sweet alone and Guy's paediatric mixture, analysis of variance did not show any statistically significant difference ($P = 0.27$).

All suspensions retained their starting physical description: opaque, light creamy-white and free from any evidence of caking. The authors suggest that with no significant differences seen in degradation between the suspending agents, the choice becomes one of tolerability, availability and cost. Guy's and HSC suspending agents are not commercially available and so may be less appropriate for some pharmacies. Syrup is known to have dental health issues and its efficacy as a suspending agent is generally regarded as poor. The authors consider the Ora-based formulations to be preferable due to their commercial availability and stability.

The data produced by Walker *et al.* (2005) show considerably better recovery of clozapine from suspension than that described by Ramuth *et al.* (1996). This may indicate that the bioavailability will be higher and/or more reproducible from the Ora-based formulations.

Ramuth *et al.* (1996) investigated the stability of a 20 mg/mL clozapine oral liquid preparation, using Guy's Hospital formula base for paediatric mixtures as the suspending agent. For each of the three suspensions under investigation, ten Clozaril 100 mg tablets were crushed using a pestle and mortar and suspended in the Guy's paediatric base; this is a syrup-based mixture containing carboxymethylcellulose BP, methylhydroxybenzoate BP and propylhydroxybenzoate BP. The suspensions were stored in amber glass bottles at room temperature. Assay was carried out using a published HPLC method, and the paediatric base did not contain any compounds that interfered with the assay (McCarthy *et al.*, 1995).

After manual shaking, undisclosed samples were diluted with a 1:1 mixture of methanol and deionised water, and assayed on days 1, 4, 8, 11, 15 and 18. No additional peaks developed, and the authors concluded the suspension to be stable for 18 days. However, the clozapine concentrations showed no clear trends over the 18-day test period, and initial starting concentrations varied widely from 15.06 mg/mL to 18.16 mg/mL. Inter-day and intra-day variations were considerable. The authors comment on the poor recovery of clozapine from the suspension and claim that 80–90% recovery rates are of minor clinical significance. However, the poor sample recovery may cause clinical problems if this is an indication of the bioavailability of the formulation, as clozapine has been acknowledged as possessing a narrow therapeutic index (Lowey and Jackson, 2008).

Acidification of the suspension using hydrochloric acid did not improve recovery, and Ramuth *et al.* (1996) suggest that adsorption of the drug to the pestle and mortar, glass or the stopper may be accountable for the loss.

The suspensions were reported to settle quickly but with easy redispersion. Anecdotal tolerability is reported as good, and informal studies of more than 10 patients maintained on the liquid formulation are reported to have shown acceptable plasma levels. The authors recommend a 7-day expiry for the suspension, when stored at room temperature. No microbiological testing was carried out.

While the study provides useful data on the relative stability of clozapine in suspension, the experimental error is significant, and the use of a syrup-based vehicle may not be clinically acceptable, given the concerns associated with dental health. Furthermore, the authors acknowledge that the drug settles easily, giving rise to concerns regarding a lack of dosage uniformity, should the patient or carer fail to shake the bottle.

Exact details of the suspending agent are not stated. An on-line review by Woods (2001) suggests that methylcellulose 1% could be used as an alternative suspending agent.

An unpublished quality control report (Anon, 2001; D. Ireland, personal communication, 2006) investigated the stability of clozapine (Clozaril tablets) suspended in xanthan gel (Stockport Pharmaceuticals). Two strengths were investigated: 40 mg/mL and 20 mg/mL. One hundred millilitres of each strength was made and divided into two 50 mL aliquots; one bottle of each strength was stored at 5–8°C and one at 22–25°C. Clozapine content was assayed by a stability-indicating RP-HPLC method on days 0, 7, 14, 28, 60, 96 and 122. Forced degradation of clozapine suspension with heated 1 M HCl gave reduction of the clozapine peak with the appearance of eight other peaks, none of which interfered with the main peak. The analytical solution was reported to be stable for 2 days.

The pH, appearance, clozapine concentration and hydroxybenzoates concentrations of the refrigerated suspensions remained constant throughout the study (122 days). The initial lemon yellow appearance of the room temperature suspensions was lost between 60 and 90 days, and the clozapine concentration fell. Secondary peaks developed at this time. The clozapine concentration remained above 90% of the starting concentration on day 122 for the 20 mg/mL formulation, but had dropped below this level by day 96 for the 40 mg/mL bottle (from a starting concentration of 37.9 mg/mL to 32.9 mg/mL; equivalent to 86.8% of the starting concentration). The pH and concentrations of the hydroxybenzoates remained constant. The report concludes with a proposed shelf-life of two months when stored below 25°C. Microbiological testing is not documented.

Bioavailability data

Whiskey and Taylor (2004) point out that for two brands to be considered 'bioequivalent', the bioavailability of the new preparation must fall between 80% and 125% of the reference preparation. While such a difference may not be clinically significant for most drugs, for a drug such as clozapine such a difference could have huge implications (Whiskey and Taylor, 2004). Furthermore, clozapine dose is a poor predictor of therapeutic response, and wide inter-patient variability of plasma levels is seen for the same dose. There have also been reports of psychotic relapses when Clozaril has been switched to a generic product (Mofsen and Balter, 2001; Whiskey and Taylor, 2004).

There have been suggestions that the use of the clozapine suspension in Guy's paediatric mixture has been associated with a reduction in plasma clozapine when compared with the standard tablet, despite passing stability testing as detailed above (Coker-Adeyemi and Taylor, 2002; Whiskey and Taylor, 2004). The reasons for this are discussed, and may include the lack of adequate shaking by nursing staff before administration (Coker-Adeyemi and Taylor, 2002). The Guy's paediatric mixture is based on syrup and carboxymethylcellulose, both of which are generally acknowledged as having issues with regard to their effectiveness at preventing settling of the drug. Despite the bioavailability concerns, Coker-Adeyemi and Taylor (2002) suggest that the liquid formulation does still have a place in therapy, particularly in covert or suspected non-compliance.

There have also been several papers published regarding switching between generic and branded clozapine tablets, with conflicting results (Kluznik *et al.*, 2001; Lam *et al.*, 2001; Makela *et al.*, 2003).

These potential bioavailability differences carry potentially serious clinical implications and patients switching dosage forms or brands may require more intensive monitoring (Coker-Adeyemi and Taylor, 2002; Whiskey and Taylor, 2004).

References

Anon (2001). Shelf-life assessment: clozapine oral suspension (unpublished), Quality Control North West (Liverpool Laboratory), 2000–2001.

British Pharmacopoeia (2007). *British Pharmacopoeia*, Vol. 1. London: The Stationery Office, pp. 552–553.

CHIP3 Regulations (2002). Chemicals (Hazard Information and Packaging for Supply) Regulations. SI 2002/1689.

Coker-Adeyemi F, Taylor D (2002). Clozapine plasma levels in patients switched from clozapine liquid to tablets. *Pharm J* 269: 650–652.

Kluznik JC, Walbek NH, Farnworth MG, Gonzales C (2001). Clinical effects of a randomised switch of patients from Clozaril to generic clozapine. *J Clin Psychiatry* 62(Suppl5): 14–17.

Lam YWF, Ereshekfsky L, Toney GB, Gonzales C (2001). Branded versus generic clozapine: bioavailability comparison and inter-changeability issues. *J Clin Psychiatry* 62(Suppl5): 18–22.

Lowey AR, Jackson MN (2008). A survey of extemporaneous dispensing activity in NHS trusts in Yorkshire, the North-East and London. *Hosp Pharmacist* 15: 217–219.

Makela EH, Cutlip WD, Stevenson JM, Weimer JM, Abdallah ES, Akhtar RS, Aboraya AS, Gunel E (2003). Branded versus generic clozapine for the treatment of schizophrenia. *Ann Pharmacother* 37: 350–353.

McCarthy PT, Hughes SA, Paton C (1995). Measurement of clozapine and norclozapine in plasma/serum by high performance liquid chromatography with ultraviolet detection. *Biomed Chromatogr* 9: 36–41.

Mofsen R, Balter J (2001). Case reports of the emergence of psychotic symptoms after conversion from brand-named clozapine to a generic formulation. *Clin Ther* 10: 1720–1731.

Nagiba Y, Elkaway ME, Elzeany BE, Wagieh NE (2002). Stability indicating methods for the determination of clozapine. *J Pharm Biomed Anal* 30: 35–37.

Ramuth S, Flanagan RJ, Taylor DM (1996). A liquid clozapine preparation for oral administration in hospital. *Pharm J* 257: 190–191.

Walker SE, Baker D, Law S (2005). Stability of clozapine stored in oral suspension vehicles at room temperature. *Can J Hosp Pharm* 58: 279–284.
Whiskey E, Taylor D (2004). Generic clozapine: opportunity or threat? *Pharm J* 273: 112.
Woods DJ (2001). *Formulation in Pharmacy Practice. eMixt, Pharminfotech*, 2nd edn. Dunedin, New Zealand. www.pharminfotech.co.nz/emixt (accessed 18 April 2006).

Co-careldopa oral liquid (levodopa and carbidopa)

Risk assessment of parent compound

A risk assessment survey completed by clinical pharmacists suggested that co-careldopa has a medium to narrow therapeutic index, with the potential to cause significant morbidity if inaccurate doses were to be administered (Lowey and Jackson, 2008). Current formulations used at NHS trusts generally show a low technical risk, based on the number and/or complexity of manipulations and calculations. Co-careldopa is therefore regarded as a medium-risk extemporaneous preparation (Box 14.12).

Summary

A body of evidence exists to support the extemporaneous formulation of co-careldopa. It would appear that the stability of the carbidopa is the key factor. Ora-Plus and Ora-Sweet have the advantage of a single formulation and published stability data, as well as being free from chloroform and ethanol.

If generic xanthan gum (also known as 'Keltrol') is chosen, the exact specification should be checked by a suitable competent pharmacist, as the formulation may vary between manufacturers. Some formulations may contain chloroform and/or ethanol. The shelf-life of the xanthan gum-based product may be extended in the future if stability data are forthcoming.

Clinical pharmaceutics

Points to consider

- Although most of the levodopa and almost all of the carbidopa are likely to be in solution at typical strengths, a significant proportion of the levodopa may be in suspension. Therefore it would be wise to re-disperse the drug before each dose. **Shake the bottle before use to improve dosage uniformity.**
- Levodopa is optically active. It is not known whether the D-isomer may be formed during preparation and/or storage of the extemporaneous formulation. The D-isomer has been linked to the development of granulocytopaenia (FDA, 2005).

***Box 14.12** Preferred formulae*

Using a typical strength of 5 mg/mL levodopa and 1.25 mg/mL carbidopa as an example:

- levodopa and carbidopa tablets (100/25 mg) × 5
- Ora-Plus to 50 mL
- Ora-Sweet to 100 mL.

Method guidance

Tablets can be ground to a fine, uniform powder in a pestle and mortar. A small amount of Ora-Plus may be added to form a paste, before adding further portions of Ora-Plus up to 50% of the final volume. Transfer to a measuring cylinder. The Ora-Sweet can be used to wash out the pestle and mortar of any remaining drug before making the suspension up to 100% volume. Transfer to an amber medicine bottle.

Shelf-life

28 days refrigerated in amber glass. **Shake the bottle.**

Alternative

An alternative formula would be:

- levodopa and carbidopa tablets (100/25 mg) × 5
- xanthan gum 0.5% suspending ('Keltrol' or similar) to 100 mL.

Shelf-life

7 days at refrigerated in amber glass. **Shake the bottle.**

Note: 'Keltrol' xanthan gum formulations vary slightly between manufacturers, and may contain chloroform and ethanol. The individual specification should be checked by the appropriate pharmacist.

- Co-beneldopa dispersible tablets may be a useful alternative in some patients.
- Ora-Plus and Ora-Sweet are free from chloroform and ethanol.
- Ora-Plus and Ora-Sweet contain hydroxybenzoate preservative systems.
- Ora-Sweet has a cherry flavour.
- 'Keltrol' is made to slightly different specifications by several manufacturers, but typically contains ethanol and chloroform. The exact specification should be assessed before use by a suitably competent pharmacist. Ethanol is a CNS depressant. Chloroform has been classified

as a class 3 carcinogen (i.e. substances that cause concern owing to possible carcinogenic effects but for which available information is not adequate to make satisfactory assessments) (CHIP3 Regulations, 2002).

- Levodopa exhibits little or no taste.
- There are no data available with regard to bioavailability of the extemporaneous formulations.

Technical information

Table 14.1

Stability profile

Physical and chemical stability

Levodopa is a white or slightly cream-coloured, crystalline powder, slightly soluble in water, practically insoluble in alcohol and ether (British Pharmacopoeia, 2007). It is almost tasteless and darkens on exposure to air (Lund, 1994). Levodopa is freely soluble in 1 M hydrochloric acid and sparingly soluble in 0.1 M hydrochloric acid, and should be stored protected from light (British Pharmacopoeia, 2007).

Table 14.1 Technical information for co-careldopa oral liquid

	Levodopa	**Carbidopa**
Structures	HO, HO, H, NH_2, COOH	COOH, HN, CH_3, NH_2, HO, OH, , H_2O
Empirical formulae (British Pharmacopoeia, 2007)	$C_9H_{11}NO_4$	$C_{10}H_{14}N_2O_4,H_2O$
Molecular weight (British Pharmacopoeia, 2007)	197.2	244.2
Solubility	3.0 mg/mL (Moffat *et al.*, 2004) to 3.33 mg/mL (Gomez *et al.*, 1976)	2 mg/mL (Moffat *et al.*, 2004)
pK_a	2.3, 8.7, 9.7, 13.4 at 25°C (Moffat *et al.*, 2004)	Not known
Optimum pH stability	Not known	Not known

In the presence of moisture, oxidation of levodopa by atmospheric oxygen causes a rapid discoloration (Lund, 1994). Levodopa powder discolours after heating at 105°C for 24 hours. The discoloration increases with heating time. Despite this discoloration, thin-layer chromatography could not detect any drug decomposition (Lund, 1994).

In alkaline solution, oxidation of levodopa leads to the formation of inactive compounds such as melanin and related intermediates (Newton, 1978; Lund, 1994). It has been suggested that levodopa should be prepared either in an acidic or non-aqueous vehicle (Anon, 2002). The pH of a 1% aqueous solution of levodopa lies between 4.5 and 7.0 (Lund, 1994).

Carbidopa is a white or yellowish-white powder, very slightly soluble in water, very slightly soluble in alcohol, practically insoluble in methylene chloride (British Pharmacopoeia, 2007). It dissolves in dilute solutions of mineral acids and should be protected from light (British Pharmacopoeia, 2007).

Degradation products

In alkaline solution, oxidation of levodopa leads to the formation of inactive compounds such as melanin and related intermediates (Newton, 1978; Lund, 1994).

The degradation products of carbidopa are not known.

Stability in practice

There is a small but useful amount of information available with regard to the stability of co-careldopa formulation, providing evidence for shelf-lives of between 7 and 28 days.

Nahata *et al.* (2000) have developed two oral suspensions of co-careldopa for use in children with amblyodopia. Ten tablets (levodopa 100 mg and carbidopa 25 mg) were ground in a pestle and mortar and used to prepared two suspensions using the following vehicles:

- Ora-Plus 1:1 Ora-Sweet
- Ora-Plus 1:1 Ora-Sweet plus ascorbic acid (2 mg/mL).

The final concentrations were 5 mg/mL levodopa and 1.25 mg/mL carbidopa. Each of the suspensions were stored in 10 amber plastic prescription bottles and split between storage at 4°C and 25°C. Samples were removed in triplicate from every bottle on days 0, 7, 14, 28, 42, 56, 70 and 91. Analysis was via a stability-indicating HPLC method.

The starting pH values were 4.41 for the suspensions without ascorbic acid and 3.56 for those with ascorbic acid. For the suspension to be considered stable, both levodopa and carbidopa had to be maintained at over 90% of the starting concentration.

At 25°C, both drugs were stable in Ora-Plus and Ora-Sweet alone after 28 days at 25°C. The lower level of the standard deviation would take the level below 90% at this time, however. For the same suspensions stored at 4°C, stability was maintained for 56 days, though again the standard deviation lower limit would take the value below 90% at this time. The suspensions containing ascorbic acid showed poorer stability. This formulation approach is therefore not recommended.

The physical appearance was reported to remain largely unchanged when the drugs were stable. However, the authors report that the 25°C samples demonstrated a darker yellow colour during storage. In conclusion, Nahata *et al.* (2000) suggest that the Ora-Plus and Ora-Sweet formulation can be stored for at least six weeks when refrigerated and for four weeks at room temperature without substantial loss of potency. The authors recommend that the suspensions be stored in a fridge to limit microbial growth.

Further aqueous and non-aqueous formulae have been suggested for a 2 mg/mL carbidopa and 20 mg/mL levodopa liquid preparation (Anon, 2002) in Table 14.2.

The paper suggests that the aqueous formulation will be stable for 7 days in the refrigerator and that the non-aqueous formulation should be stable for at least 30 days (Anon, 2002). However, these formulations are not directly supported or validated by any data, and the shelf-life of the non-aqueous formulation cannot be endorsed due to a lack of related data.

Allwood (1987) has studied the stability of four drugs for Parkinson's disease in liquid formulation, using stability-indicating HPLC techniques. The four drugs were levodopa (Bracodopa), co-careldopa (Sinemet), co-beneldopa (Madopar) and selegiline (Eldepryl).

Table 14.2 Further suggested aqueous and non-aqueous formulae for a 20 mg/mL levodopa liquid preparation

	Aqueous	Non-aqueous
Carbidopa	200 mg	200 mg
Levodopa	2 g	2 g
Propylene glycol	5 mL	n/a
Ora-Plus	50 mL	n/a
Ora-Sweet or Ora-Sweet SF	to 100 mL	n/a
Saccharin	n/a	100 mg
Flavour (tangerine or other)	n/a	qs
Almond oil	n/a	to 100 mL

The co-careldopa suspension (100 mg levodopa and 10 mg carbidopa in 5 mL, or 20 and 2 mg/mL) was prepared using 'Keltrol' 0.4% w/v. Tests were carried out in buffered (pH 4 to 4.5) and unbuffered formulations, although the exact details of the formulae are not provided, and only the results of the buffered preparation are presented.

The degradation of the carbidopa is quicker than that of levodopa, and so becomes the important factor. Ninety-one per cent and 92% of the initial concentration of carbidopa remained after 20 days storage at 2–6°C and at ambient temperature, respectively. Allwood (1987) states that the degradation of the carbidopa in the unbuffered formulation was 'slightly more rapid', but no details are given. The author observed that the blue coloration from the tablet dyes faded rapidly in the mixtures stored at room temperature.

The results are limited by the lack of detail supplied in the number of samples, sampling procedures and exact formulation. The accelerated degradation of carbidopa when compared with levodopa may be partially explained by the higher solubility and lower strength of the carbidopa in the formulation. Therefore, small amounts of degradation in real terms appear greater when presented in terms of percentage degradation.

Microbial stability was not measured. Allwood (1987) suggests that levodopa is stable in the recommended formula for at least four weeks. Carbidopa degrades slowly and the rate of degradation may be pH dependent; Allwood (1987) also recommends the use of a 14-day shelf-life for the buffered co-careldopa preparation if stored in the refrigerator.

The use of chloroform as a preservative cannot be endorsed as this agent is known to be a class 3 carcinogen (CHIP Regulations, 2002). 'Keltrol' in this research refers to plain xanthan gum. However, most commercially available xanthan gum suspending agents, often termed 'Keltrol', now contain preservative systems. These may or may not include ethanol and/or chloroform. The specification of the suspending agent chosen should be checked carefully.

Kurth *et al.* (1993) have described the use of a levodopa/carbidopa/ascorbic acid solution (LCAS). Ten tablets of Sinemet 25/100 and 2 g of crystalline ascorbic acid were added to 1 L of tap water to give a solution of 1 mg/mL levodopa, 0.25 mg/mL carbidopa and 2 mg/mL ascorbic acid. The authors report that the solution is stable for 24 hours but the supported HPLC analytical data are not shown. Although Kurth *et al.* (1993) argue that LCAS may have advantages over the use of standard tablets, no further supporting evidence of stability is available, and the use of ascorbic acid must be questioned considering the data published by Nahata *et al.* (2000) (see above). The data produced by Nahata *et al.* suggested that ascorbic acid may have a deleterious effect on co-careldopa stability.

Metman *et al.* (1994) report dissolving 10.5 tablets of 100/25 mg co-careldopa, 2 g of ascorbic acid and 5 g of sodium chloride in 1 L of water, but no stability data are provided. The Pharmaceutical Codex (Lund, 1994) reports that excipients that have been used in formulation of suspensions include syrup.

Related products

den Hartigh *et al.* (1991) have investigated the stability of levodopa 1 mg/mL injection concentrate in water for injections (pH 3.7) that was stored in the dark at 0–4°C, and also of a ready-to-use infusion containing ascorbic acid as a stabilising agent. The ready-to-use solution was stored at room temperature and protected from light (Lund, 1994). The authors suggest the injection concentrate could be stored for at least three months under the conditions studied and the infusion was reported to be stable for 24 hours (den Hartigh *et al.*, 1991; Lund, 1994).

Stennet *et al.* (1986) have studied the stability of levodopa 1 mg/mL in 5% dextrose injection, adjusted to pH 5 or 6. Three samples of pH 5 solutions were stored at each of three temperatures (4, 25 and 45°C) and three samples of the pH 6 solutions were stored at 25°C only. HPLC analysis was carried out over 21 days.

All solutions maintained at least 90% of the initial concentration for 7 days, and all the solutions except for those stored at 4°C showed discoloration. The pH 5 samples remained clear and colourless and maintained stability for the full 21-day period, although the final day analysis showed more than 100% drug remaining. The solutions stored at 45°C showed discoloration in less than 12 hours. pH values remained unchanged for all solutions.

A rapidly dissolving formulation of carbidopa and levodopa (Parcopa, Schwarz) has become available in some countries (Anon, 2005). It can be taken with or without water. An intestinal gel in a 100 mL cassette is available in the UK for use with an enteral tube, via a Duodopa portable pump (BNF, 2006).

Bioavailability data

No bioavailability data are available. Nahata *et al.* (2000) report a small-scale trial of the Ora-Plus and Ora-Sweet trial in a limited number of children. Kurth *et al.* (1993) and Metman *et al.* (1994) have reported limited studies of oral levodopa/carbidopa solutions versus tablets.

References

Allwood MC (1987). Liquid formulations of drugs for Parkinson's disease. *Br J Pharm Pract* 9:34–36.

Anon (2002). Carbidopa 2 mg/ml and levodopa 20 mg/ml oral suspension. *Int J Pharm Compound* 6: 202.

Anon (2005). Parcopa: a rapidly dissolving formulation of carbidopa/levodopa. *Med Lett* 47 (1201): 12.

British Pharmacopoeia (2007). *British Pharmacopoeia*, Vol. 1. London: The Stationery Office, p. 374 and Vol. 2, p. 1234.

CHIP3 Regulations (2002). Chemicals (Hazard Information and Packaging for Supply) Regulations. SI 2002/1689.

den Hartigh J, Twiss I, Vermeij P, (1991). *Pharm Weekbl Sci* 13: J3.

FDA (Food and Drug Administration) (2005). FDA policy statement for the development of new stereoisomer drugs. www.fda.gov/cder/guidance/stereo.htm. (accessed 25 September 2006).

Gomez R, Hagel RB, MacMullan EA (1976). Levodopa. In: Florey K, ed. *Analytical Profiles of Drug Substances*; Vol. 13. London: Academic Press, pp. 189–223.

Joint Formulary Committee (2010). *British National Formulary*, No. 59. London: British Medical Association and Royal Pharmaceutical Society of Great Britain.

Kurth MC, Tetrud JW, Irwin BA, Lyness WH, Langston MD (1993). Oral levodopa/carbidopa solution versus tablets in Parkinson's patients with severe fluctuations: a pilot study. *Neurology* 43:1036–1039.

Lund W, ed. (1994). *The Pharmaceutical Codex: Principles and Practice of Pharmaceutics*, 12th edn. London: Pharmaceutical Press.

Lowey AR, Jackson MN (2008). A survey of extemporaneous dispensing activity in NHS trusts in Yorkshire, the North-East and London. *Hosp Pharmacist* 15:217–219.

Metman LV, Hoff J, Mouradian MM, Chase TN (1994). Fluctuations in plasma levodopa and motor responses with liquid and tablet. *Mov Disord* 9:463–465.

Moffat AC, Osselton MD, Widdop B (2003). *Clarke's Analysis of Drugs and Poisons*, 3rd edn. London: Pharmaceutical Press. ISBN 978 0 85369 473 1

Nahata MC, Morosco RS, Leguire LE (2000). Development of two stable oral suspensions of levodopa-carbidopa for children with amblyodopia. *J Pediatr Ophthalmol Strabismus* 37:333–337.

Newton DW (1978). Physicochemical determinants of incompatibility and instability in injectable drug solutions and admixtures. *Am J Hosp Pharm* 35:1213–1222.

Stennett DJ, Christensen JM, Anderson JL, Parrott KA (1986). Stability of levodopa in 5% dextrose injection at pH 5 or 6. *Am J Hosp Pharm* 43:1726–1728.

Co-enzyme Q10 oral liquid

Synonyms

Ubiquinone, ubidecarenone, mitoquinone.

Risk assessment of parent compound

A risk assessment survey completed by clinical pharmacists suggested that co-enzyme Q10 has a wide therapeutic index, with the potential to cause generally mild side-effects if inaccurate doses were to be administered (Lowey and Jackson, 2008). Current formulations used at NHS trusts generally show a low technical risk, based on the number and/or complexity of manipulations and calculations. Co-enzyme Q10 is therefore generally regarded as a low-risk extemporaneous preparation (Box 14.13).

Summary

A preparation with a marketing authorisation in Italy is known to be available for import. Other imports may also be available. The use of preparations

> ***Box 14.13*** *Preferred formula*
>
> **Note:** An oral liquid preparation is available as an import from Italy, where it carries a marketing authorisation.
>
> Using a typical strength of 10 mg/mL as an example:
>
> - ubiquinone 25 mg tablets × 40
> - methylcellulose solution 2% to 100 mL.
>
> ### Method guidance
>
> Tablets can be ground to a fine, uniform powder in a pestle and mortar. A small amount of methylcellulose may be added to form a paste, before adding further portions of methylcellulose up to 75% of the final volume. Transfer to a measuring cylinder. The remaining suspending agent can be used to wash out the pestle and mortar of any remaining drug before making the suspension up to 100% volume. Transfer to an amber medicine bottle.
>
> ### Shelf-life
>
> 7 days refrigerated in amber glass.

licensed abroad should be considered as a priority. The exact specification of these products should be assessed by a suitably competent pharmacist before use.

Pharmacists are referred to the document 'Guidance for the purchase and supply of unlicensed medicinal products – notes for prescribers and pharmacists' for further information (NHS Pharmaceutical Quality Assurance Committee, 2004).

The preparation above was made in a very limited number of centres in a recent survey of NHS trusts in Yorkshire, the North-East and London. There are no direct data to support the stability of the preparation, and so it must be used with caution. The quality of the active ingredients used in co-enzyme Q10 should be assessed carefully, as many preparations are marketed as food and nutritional supplements rather than as pharmaceuticals. It may also be possible to round doses to equate to whole tablet/capsules, which can then be dispersed in water.

Clinical pharmaceutics

Points to consider

- Co-enzyme Q10 is practically insoluble and so will be held in suspension rather than solution. **The suspension should be shaken well to ensure adequate dosage uniformity.**
- Co-enzyme Q10 is reported to be odourless and tasteless (Anon, 2000).
- There are few bioavailability data available for liquid forms of co-enzyme Q10. Absorption from solid dosage forms is reported to be poor.

Other notes

Thirty milligrams strength fast-melting and effervescent preparations have been reported to have been used (Joshi, 2003).

The formulation of chewable troches has been reported in the literature (Anon, 2000). However, no stability data are presented.

Technical information

Structure

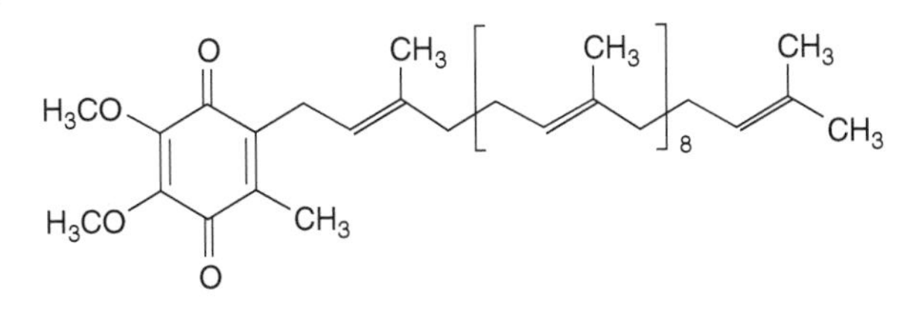

Empirical formula

$C_{59}H_{90}O_4$ (British Pharmacopoeia, 2007).

Molecular weight

863 (British Pharmacopoeia, 2007).

Solubility

Practically insoluble in water (<100 micrograms/mL) (Lund, 1994).

pK_a

Not known.

Optimum pH stability

Not known.

Stability profile

Physical and chemical stability

Co-enzyme Q10 is a yellow or orange, crystalline powder, practically insoluble in water and very slightly soluble in ethanol (British Pharmacopoeia, 2007). It is a fat-soluble benzoquinone derivative (Zhiri and Belichard, 1994; Joshi *et al.*, 2003).

It gradually decomposes and darkens on exposure to light (British Pharmacopoeia, 2007). The darkening of the crystals is attributed to change in the surface condition in addition to the structural changes of co-enzyme Q10 molecules. These physico-chemical changes have been shown to affect the UV absorption properties of solid-state samples (Kommuru *et al.*, 1999).

Antioxidants have been used to improve the stability of solid dose drug formulations (Kommuru *et al.*, 1999). Solution preparations of co-enzyme Q10 are reported to be far more prone to degradation than in the solid state (Kommuru *et al.*, 1999).

Co-enzyme Q10 should be stored in an airtight container and protected from light (British Pharmacopoeia, 2007).

Stability in practice

Soonswang *et al.* (2005) used either a soft gelatin capsule or a soya bean suspension to investigate the effect of co-enzyme Q10 on idiopathic chromic dilated cardiomyopathy in children. Details of the suspension were requested but not received; no indication of stability is provided in the paper.

Related preparations – solid dose

The stability of co-enzyme Q10 as a powder has been investigated at 25, 37, 45 and 55°C and at various humidities (Kommuru *et al.*, 1999). Various antioxidants were added to samples of 750 mg in glass vials that were wrapped in aluminium foil. The degradation profiles were first order in nature, and increased with increasing temperature. The effect of humidity was insignificant.

Interpretation of data suggested a predicted shelf-life at room temperature of 6.3 years. Two commercially marketed preparations were investigated; Kommuru *et al.* (1999) suggested that some marketed preparations may not be stable during their shelf-life.

Among the antioxidants studied, mixtures of ascorbic acid (5%) and EDTA (0.1%) offered better protection than phenolic antioxidants such as butylated anisole (BHA), butylated toluene (BHT) or propyl gallate (PG) when the powder was exposed to light or a higher temperature (55°C).

Bioavailability data

Commercial co-enzyme Q10 formulations have been associated with poor intestinal absorption, given its relatively large molecular weight and hydrophobic nature (Weis *et al.*, 1994; Chopra *et al.*, 1998; Joshi *et al.*, 2003; Ullman *et al.*, 2005). Various approaches have been investigated to improve absorption by creating a more water-miscible preparation, mixing with other oils such as rice bran oil and soya bean oil, and modifying particle size.

Maggi and Orsenigo (1993) investigated the bioavailability of two different formulations of ubidecarenone for oral use. Eighteen healthy volunteers received 50 mg three times daily from an extemporaneous standard aqueous suspension and an aqueous solution in a cross-over study. The concentration of co-enzyme Q10 in plasma was assayed using an HPLC assay. The formulations were reported to show comparable bioavailability. Details of the formulations are not known.

References

Anon (2000). Coenzyme Q10 200 mg chewable troches. *Int J Pharm Compound* 4: 129.

British Pharmacopoeia (2007). *British Pharmacopoeia*, Vol. 2. London: The Stationery Office, pp. 2127–2128.

Chopra RK, Goldman R, Sinatra ST, Bhagavan HN (1998). Relative bioavailability of coenzyme Q10 formulations in human subjects. *Int J Vitam Nutr Res* 68: 109–113.

Joshi SS, Sawant SV, Shedge A, Halpner AD (2003). Comparative bioavailability of two novel coenzyme Q10 preparations in humans. *Int J Clin Pharmacol Ther* 41(4): 42–48.

Kommuru TR, Ashraf M, Khan MA, Reddy IK (1999). Stability and bioequivalence studies of two marketed formulations of coenzyme Q10 in beagle dogs. *Chem Pharm Bull* 47: 1024–1028.

Lund W, ed. (1994) *The Pharmaceutical Codex: Principles and Practice of Pharmaceutics*, 12th edn. London: Pharmaceutical Press.

Lowey AR, Jackson MN (2008). A survey of extemporaneous dispensing activity in NHS trusts in Yorkshire, the North-East and London. *Hosp Pharmacist* 15: 217–219.

Maggi GC, Orsenigo F (1993). Comparative evaluation of bioavailability of two different formulations of ubidecarenone for oral use. Steady-state in cross-over study in healthy volunteers. *Cuore* 10: 519–527.

NHS Pharmaceutical Quality Assurance Committee (2004). Guidance for the purchase and supply of unlicensed medicinal products. Unpublished document available from regional quality assurance specialists in the UK or after registration from the NHS Pharmaceutical Quality Assurance Committee website: www.portal.nelm.nhs.uk/QA/default.aspx.

Ullmann U, Metzner J, Schulz C, Perkins J, Leuenberger (2005). A new coenzyme Q10 tablet-grade formulation (all-Q) is bioequivalent to Q-Gel and both have better bioavailability properties than Q-SorB. *J Med Food* 8: 397–399.

Soonswang J, Santawesin C, Durongpisitkul K, Laohaprasitporn D, Nana A, Punlee K, Kangkagate C (2005). The effect of coenzyme Q10 on idiopathic chronic dilated cardiomyopathy in children. *Pediatr Cardiol* 26: 361–366.

Weis M, Mortensen SA, Rassing MR, Mooler-Sonnergaard J, Poulsen G, Rasmussen SN (1994). Bioavailability of four oral coenzyme Q10 formulations in healthy volunteers. *Mol Aspects Med* 15: S273–S280.

Zhiri A, Belichard P (1994). Reversed-phase liquid chromatographic analysis of coenzyme Q10 and stability study in human plasma. *J Liq Chromatogr* 17: 2633–2640.

Dexamethasone oral liquid and dexamethasone sodium phosphate oral liquid

Note: Dexamethasone sodium phosphate 1.3 mg is approximately equivalent to dexamethasone base 1 mg.

Risk assessment of parent compound

A risk assessment survey completed by clinical pharmacists suggested that dexamethasone has a medium therapeutic index. Inaccurate doses were thought to be associated with significant morbidity (Lowey and Jackson, 2008). Current formulations used at NHS trusts were generally of a moderate technical risk, based on the number and/or complexity of manipulations and calculations. Dexamethasone is therefore regarded as a medium- to high-risk extemporaneous preparation (Box 14.14).

Summary

NHS experience is divided between dexamethasone (base) tablets in 'Keltrol' and dilution of the sodium phosphate to 100–200 micrograms/mL.

Several formulations of dexamethasone oral liquid are available from NHS and non-NHS 'Specials' manufacturers. The formulations and their supporting evidence should be assessed by a suitably competent pharmacist before use. The purchase of a 'Special' product should be considered in preference to an extemporaneous preparation under normal circumstances.

Clinical pharmaceutics

Points to consider

- Dexamethasone sodium phosphate displays polymorphism. In theory, changes in polymorphic form could alter particle size (Han *et al.*, 2006). Whether this occurs with oral liquid forms of dexamethasone is not known.

***Box 14.14** Preferred formulae*

Note: 'Specials' are now available from reputable manufacturers. Extemporaneous preparation should only occur where there is a demonstrable clinical need.

Strengths of the order of 400–500 micrograms have been largely superseded by the advent of a licensed product of 400 micrograms/mL.

Using a strength of 100 micrograms/mL dexamethasone base (equivalent to approximately 130 micrograms/mL dexamethasone sodium phosphate) as an example:

- dexamethaxone (as sodium phosphate) injection (4 mg/mL) 2.5 mL
- Ora-Plus 50 mL
- Ora-Sweet to 100 mL.

Method guidance

Draw up the required amount of injection using a filter needle or filter straw and transfer to a measuring cylinder. Add Ora-Plus up to 50% of the final volume. Make up to 100% volume with the Ora-Sweet. Mix thoroughly and transfer to an amber medicine bottle.

Shelf-life

28 days at room temperature in amber glass.

- Ora-Plus and Ora-Sweet are free from chloroform and ethanol.
- Ora-Plus and Ora-Sweet contain hydroxybenzoate preservative systems.
- Ora-Sweet SF has a cherry flavour which may help to mask the bitter taste of the drug.
- There are no known bioavailability data available for the extemporaneous formulations.

Other notes

- Licensed preparations are available in the UK. The Dexsol brand (400 micrograms/mL) contains benzoic acid, propylene glycol, citric acid monohydrate, liquid maltitol, garden mint flavour (containing isopropanol and propylene glycol), liquid sorbitol non-crystallising, sodium citrate and purified water (Summary of Product Characteristics, 2006). The product currently has a 2-year shelf-life but should be used within three months of opening. Storing the Dexsol formulation at

temperatures above 25°C may lead to precipitation (Summary of Product Characteristics, 2006). The product should not be used if this occurs.
- There are some preparations that have been licensed abroad. However, they are known to contain up to 30% ethanol (Woods, 2001).
- Some tablet formulations disperse readily in water and are often scored, allowing administration of fractional doses (Woods, 2001).

Technical information

Structure of dexamethasone as the free base

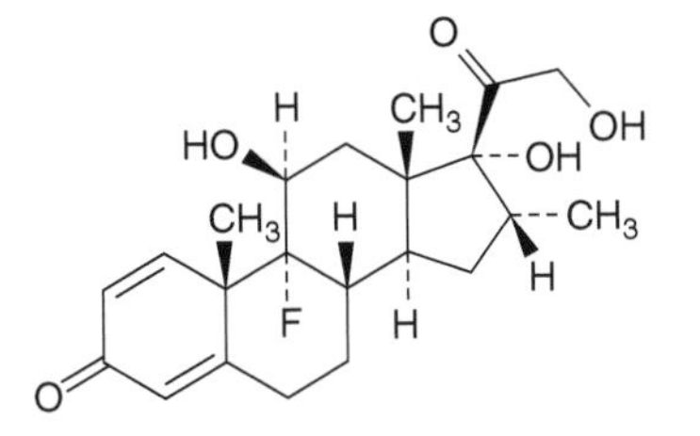

Empirical formula

$C_{22}H_{29}FO_5$ (British Pharmacopoeia, 2007).

Molecular weight

392.5 (British Pharmacopoeia, 2007).

Solubility

- Dexamethasone base is practically insoluble in water (British Pharmacopoeia, 2007). Approximately 100 micrograms/mL (Trissel, 2000).
- Dexamethasone sodium phosphate is freely soluble in water (British Pharmacopoeia, 2007). Approximately 500 mg/mL (Trissel, 2000).

pK_a

Not known.

Optimum pH stability

Not known.

Stability profile

Physical and chemical stability

Although dexamethasone displays poor solubility, the low strength of typical extemporaneous preparations (e.g. 100 micrograms/mL) means that most or all of the drug is likely to be in solution, and therefore open to aqueous degradation mechanisms. At higher strengths, such as 400–500 micrograms/mL, the majority of drug may be in suspension rather than solution, thus potentially limiting degradation. Dexamethasone sodium phosphate is highly soluble, so the strength of the preparation will not be an issue.

Dexamethasone is subject to degradation following exposure to ultraviolet light, oxygen and alkali (Accordino *et al.*, 1994). In the D-ring, the C-17 dihydroxyacetate side-chain is quite susceptible to aerobic and anaerobic transformations. Accordino *et al.* (1994) suggest that the predominant reaction, under base catalysed aerobic conditions, appears to be oxidative cleavage of the side-chain to the corresponding etianic acid. Solutions of dexamethasone lose approximately 50% of the C-17 α-ketol side-chain within 6–8 minutes in the presence of a base catalyst (Cohen, 1973; Lund, 1994). It has been suggested that formation of the C-17 ketone is only a minor reaction product (Cohen, 1973; Accordino *et al.*, 1994).

Trissel (2000) suggests that solutions and suspensions of dexamethasone and its salts are heat labile, and therefore should not be autoclaved for sterilisation. Injections should be protected from light and freezing (Trissel, 2000; United States Pharmacopeia, 2003).

Dexamethasone is a white or almost white crystalline powder, practically insoluble in water and sparingly soluble in ethanol (British Pharmacopoeia, 2007). In the solid state, it is stable in air but should be protected from light (Lund, 1994; British Pharmacopoeia, 2007). Dexamethasone oral solution has a pH of 3–5, while the oral elixir (Merck Sharp and Dohme) is reported to have a pH of 3.3 (Cutie, 1983; Trissel, 2000).

Dexamethasone sodium phosphate is a white or almost white powder, very hygroscopic, freely soluble in water and slightly soluble in alcohol. It displays polymorphism (British Pharmacopoeia, 2007); two crystal forms have been reported (Cohen, 1973).

A 1% dexamethasone sodium phosphate aqueous solution has a pH between 7.5 and 10.5 (Trissel, 2000). Dexamethasone sodium phosphate injection has a pH of 7.0–8.5 (Trissel, 2000). Dexamethasone sodium phosphate should be stored in an airtight container, protected from light (British Pharmacopoeia, 2007).

Degradation products

May include the oxidative products 16α- and 16β-methyl-17-ketone (Lund, 1994).

Stability in practice

Several studies suggest that dexamethasone may be formulated with satisfactory stability data to support a reasonable shelf-life.

The use of the injection (dexamethasone sodium phosphate) diluted in flavoured syrups and vehicles has been suggested (Canadian Society of Hospital Pharmacists, 1988; Woods, 2001). Wen-Lin Chou *et al.* (2001) have studied the stability of dexamethasone in extemporaneously prepared oral suspensions. Concentrations of 0.5 mg/mL and 1 mg/mL suspensions were prepared using a 1:1 mixture of Ora-Sweet and Ora-Plus as the vehicle.

Dexamethasone sodium phosphate injection solution 4 mg/mL was used to prepare six 40 mL suspensions of each of the two strengths. Three of these suspensions were stored at 4°C and a further three suspensions were stored at 25°C. All samples were exposed only to fluorescent light in the laboratory. Over a period of 91 days, all suspensions were evaluated for changes in colour, pH, viscosity, presence of precipitates and ease of resuspension. Following these physical checks, each bottle was shaken manually for 10 seconds before 0.8 mL was removed at stored at –84°C until analysis by a stability-indicating HPLC method.

No significant changes in physical appearance or odour were noticed over the 91 days. The suspensions were cloudy white in appearance, with a sweet smell. Viscosity did not change and resuspension did not prove problematic. The mean pH values were:

- 5.18 for the 0.5 mg/mL suspensions stored at 4°C
- 5.09 for the 0.5 mg/mL suspensions stored at 25°C
- 5.63 for the 1 mg/mL suspensions stored at 4°C
- 5.56 for the 1 mg/mL suspensions stored at 25°C.

The assay results showed that all suspensions maintained over 90% of the original concentration over the 91 days. The authors conclude that the dexamethasone suspension remained stable for 91 days at both temperatures and at both concentrations.

Lugo and Nahata (1994) also studied the stability of diluted dexamethasone sodium phosphate injection, for use in the treatment of premature neonates with bronchopulmonary dysplasia. Ten vials of 4 mg/mL injection were diluted to 1 mg/mL with bacteriostatic sodium chloride and divided equally between storage at 4°C and 22°C. The samples were exposed to fluorescent light. Drug content was assayed on days 0, 1, 3, 7, 14, 21 and 28 using a stability-indicating HPLC method. Visual inspection for precipitation and discoloration was also undertaken on the study days.

Less than 5% degradation occurred at both temperatures over 28 days. Lugo and Nahata (1994) suggest that dilution of the licensed injection may help to minimise errors in measuring the small dexamethasone doses required in this patient group. Bulk preparation may also help to reduce labour and drug wastage costs. The authors conclude that dexamethasone sodium phosphate injection 1 mg/mL in bacteriostatic sodium chloride 0.9% was found to be stable for 28 days at 4°C and 22°C.

Accordino *et al.* (1994) have published a short-term stability study of an oral solution of dexamethasone, using the following formula:

- dexamethasone BP 100 mg
- ethanol 95% BP 1973 15 mL
- propylene glycol BP 20 mL
- glycerol BP 50 mL
- raspberry essence HC 417 0.5 mL
- saccharin sodium BP 0.3 g
- purified water BP to 100 mL.

Accordino *et al.* (1994) suggested that the use of the free base was preferable to that of the sodium phosphate due to previous concerns that had been raised regarding the bioavailability of the sodium phosphate ester (Fleisher *et al.*, 1986). The propylene glycol content in the formula allowed the overall alcohol content to be reduced to 15%. The glycerol was added to reduce the water content and to act as a sweetener. Palatability and flavour were improved via the use of raspberry essence and saccharin. Accordino *et al.* (1994) suggest that preservatives were not necessary as the propylene glycol, ethanol and glycerol can individually act as preservatives at the listed concentrations (Block, 1991).

Prednisolone was used as an internal standard in a stability-indicating HPLC assay. Samples were duplicated and the mean result used. Three separate batches of dexamethasone oral solution were prepared and stored in 50 mL amber glass bottles with polyethylene closures. Bottles from each batch were stored at 2–8°C, 25°C in the dark, 25°C when exposed to sunlight, and 37°C. Analysis took place periodically over a period of 26 weeks.

Throughout the study period, the mean dexamethasone concentration remained above 95% in samples of all batches, except for those stored at 37°C. However, even at this elevated temperature, the mean drug concentration remained above 90% (91.4%, 95.0% and 93.4% remaining in the three batches at 26 weeks). Small amounts of degradation (<3%) were noted by the authors after 13 weeks at 37°C. pH values were not deemed to have changed significantly, and no physical stability problems were noted.

Although the researchers consider that the solution would not support microbial growth, formal microbiology tests were not carried out. Therefore, a maximum expiry of 30 days is recommended (Accordino *et al.*, 1994). The use of the formula, however, may well be limited by the alcohol content.

Excipients that have been used in presentations of dexamethasone as an oral solution include benzoic acid solution, wild cherry-flavoured syrup, chloroform water and ethanol (Lund, 1994; Bjornson *et al.*, 2004).

Bioavailability data

There are no available data for the extemporaneous preparations. The oral bioavailability of dexamethasone sodium phosphate may be less than that of dexamethasone (Fleisher *et al.*, 1986).

References

Accordino A, Chambers R, Thompson B (1994). A short-term study of an oral solution of dexamethasone. *Aust J Hosp Pharm* 24: 312–316.

Bjornson CL, Klassen TP, Williamson J, Brant L, Mitton C, Plint A *et al.* (2004). A randomized trial of a single dose of oral dexamethasone for mild croup. *N Engl J Med* 351: 1306–1313.

Block SS, ed. (1991). *Disinfection, Sterilisation, and Preservation*, 4th edn. Philadelphia: Lea and Febiger, pp. 882–883.

British Pharmacopoeia (2007). *British Pharmacopoeia*, Vol. 1. London: The Stationery Office, pp. 783–785 and Vol. 2, pp. 2090–2092.

Canadian Society of Hospital Pharmacists (1988). *Extemporaneous Oral Liquid Dosage Preparations*. Toronto, Ontario: CSHP.

Cohen E (1973). Dexamethasone In: Florey K, ed. *Analytical Profiles of Drug Substances*, Vol, 2. London: Academic Press, pp. 163–197.

Cutie AJ, Altman E, Lenkel L (1983). Compatibility of enteral products with commonly-employed drug additives. *J Parenter Enter Nutr* 7: 186–191.

Fleisher D, Johnson KC, Stewart BH, Amidon GL (1986). Oral absorption of 21-corticosteroid esters: a function of aqueous stability and intestinal enzyme activity and distribution. *J Pharm Sci* 75: 934–939.

Han J, Beeton A, Long PF, Wong I, Tuleu C (2006). Physical and microbiological stability of an extemporaneous tacrolimus suspension for paediatric use. *J Clin Pharm Ther* 31: 167–172.

Lowey AR, Jackson MN (2008). A survey of extemporaneous dispensing activity in NHS trusts in Yorkshire, the North-East and London. *Hosp Pharmacist* 15: 217–219.

Lugo RA, Nahata MC (1994). Stability of diluted dexamethasone sodium phosphate injection at two temperatures. *Ann Pharmacother* 28: 1018–1019.

Lund W, ed. (1994). *The Pharmaceutical Codex: Principles and Practice of Pharmaceutics*, 12th edn. London: Pharmaceutical Press.

Summary of Product Characteristics (2006). Dexsol. www.medicines.co.uk (accessed 4 October 2006).

Trissel LA, ed. (2000). *Stability of Compounded Formulations*, 2nd edn. Washington DC: American Pharmaceutical Association.

United States Pharmacopeia (2003). *United States Pharmacopeia*, XXVI. Rockville, MD: The United States Pharmacopeial Convention.

Wen-Lin Chou J, Decarie D, Dumont R, Ensom MHH (2001). Stability of dexamethasone in extemporaneously prepared oral suspensions. *Can J Hosp Pharm* 54: 96–101.

Woods DJ (2001). *Formulation in Pharmacy Practice. eMixt, Pharminfotech*, 2nd edn. Dunedin, New Zealand. www.pharminfotech.co.nz/emixt (accessed 18 April 2006).

Diazoxide oral liquid

Risk assessment of parent compound

A survey of clinical pharmacists suggested that diazoxide has a moderate therapeutic index, with the potential to cause serious morbidity if inaccurate doses are administered (Lowey and Jackson, 2008). Current formulations used at NHS trusts show a moderate technical risk, based on the number and complexity of manipulations and calculations. Diazoxide oral liquid is therefore regarded as a high-risk extemporaneous preparation (Box 14.15).

Summary

Importation of a licensed product should be considered if the formulation is appropriate for the intended patient(s). By using the product licensed in a mutually recognised country, there will be a greater level of quality assurance, as there will have been both stability and bioavailability data presented to the licensing authority as part of the licensing process. The appropriateness of the formulation should be assessed on an individual patient basis by a suitably competent pharmacist.

Pharmacists are referred to the document 'Guidance for the purchase and supply of unlicensed medicinal products – notes for prescribers and pharmacists' for further information (NHS Pharmaceutical Quality Assurance Committee, 2004).

Most of the experience regarding extemporaneous formulations of diazoxide is based on xanthan gum preparations. Given the very limited solubility of diazoxide, extensive drug breakdown is unlikely. However, no chemical stability data are available to support these preparations; a conservative shelf-life of 7 days is therefore suggested.

Alternative suspending agents could be used (e.g. carboxymethylcellulose, methylcellulose).

Clinical pharmaceutics

Points to consider

- Given the typical preparation strengths of 10–50 mg/mL, the majority of drug is likely to be in suspension rather than solution. **Shake the bottle before use to improve dosage uniformity.**

Box 14.15 *Preferred formula*

Note: A diazoxide oral liquid is now also available as a import from mutually recognised countries, where it is known to carry a product licence.

Using a typical strength of 10 mg/mL as an example:

- diazoxide 50 mg tablets × 20
- Ora-Plus 50 mL
- Ora-Sweet/Ora-Sweet SF to 100 mL.

Method guidance

Tablets can be ground to a fine, uniform powder in a pestle and mortar. A small amount of Ora-Plus may be added to form a paste, before adding further portions of Ora-Plus up to 50% of the final volume. Transfer to a measuring cylinder. The Ora-Sweet SF can be used to wash out the pestle and mortar of any remaining drug before making the suspension up to 100% volume. Transfer to an amber medicine bottle.

Shelf-life

7 days at room temperature in amber glass. **Shake the bottle.**

Alternative

An alternative formula would be:

- diazoxide 50 mg tablets × 20
- xanthan gum 0.4% ('Keltrol' or similar) to 100 mL.

Shelf-life

7 days at room temperature in amber glass. **Shake the bottle.**

Note: 'Keltrol' xanthan gum formulations vary slightly between manufacturers, and may contain chloroform and ethanol. The individual specification should be checked by the appropriate pharmacist.

- 'Keltrol' is made to slightly different specifications by several manufacturers, but typically contains ethanol and chloroform. The exact specification should be assessed before use by a suitably competent pharmacist. Ethanol is a CNS depressant. Chloroform has been classified as a class 3 carcinogen (i.e. substances that cause concern owing to possible carcinogenic effects but for which available information is not adequate to make satisfactory assessments) (CHIP3 Regulations, 2002).

- Alternative formulations may contain excipients that are not suitable for certain neonates and young children (e.g. alcohol, hydroxybenzoates).
- Higher strength preparations may be available for import. A suitably competent pharmacist should assess the suitability of any imported products before use.
- There are no data available with regard to the bioavailability of these preparations.

Technical information

Structure

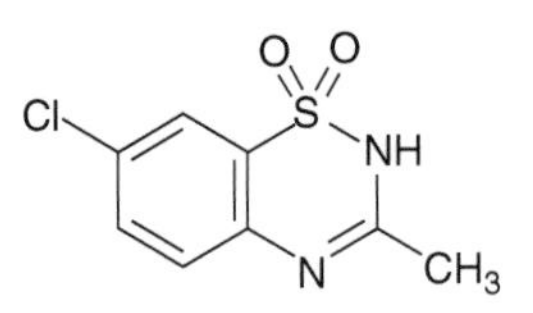

Empirical formula

$C_8H_7ClN_2O_2S$ (British Pharmacopoeia, 2007).

Molecular weight

230.7 (British Pharmacopoeia, 2007).

Solubility

Practically insoluble in water (more than 10,000 mL of solvent required for each gram of solute; equivalent to less than 100 micrograms/mL) (British Pharmacopoeia, 2007).

pK_a

8.5 (Moffat, 1986).

Optimum pH stability

Not known.

Stability profile

Physical and chemical stability

Diazoxide is a white or almost white, fine or crystalline powder. It is practically insoluble in water, slightly soluble in alcohol and very soluble in dilute solutions of alkali hydroxides (British Pharmacopoeia, 2007).

Diazoxide injection contains sodium hydroxide (pH approximately 11.6). It can be stored at controlled room temperature or in the refrigerator. The clear, colourless solution darkens on exposure to light and should be protected from light, heat and freezing. Darkened solutions may be subpotent and should not be used (Trissel, 2005).

Diazoxide is reported to be physically compatible with 'Keltrol' (Nova Laboratories Ltd, 2003).

The limited solubility of diazoxide (less than 100 micrograms/mL) should limit the amount of drug in solution, and therefore limit the amount of drug open to degradation mechanisms.

Degradation products

No data.

Stability in practice

There are no published stability data to support the stability of an extemporaneously prepared diazoxide suspension. Woods (2001) suggests the use of the following formula for a 5 mg/mL preparation:

- diazoxide tablets 25 mg 20
- sodium carboxymethylcellulose* 400 mg
- parabens 0.1%
- ethanol (95% or 90%) 7 mL
- glycerol 40 mL
- water to 100 mL.

Similar formulations based on carboxymethycellulose have been used in some NHS centres (Lowey and Jackson, 2008). Woods (2001) suggests that similar formulations are available in other countries, although the lack of formal stability studies is acknowledged. The tablets or capsule contents should be triturated with around 10 mL of glycerol. The sodium carboxymethylcellulose should be hydrated in approximately 25 mL of water and added to the crushed tablets when cooled. The resulting liquid should be

*Alternatively, use methylcellulose.

mixed thoroughly before the remaning excipients are added. A shelf-life of 7 days in the refrigerator (when protected from light) is suggested, although Woods suggests that the preparation is likely to be stable for at least 30 days.

The use of ethanol in the preparation is not ideal and may lead to pharmacological effects. Parabens (also known as hydroxybenzoates) are a group of widely used antimicrobial preservatives in cosmetics and oral and topical pharmaceutical formulations. Concern has been expressed over their use in infant parenteral products as bilirubin binding may be affected (Rieger, 2003). This is potentially hazardous in hyperbilirubinaemic neonates (Loria, 1976). However, there is no proven link to similar effects when used in oral products. The World Health Organization (WHO/FAO, 1974) has suggested an estimated total acceptable daily intake for methyl-, ethyl- and propyl-hydroxybenzoates at up to 10 mg/kg body weight.

Bioavailability data

No information available.

References

British Pharmacopoeia (2007). *British Pharmacopoeia*, Vol. 1. London: The Stationery Office, pp. 783–785 and Vol. 2, pp. 2090–2092.

CHIP3 Regulations (2002). Chemicals (Hazard Information and Packaging for Supply) Regulations. SI 2002/1689.

Loria CJ, Excehverria P, Smith AL (1976). Effect of antibiotic formulations in serum protein: bilirubin interaction of newborn infants. *J Pediatr* 89:479–482.

Lowey AR, Jackson MN (2008). A survey of extemporaneous dispensing activity in NHS trusts in Yorkshire, the North-East and London. *Hosp Pharmacist* 15:217–219.

Moffat AC, Osselton MD, Widdop B (2003). *Clarke's Analysis of Drugs and Poisons*, 3rd edn. London: Pharmaceutical Press.

NHS Pharmaceutical Quality Assurance Committee (2004). Guidance for the purchase and supply of unlicensed medicinal products. Unpublished document available from regional quality assurance specialists in the UK or after registration from the NHS Pharmaceutical Quality Assurance Committee website: www.portal.nelm.nhs.uk/QA/default.aspx.

Nova Laboratories Ltd (2003). Information Bulletin – Suspension Diluent Used at Nova Laboratories.

Rieger MM (2003). Ethylparabens. In: Rowe RC, Sheskey PJ, Weller PJ, eds. *Handbook of Pharmaceutical Excipients*, 4th edn. London: Pharmaceutical Press and Washington DC: American Pharmaceutical Association.

Trissel LA (2005). *Handbook on Injectable Drugs*, 13th edn. Bethesda, MD: American Society of Health System Pharmacists.

WHO/FAO (1974). Toxicological evaluation of certain food additives with a review of general principles and of specifications. Seventeenth report of the Joint FAO/WHO Expert Committee on Food Additives. World Health Organ. Tech. Rep. Ser. No. 539.

Woods DJ (2001). *Formulation in Pharmacy Practice. eMixt, Pharminfotech*, 2nd edn. Dunedin, New Zealand. www.pharminfotech.co.nz/emixt (accessed 18 April 2006).

Dinoprostone (prostaglandin E2) oral liquid

Risk assessment of parent compound

A recent survey of clinical pharmacists suggested that dinoprostone has a moderate therapeutic index (Lowey and Jackson, 2008). However, given its use in neonatal cardiology centres, it carries the potential to cause serious morbidity or death if inaccurate dosages are administered. The formulation seen in the survey shows a moderate technical risk, based on the number and complexity of manipulations and calculations. Dinoprostone is therefore regarded as a high-risk extemporaneous preparation. (Box 14.16).

Summary

This preparation is made in a limited number of NHS centres. Some centres report the preferential routine use of the intravenous route, due to the reliable and reproducible plasma levels associated with the intravenous method of administration (S. Jarvis, Paediatric Clinical Pharmacist, United Bristol Healthcare NHS Trust, personal communication, 2007).

Box 14.16 *Preferred formula*

Using a typical strength of 100 micrograms/mL as an example:

- dinoprostone injection 1 mg/mL 0.75 mL
- water for injection to 7.5 mL (or adjust values for larger volumes).

Method guidance

Draw up 0.75 mL of dinoprostone using a filter needle and 1 mL syringe. Open the water ampoule and draw up 6.75 mL. Add to an amber medicine bottle. Remove the filter needle and add the dinoprostone to the bottle. Mix well.

Shelf-life

7 days refrigerated in amber glass.

There is a paucity of published data to support the stability or bioavailability of the oral preparation. The available unpublished stability data appears to suggest that a shelf-life of 7 days is justifiable for an unpreserved system stored in the fridge.

The oral solution contains approximately 10% alcohol (w/v).

Clinical pharmaceutics

Points to consider

- Dinoprostone is formulated as a solution for intravenous use; this is diluted for oral use.
- The licensed injection is formulated with dehydrated alcohol; the 1 mg/mL ampoule contains 99.86% alcohol (798.9 mg) and the 10 mg/mL ampoule contains 98.63% alcohol (800 mg) (A. Wdowiak, Pfizer Medical Information, personal communication, 2006). Therefore, the alcohol content of the suggested formulation is approximately 10% (w/v).
- There are no data available regarding bioavailability.
- The use of hydroxybenzoate preservative compounds such as Nipasept (see below) may not be advisable in neonates, particularly those with jaundice, kernicterus or hyperbilirubinaemia.

Technical information

Structure

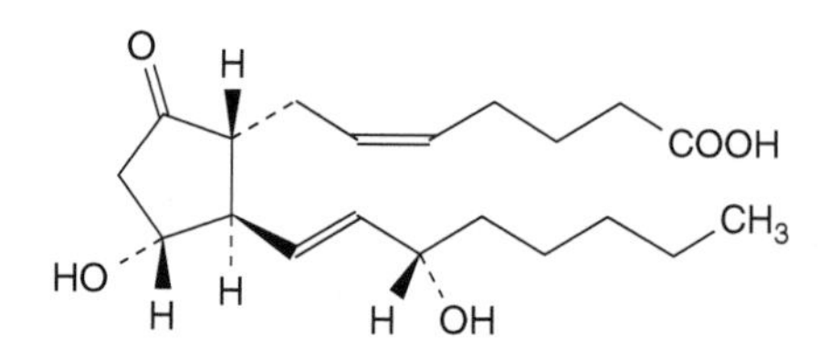

Empirical formula

$C_{20}H_{32}O_5$ (British Pharmacopoeia, 2007).

Molecular weight

352.5 (British Pharmacopoeia, 2007).

Solubility

1.3 mg/mL (Trissel, 2000).

pK_a

4.6 (Trissel, 2000).

Optimum pH stability

Various citations: 3.5 (Thompson, 1973; Cho *et al.*, 1977), 4–8 (Granstrom *et al.*, 1980; Budavari, 1989), 3–8.5 (Karim *et al.*, 1968).

Stability profile

Physical and chemical stability

Dinoprostone exists as a white, or almost white, crystalline powder or colourless crystals. It is practically insoluble in water (British Pharmacopoeia, 2007). However, the low concentration of drug in this preparation and the formulation of the licensed injection mean that the drug will be in solution, and therefore be vulnerable to aqueous degradation mechanisms.

The β-hydroxy ketone system is unstable in acidic and basic conditions, readily undergoing dehydration to form prostaglandin A2 (Stehle and Oesterling, 1977). Loss of protons at C-8 or C-10 may result in a negative charge, explusion of the hydroxyl group at C-11, and formation of a Δ10 double bond (PGA compound) (Monkhouse *et al.*, 1973). Under basic conditions, the A prostaglandins can isomerise further to B prostaglandins such as prostaglandin B2 (Stehle and Oesterling, 1977). Lee (1982) has also commented on the problems of dehydration and isomerisation.

In addition to the main degradation pathway, the formation of epimers at C-15 may occur under strongly acidic conditions (Stehle and Oesterling, 1977). In mildly to strongly basic media, E prostaglandins can form an equilibrium mixture with their C-8 isomer, and may also undergo 13,15 rearrangement in acidic conditions.

In aqueous media, dehydration occurs rapidly at very high or very low pH (<4 to >8) (Granstrom *et al.*, 1980; Budavari, 1989). Degradation follows first-order kinetics for both hydrogen ion and hydroxide ion concentrations. Rate of degradation would appear to be relatively independent of drug concentration (Stehle and Smith, 1976).

Monkhouse *et al.* (1973) studied the degradation of prostaglandins E1 and E2 at 60°C at a pH range of 1–10. It was assumed that the increased temperature did not alter the route of degradation. A consecutive first-order reaction appeared to occur over pH 4 for the dehydration and rearrangement reactions. At a pH below 3, the dehydration appeared to be first order with respect to hydrogen ion concentration. The rate of dehydration of prostaglandin E2 (PGE2) was 2–3 times faster than that of PGE1 under the same conditions at pH 4–10. The rate of rearrangement of PGE2 progressively exceeded that of PGE1 as pH decreased (e.g. at pH 6, PGE2 rearranged 15 times faster than PGE1). The greater reactivity of PGE2 when compared with PGE1 was attributed to the *cis*-Δ5 double bond, which may be involved in the rearrangement process.

Following the work by Monkhouse *et al.* (1973), Thompson *et al.* (1973) produced a log rate-pH profile for the decomposition of PGE2 in water/methanol (95:5 v/v) at 25°C. The data suggest that in aqueous systems, dinoprostone shows maximum stability at pH 3.5, with a half-life reported to be around 40 days. This equates to a shelf-life (t_{90}) of approximately 6 days at 25°C. The rate profile suggests that degradation is exponentially increased at pH values less than 2 and more than 8.

Stehle and Oesterling (1977) found typical first-order time data for the disappearance of dinoprostone from solution at pH 2.05. The rate of total loss of dinoprostone was proportional to the hydrogen ion concentration in the pH range 1–2 but not around pH 3. The authors suggest that above a pH of around 2.5, specific hydrogen-ion catalysis decreases in relative importance, a significant part of the total loss of prostaglandin E being due to water or buffer catalysis. The authors suggest that under the most severe conditions (pH 1–1.5, 37°C) encountered during oral administration of a prostaglandin E, approximately 5% of the prostaglandin would be lost per hour from non-enzymatic causes.

Although first-order degradation kinetics are also seen in alkaline conditions, the activation energies are lower, possibly suggesting that different mechanisms of degradation operate under these conditions (Stehle and Oesterling, 1977).

Degradation products

These may include A and B-type prostaglandins (Stehle and Oesterling, 1977).

Stability in practice

There are no known published stability studies of oral formulations of dinoprostone.

Unpublished work carried out by Davidson (1993) at the Regional Pharmaceutical Laboratory at Belfast City Hospital suggests that dinoprostone 100 micrograms/mL in Nipasept 0.2% (methyl-, ethyl- and propyl-hydroxybenzoate) shows less than 6% decomposition when stored at 4°C in amber glass for 7 days. Standard deviations are not quoted.

The solution was reported to be prepared as follows:

1 Prepare 0.2% Nipasept solution:

   a Dissolve 100 mg well-mixed Nipasept powder in approximately 2 mL 96% alcohol.
   b Make up to 50 mL with sterile water.
2. Dilute 0.3 mL of Prostin E2 sterile solution (100 micrograms/mL) to 30 mL with the freshly prepared Nipasept 0.2% solution.

Details of the sampling and HPLC method have been supplied. Three 30 mL solutions were prepared and stored in 50 mL amber glass bottles in the fridge. At each assay time, a 3 mL aliquot was removed from each bottle. A 1 mL volume was used for the assay and the remaining 2 mL for pH testing. Starting pHs were found to be 4.531, 4.542 and 4.628. After an initial small drop in pH over the first 2 days (<0.2 pH units), the values rose to finish at 4.621, 4.680 and 4.725, respectively.

The solution also passed the BP Antimicrobial Preservative Effectiveness Test. The author concludes that this formulation is chemically stable for 7 days if stored in the fridge, and will be adequately preserved for this time.

This formulation also contains approximately 10% alcohol (w/v) and Nipasept. Alcohol is known to possess pharmacological activity as a CNS depressant (Owen, 2003). Nipasept is a mixture of methyl, ethyl and propyl hydroxybenzoates (also called 'parabens'). Concern has been expressed about the use of methyl hydroxybenzoate in infant parenteral products because bilirubin binding may be affected (Rieger, 2003). This effect has not been demonstrated *in vivo*, and it is unclear whether the amounts of hydroxyl benzoates in oral formulations will cause such problems. However, it may be a prudent risk management step to consider avoiding the use of such a preservative system, particularly in neonates with jaundice, kernicterus or hyperbilirubinaemia.

The World Health Organization has set an estimated total acceptable daily intake for methyl, ethyl, and propyl hydroxybenzoates at up to 10 mg/kg body weight (WHO/FAO, 1974).

Related preparations

Research on topical delivery of dinoprostone by Lee (1982) investigated stability of the drug in methylhydroxyethylcellulose gel. Degradation, measured by gas chromatography, adhered to first-order principles, with 5% breakdown occurring after 7 days storage in the refrigerator, and after 32 hours at room temperature.

Bioavailability data

No information.

References

British Pharmacopoeia (2007). *British Pharmacopoeia*, Vol. 1. London: The Stationery Office, pp. 704–706.

Budavari S, ed. (1989). *Merck Index: An Encyclopedia of Chemicals, Drugs and Biologicals*, 11th edn. Rahway, NJ: Merck and Co.

Cho MJ, Krueger WC, Oesterling TO (1977). Nucleophilic addition of bisulfite ion to prostaglandins E2 and A2: implication in aqueous stability. *J Pharm Sci* 66: 149–154.

Davidson H (1993). Dinoprostone stability study (unpublished). Regional Pharmaceutical Laboratory, Belfast City Hospital.

Granstrom E, Hamberg M, Hansson G, Kindahl H (1980). Chemical instability of 15-keto-13,14-dihydro-PGE2: the reason for low assay reliability. *Prostaglandins* 19: 933–957.

Karim SMM, Devlin J, Hillier K (1968). The stability of dilute solutions of prostaglandins E1, E2 F1α, F2α. *Eur J Pharmacol* 4: 416–420.

Lee MG (1982). A study on the stability of prostaglandin E2 in methylhydroxyethylcellulose gel by gas chromatography. *J Hosp Clin Pharm* 7: 67–70.

Lowey AR, Jackson MN (2008). A survey of extemporaneous dispensing activity in NHS trusts in Yorkshire, the North-East and London. *Hosp Pharmacist* 15: 217–219.

Monkhouse DC, Van Campen L, Aguiar AJ (1973). Kinetics of dehydration and isomerisation of prostaglandins E1 and E2. *J Pharm Sci* 62: 576–580.

Owen SC (2003). Alcohol. In: Rowe RC, Sheskey PJ, Weller PJ, eds. *Handbook of Pharmaceutical Excipients*, 4th edn. London: Pharmaceutical Press and Washington DC: American Pharmaceutical Association, pp. 13–15.

Rieger MM (2003). Methylparaben. In: Rowe RC, Sheskey PJ, Weller PJ, eds. *Handbook of Pharmaceutical Excipients*, 4th edn. London: Pharmaceutical Press and Washington DC: American Pharmaceutical Association, pp. 390–394.

Stehle RG, Oesterling TO (1977). Stability of prostaglandin E1 and dinoprostone (prostaglandin E2) under strongly acidic and basic conditions. *J Pharm Sci* 66: 1590–1595.

Stehle RG, Smith RW (1976). Relative aqueous stabilities of dinoprostone free acid (prostaglandin E2) and its carbamoylmethyl ester [letter]. *J Pharm Sci* 65: 1844–1845.

Thompson GF, Collins JM, Schmalzfried LM (1973). Total rate equation for decomposition of prostaglandin E2. *J Pharm Sci* 62: 1738–1739.

Trissel LA, ed. (2000). *Stability of Compounded Formulations*, 2nd edn. Washington DC: American Pharmaceutical Association.

WHO/FAO (1974). Toxicological evaluation of certain food additives with a review general principles and of specifications. Seventeenth report of the Joint FAO/WHO Expert Committee on Food Additives. World Health Organ. Tech. Rep. Ser. No. 539.

Ergocalciferol oral liquid

Synonym

Vitamin D2.

Risk assessment of parent compound

A risk assessment survey completed by clinical pharmacists suggested that ergocalciferol has a wide to moderate therapeutic index, with the potential to cause minor side-effects if inaccurate doses were to be administered (Lowey and Jackson, 2008). Current formulations used at NHS trusts generally show a low technical risk, based on the number and/or complexity of manipulations and calculations. Ergocalciferol is therefore regarded as a low- to moderate-risk extemporaneous preparation (Box 14.17).

Summary

There is a lack of data in the public domain to support this formulation. It should be used under specialist supervision with careful patient monitoring.

***Box 14.17** Preferred formula*

Note: This product is known to be available from some 'Specials' manufacturers. Extemporaneous preparation should only occur where there is a demonstrable clinical need.

Using a typical strength of 3,000 IU/mL:

- ergocalciferol injection 300 000 IU/mL × 0.5 mL
- olive oil to 50 mL.

Method guidance

Using an appropriate-sized syringe and 19G needle measure the required volume of ergocalciferol. Change the needle for a filter needle and transfer to a measure. Make up to volume with olive oil. Stir well before packing.

Shelf-life

7 days at room temperature in amber glass.

Note: Ergocalciferol is known to bind to plastic. It has been recommended that plastic syringes are not used to draw up ergocalciferol injection (Summary of Product Characteristics, 2005).

An oral liquid is known to be available as a 'Special'. The quality and appropriateness of such a preparation should be assessed by a suitably competent pharmacist before use. Some 'Specials' may contain peanut oil (BNF for Children, 2007).

Clinical pharmaceutics

Points to consider

- Ergocalciferol is soluble in fatty oils. Dosage uniformity should not be problematic.
- There are no known data for bioavailability.
- The strength of ergocalciferol formulations may also be expressed in terms of units of weight or anti-rachitic activity. Each microgram of ergocalciferol is equivalent to 40 IU of anti-rachitic activity (British Pharmacopoeia, 2007).
- Ergocalciferol injection is formulated in ethyl oleate (Summary of Product Characteristics, 2005). It should not be administered with a plastic syringe, as ergocalciferol binds strongly to plastic (Summary of Product Characteristics, 2005; Martindale, 2005). The type of plastic that is incompatible is not known. Anecdotal reports suggest that some brands of plastic syringes have been used successfully (UCB Pharma Medical Information, personal communication, 2007). However, it is not known which particular plastic(s) is/are problematic.
- The active ingredient has also been anecdotally reported to stick to some rubber stoppers (UCB Pharma Medical Information, personal communication, 2007).

Other notes

- Alternative treatment strategies may include the use of colecalciferol preparations, depending on the needs of the individual patient.

Technical information

Structure

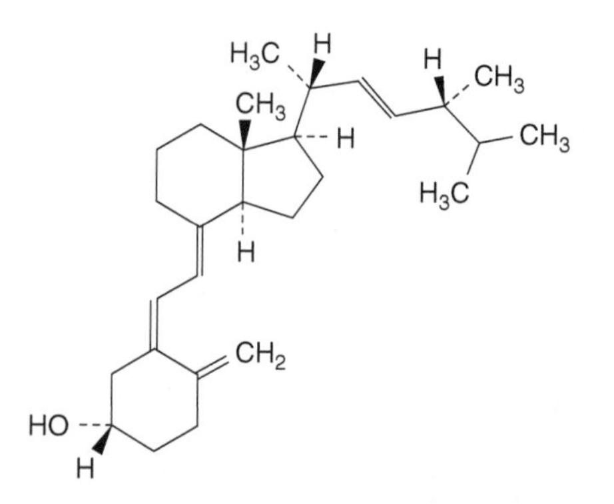

Empirical formula

$C_{28}H_{44}O$ (British Pharmacopoeia, 2007).

Molecular weight

396.7 (British Pharmacopoeia, 2007).

Solubility

Practically insoluble in water but soluble in fatty oils (Lund, 1994).

pK_a

Not known.

Optimum pH stability

Not known.

Stability profile

Physical and chemical stability

Ergocalciferol is a white or slightly yellowish, crystalline powder or white or almost white crystals. It is practically insoluble in water, freely soluble in alcohol and soluble in fatty oils (British Pharmacopoeia, 2007).

Ergocalciferol is known to be sensitive to air, heat and light. The oxidative degradation of ergocalciferol has been recognised for over 60 years (Stewart *et al.*, 1984). Solutions in volatile solvents are unstable and must be used immediately (British Pharmacopoeia, 2007). When in solution, a reversible isomerisation to pre-ergocalciferol may occur dependent on temperature and time (British Pharmacopoeia, 2007). The activity is due to both compounds.

The degradation of crystalline ergocalciferol has been studied by Stewart *et al.* (1984). The authors found that the degradation of ergocalciferol in ordinary fluorescent light at average room temperature and humidity results in the oxidation and fragmentation of the triene functionality. Previous research has also indicated that crystalline ergocalciferol degrades in the presence of air and light, while storage under vacuum improves stability (Fuchs and Van Niekirk, 1935; Huber and Barlow, 1943; Kanzawa and Kolak, 1953; Amer *et al.*, 1970). Although Stewart *et al.* (1984) did not rule out other forms of degradation, oxidation is suggested to be the primary route. While the nucleus of the molecule is thought to undergo extensive oxidation, the side-chain appears to remain intact.

Ergocalciferol oral solution has appeared in previous editions of the British Pharmacopoeia, including the 1998 edition, under the monograph for calciferol oral solution BP. Calciferol oral solution was described as 0.0075% solution of calciferol or ergocalciferol in a suitable vegetable oil.

The 1998 British Pharmacopoeia states, 'Calciferol Oral Solution may be prepared by warming to 40° a 1% suspension of Colcalciferol or Ergocalciferol in a suitable vegetable oil, such as Arachis Oil, Carbon Dioxide being bubbled through it to facilitate solution, and adding a sufficient quantity of the oil to produce a solution containing 0.0075% of Colecalciferol or Ergocalciferol.' The characteristics of the solution are described as a pale yellow, oily liquid with a slight but not rancid odour (British Pharmacopoeia, 1998). Calciferol oral solution BP should be stored in a well-filled, well-closed container, protected from light and at a temperature not exceeding 25°C (British Pharmacopoeia, 1998). However, subsequent editions of the British Pharmacopoeia do not contain such a monograph.

An official monograph for ergocalciferol oral solution appears in the United States Pharmacopeia. It is described as a solution of ergocaliciferol in an edible vegetable oil, in Polysorbate 80, or in propylene glycol (United States Pharmacopeia, 2003). It should be preserved in tight, light-resistant containers. No more details are available.

Ergocalciferol injection BP is a sterile solution containing 0.75% w/v of ergocalciferol in ethyl oleate (British Pharmacopoeia, 2007). It is a pale yellow oily liquid which should be protected from light. It has a typical shelf-life of 36 months (Summary of Product Characteristics, 2005). Ergocalciferol tablets 0.25 mg and 1.25 mg also carry 36-month shelf-lives (Summary of Product Characteristics, 2002). They are formulated with Avicel PH 101 (microcrystalline cellulose), magnesium stearate, lactose DCL 11, acacia, sugar, talc, gelatin, titanium dioxide and opaglos AG 7350.

Ergocalciferol powder should be stored in an airtight container, under nitrogen, protected from light, and at a temperature between 2 and 8°C (British Pharmacopoeia, 2007). The contents of any open containers should be used immediately (British Pharmacopoeia, 2007).

Degradation products

No information.

Stability in practice

There is no known information available in the public domain.

Bioavailability data

No information.

References

Amer MM Ahmad AKS, Varda SP (1970). *Zette Seifen Anstrichm* s72: 1040.

British Pharmacopoeia (1998). *British Pharmacopoeia*. London: The Stationery Office, pp. 1531–1532.

British Pharmacopoeia (2007). *British Pharmacopoeia*, Vol. 1. London: The Stationery Office, pp. 773–774 and Vol. 3, p. 2546.

Fuchs L, Van Niekirk J (1935). *Biochemistry* 277: 32.

Huber W, Barlow OW (1943). Chemical and biological stability of crystalline vitamins D2 and D3 and their derivatives. *J Biol Chem* 149: 125.

Joint Formulary Committee (2009). *BNF for Children*. London: British Medical Association and Royal Pharmaceutical Society of Great Britain.

Kanzawa T, Kolak S (1953). *J Pharm Soc Jpn* 73: 1357.

Lowey AR, Jackson MN (2008). A survey of extemporaneous dispensing activity in NHS trusts in Yorkshire, the North-East and London. *Hosp Pharmacist* 15: 217–219.

Lund W, ed. (1994) *The Pharmaceutical Codex: Principles and Practice of Pharmaceutics*, 12th edn. London: Pharmaceutical Press.

Stewart BA, Midland SL, Byrn SR (1984). Degradation of crystalline ergocalciferol [vitamin D2, (3β, 5*Z*, 22*E*)-9,10,Secoergosta-5,7,10(19),22-tetraen-3-ol]. *J Pharm Sci* 73(9):1322–1323.

Summary of Product Characteristics (2002). Ergocalciferol tablets 0.25 mg, Celltech Manufacturing Services. www.medicines.org.uk (accessed 12 June 2007).

Summary of Product Characteristics (2005). Ergocalciferol injection, UCB Pharma. www.medicines.org.uk (accessed 12 June 2007).

Sweetman SC (2009). *Martindale: The Complete Drug Reference*, 36th edn. London: Pharmaceutical Press.

United States Pharmacopeia (2003). *United States Pharmacopeia*, XXVI. Rockville, MD: The United States Pharmacopeial Convention.

Ethambutol hydrochloride oral liquid

Risk assessment of parent compound

A risk assessment survey completed by clinical pharmacists suggested that ethambutol has a moderate therapeutic index, with the potential to cause significant morbidity if inaccurate doses were to be administered (Lowey and Jackson, 2008). Current formulations used at NHS trusts show a high technical risk, based on the number and complexity of manipulations and calculations. Ethambutol is therefore regarded as a very high-risk extemporaneous product (Box 14.18).

Summary

Given the concerns regarding the optical chemistry of ethambutol, further work is required to validate a suitable formulation. However, ethambutol oral liquid has been used for many years with no known reports of specific problems. The formulae have typically consisted of xanthan gum vehicles or followed the Lederle formula (see below). However, the Lederle formula contains chloroform water and may not be suitable. Chloroform has been classified as a class 3 carcinogen (substances that cause concern owing to possible carcinogenic effects but for which available information is not adequate to make satisfactory assessments) (CHIP3 Regulations, 2002).

An individual risk assessment should be carried out and it is suggested that more stringent clinical review and monitoring may be required.

Clinical pharmaceutics

Points to consider

- Ethambutol hydrochloride is freely soluble and therefore, at typical strengths (50–100 mg/mL), most or all of the drug will be dissolved in the vehicle (no uniformity of dose considerations).
- Ethambutol exhibits optical chemistry (it is the D-isomer). While conversion to the more toxic L-isomer is theoretically possible, solutions of the drug in water are thought to be dextrorotatory. Whether such a conversion may occur on storage is not known.
- Ora-Plus and Ora-Sweet SF are free from chloroform and ethanol.

***Box 14.18** Preferred formula*

Using a typical strength of 100 mg/mL as an example:

- ethambutol 400 mg tablets × 20
- ethambutol 100 mg tablets × 20
- Ora-Plus 50 mL
- Ora-Sweet SF to 100 mL.

Method guidance

Tablets can be ground to a fine, uniform powder in a pestle and mortar. A small amount of Ora-Plus may be added to form a paste, before adding further portions of Ora-Plus up to 50% of the final volume. Transfer to a measuring cylinder. The Ora-Sweet SF can be used to wash out the pestle and mortar before making the suspension up to 100% volume. Transfer to an amber medicine bottle.

Shelf-life

28 days at room temperature in amber glass.

Alternative

An alternative formula would be:

- ethambutol 400 mg tablets × 20
- ethambutol 100 mg tablets × 20
- citric acid 2 g
- orange tincture 6 mL
- syrup 50 mL
- double strength chloroform water to 100 mL.

Shelf-life

28 days at room temperature in amber glass.

- Ora-Plus and Ora-Sweet SF contain hydroxybenzoate preservative systems.
- Ora-Sweet SF has a cherry flavour which may help to mask the bitter taste of the drug.
- No bioavailability data are available.
- The alternative formula contains chloroform, which has been classified as a class 3 carcinogen (i.e. substances that cause concern owing to possible

carcinogenic effects but for which available information is not adequate to make satisfactory assessments) (CHIP3 Regulations, 2002).

Technical information

Structure

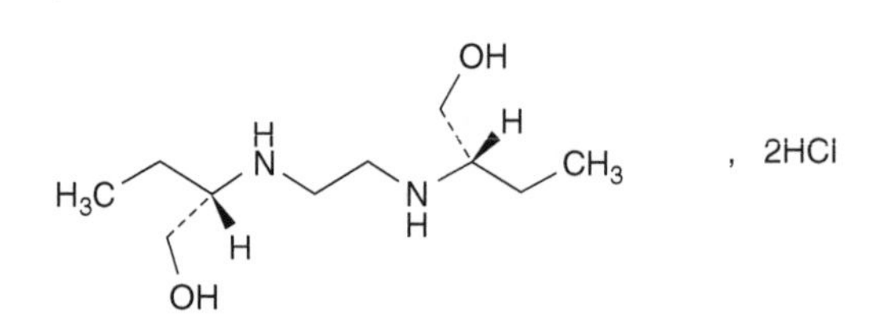

Empirical formula

$C_{10}H_{24}N_2O_2.2HCl$ (British Pharmacopoeia, 2007).

Molecular weight

277.2 (British Pharmacopoeia, 2007).

Solubility

1 in 1 of water (1 g in 1 mL) (Lund, 1994) (ethambutol base is sparingly soluble).

pK_a

6.3 and 9.5 (20°C) (Moffat, 1986; Lund, 1994).

Optimum pH stability

Not known.

pH

A 2% solution of ethambutol hydrochloride has a pH of 3.7–4.0 (Lund, 1994).

Stability profile

Physical and chemical stability

Ethambutol base is reported to be a white, scentless, bitter tasting, thermostable crystalline powder (Thomas *et al.*, 1961; Chings-San and Benet, 1978).

Ethambutol hydrochloride, a weak base, is freely soluble. Therefore, degradation reactions in solution would theoretically apply. However, the structure suggests that there are no obvious vulnerable groups for degradation under mild conditions (Dekker, 2005).

Ethambutol hydrochloride has two chiral carbon atoms and is optically active (Ching-San and Benet, 1978; Dekker, 2005). Ethambutol is the name given to the D-isomer of the synthetically constructed compound, 2,2′-(ethylenediimino)-di-1-butanol (Stowe and Jacobs, 1999; Margolin, 1993). It is possible that a racemate could form during extemporaneous preparation. However, a solution of ethambutol in water is thought to be dextrorotatory (Reynolds, 1982). Whether isomeric conversion may occur on storage is not known.

It is widely known that stereospecific biotransformation of a drug can affect its clinical properties (Wainer, 2001). *In vivo* studies of the stereoisomers in mice indicated that at maximal tolerated doses the L form of the compound was inactive, and that the D-isomer accounted for the greater amount of the tuberculosis activity (Thomas *et al.*, 1961; Peets and Busyke, 1964). The L-isomer has been reported to be associated with a greater degree of toxicity than the D-isomer (Peets and Busyke, 1964), and has been reported to be associated with an increased incidence of blindness (Aboul-Enein and Abou-Basha, 1997; Blessington, 1997). A third isomer, the optically inactive meso form, also exists. Neither the meso nor the levorotatory forms are used commercially (Blessington and Beiragh, 1991). The meso form displays approximately one-sixteenth of the activity of the dextro form. The levo form is inactive. There are also major differences in the isomers' distribution and elimination processes (Peets and Busyke, 1964).

Solutions of ethambutol hydrochloride are reported to be stable when heated at 121°C for 10 minutes (Lund, 1994).

Degradation products

The degradation products are not known.

Stability in practice

The only published formula available was formerly provided by Lederle (Martindale, 1989):

- ethambutol powder 500 mg
- citric acid 100 mg
- orange tincture 0.3 mL
- syrup 2.5 mL
- double strength chloroform water to 5 mL.

Although this formulation is reported to be stable for one month at room temperature, there are no data available to validate this claim. The use of the Lederle formula is no longer recommended as first line due to the presence of double strength chloroform water in the preparation. Chloroform has been classified as a class 3 carcinogen (substances that cause concern owing to possible carcinogenic effects but for which available information is not adequate to make satisfactory assessments) (CHIP3 Regulations, 2002). Syrup is known to adversely affect dental health.

The manufacturers of Myambutol recommend that tablets can be ground to a fine powder and mixed with about 100 mL of apple juice or a small quantity of apple sauce immediately before administration to the patient. This approach is endorsed by Van Scoy and Wilkowski (1983).

Ethambutol powder is available in the UK and is used in the preparation of oral liquids by some NHS trusts (Lowey and Jackson, 2008). The use of powder will avoid any issues associated with the excipients in tablet preparations. However, the powder should be subject to quality control testing before use; many trusts will not have the necessary facilities and/or volume of work needed to justify the associated costs of using pharmaceutical grade powder.

The Pharmaceutical Codex (Lund, 1994) suggests that the following excipients have been used in oral liquids of ethambutol hydrochloride: acacia, chloroform, citric acid, hydroxybenzoate esters, methylcellulose, orange tincture, saccharin sodium, sodium benzoate, sorbitol, starch, sucrose, syrup and tragacanth. The success or otherwise of the use of these excipients is not stated.

Further stability work is required for ethambutol oral liquid due to the paucity of data associated with the available formulations.

Bioavailability data

No data. See optical chemistry concerns.

References

Aboul-Enein HY, Abou-Basha LI (1997). Chirality and drug hazards. In: Aboul-Enein HY, Wainer IW, eds. *The Impact of Stereochemistry on Drug Development and Use*. New York: John Wiley, pp. 1–20.

Blessington B (1997). Ethambutol and tuberculosis, a neglected and confused puzzle. In: Aboul-Enein HY, Wainer IW, eds. *The Impact of Stereochemistry on Drug Development and Use.* New York: John Wiley, pp. 235–262.

Blessington B, Beiragh A (1991). A method for the quantitation enantioselective HPLC analysis of ethambutol and its stereoisomers. *Chirality* 3: 139–144.

British Pharmacopoeia (2007). *British Pharmacopoeia*, Vol. 1. London: The Stationery Office, p. 806.

Chings-San L, Benet LZ (1978) In: Florey K, ed. *Analytical Profiles of Drug Substances*, Vol. 7. London: Academic Press, pp. 231–249.

CHIP3 Regulations (2002). Chemicals (Hazard Information and Packaging for Supply) Regulations. SI 2002/1689.

Dekker T (2005). Presentation at the Workshop on GMP and Quality Assurance of Multisource Tuberculosis Medicines, Kuala Lumpur, Malaysia, 21–25 February 2005.

Lowey AR, Jackson MN (2008). A survey of extemporaneous dispensing activity in NHS trusts in Yorkshire, the North-East and London. *Hosp Pharmacist* 15: 217–219.

Lund W, ed. (1994). *The Pharmaceutical Codex: Principles and Practice of Pharmaceutics*, 12th edn. London: Pharmaceutical Press.

Margolin AL (1993). Enzymes in the synthesis of chiral drugs. *Enzyme Microb Technol* 15: 266–280.

Moffat AC, Osselton MD, Widdop B (2003). *Clarke's Analysis of Drugs and Poisons*, 3rd edn. London: Pharmaceutical Press. ISBN 978 0 85369 473 1.

Peets EA, Buyske DA (1964). Comparative metabolism ethambutol and its l-isomer. *Biochem Pharmacol* 13: 1403–1419.

Reynolds JEF, ed. (1982). *Martindale: The Extra Pharmacopoeia*, 28th edn. London: Pharmaceutical Press, pp. 1569–1570.

Stowe CD, Jacobs RF (1999). Treatment of tuberculosis infection and disease in children: the North American perspective. *Paediatr Drugs* 1: 299–312.

Sweetman SC (2009). *Martindale: The Complete Drug Reference*, 36th edn. London: Pharmaceutical Press.

Thomas JP, Baughin CO, Wilkinson RG, Shepherd RG (1961). A new synthetic compound with antituberculous activity in mice: ethambutol (dextro-2,2′-[ethylenediimino]-di-1-butanol). *Am Rev Respir Dis* 83: 891.

Van Scoy RE, Wilkowski CJ (1983). Antituberculous agents isoniazid, rifampicin, streptomycin, ethambutol and pyrazinamide. *Mayo Clin Proc* 58: 233–240.

Wainer IW (2001). The therapeutic promise of single enantiomers: introduction. *Hum Psychopharmacol Clin Exp* 16: S73–S77.

Woods DJ (2001). *Formulation in Pharmacy Practice. eMixt, Pharminfotech*, 2nd edn. Dunedin, New Zealand. www.pharminfotech.co.nz/emixt (accessed 18 April 2006).

Gabapentin oral liquid

Risk assessment of parent compound

A risk assessment survey completed by clinical pharmacists suggested that gabapentin has a medium therapeutic index, with the potential to cause significant morbidity if inaccurate doses were to be administered (Lowey and Jackson, 2008). Current formulations used at NHS trusts generally show a low technical risk, based on the number and/or complexity of manipulations and calculations. Gabapentin is therefore regarded as a medium-risk extemporaneous preparation (Box 14.19).

Summary

Importation of the licensed product from the USA should be considered as a priority, if the formulation is appropriate for the intended patient(s). By using the product licensed in USA, there will be a greater level of quality assurance, as there will have been both stability and bioavailability data presented to the licensing authority as part of the licensing process. The inactive ingredients are glycerin, xylitol, purified water and artificial cool strawberry anise flavour (Summary of Product Characteristics, 2006). The appropriateness of the formulation should be assessed on an individual patient basis by a suitably competent pharmacist.

Pharmacists are referred to the document 'Guidance for the purchase and supply of unlicensed medicinal products – notes for prescribers and pharmacists' for further information (NHS Pharmaceutical Quality Assurance Committee, 2004).

The research on the extemporaneous formulations suggests that gabapentin is relatively stable, and can be formulated in a range of suitable suspending vehicles. Ward-based manipulations have been used in some centres. For whole-capsule doses, capsules have been opened and the contents added to water and administered immediately (Medicines for Children, 2003; P. Dale, Paediatric Pharmacist, Royal Cornwall Hospital, personal communication, 2007).

Clinical pharmaceutics

Points to consider

- Most or all of the drug is liable to be in solution. Dosage uniformity should not be a major issue.

Box 14.19 *Preferred formula(e)*

Note: A brand-leading gabapentin oral liquid is available as a import from the USA, where it carries a product licence (Neurontin).

Using a typical strength of 100 mg/mL as an example:

- gabapentin 400 mg capsules × 25
- Ora-Plus to 50 mL
- Ora-Sweet to 100 mL.

Method guidance

Tablets can be ground to a fine, uniform powder in a pestle and mortar. A small amount of Ora-Plus may be added to form a paste, before adding further portions of Ora-Plus up to 50% of the final volume. Transfer to a measuring cylinder. The Ora-Sweet can be used to wash out the pestle and mortar of any remaining drug before making the suspension up to 100% volume. Transfer to an amber medicine bottle.

Shelf-life

28 days at room temperature in amber glass. **Shake the bottle.**

Alternatives

An alternative formula would be:

- gabapentin 400 mg capsules × 25
- simple syrup to 50 mL
- methylcellulose 1% to 100 mL.

Shelf-life

7 days at room temperature in amber glass; may be extended to 30 days if a suitable preservative is added. (Note: Syrup is not 'self-preserving' when diluted to this concentration.)

Another alternative formula would be:

- gabapentin 400 mg capsules × 25
- xanthan gum 0.5% suspending agent (or similar) to 100 mL.

Shelf-life

28 days at room temperature in amber glass.

Note: 'Keltrol' xanthan gum formulations vary slightly between manufacturers, and may contain chloroform and ethanol. The individual specification should be checked by the appropriate pharmacist.

- Ora-Plus and Ora-Sweet are free from chloroform and ethanol.
- Ora-Plus and Ora-Sweet contain hydroxybenzoate preservative systems.
- Ora-Sweet has a cherry flavour which may help to mask the extremely bitter taste of the drug.
- 'Keltrol' is made to slightly different specifications by several manufacturers, but typically contains ethanol and chloroform. The exact specification should be assessed before use by a suitably competent pharmacist. Ethanol is a CNS depressant. Chloroform has been classified as a class 3 carcinogen (i.e. substances that cause concern owing to possible carcinogenic effects but for which available information is not adequate to make satisfactory assessments) (CHIP3 Regulations, 2002).
- There are no bioavailability data available for the extemporaneous products.

Technical information

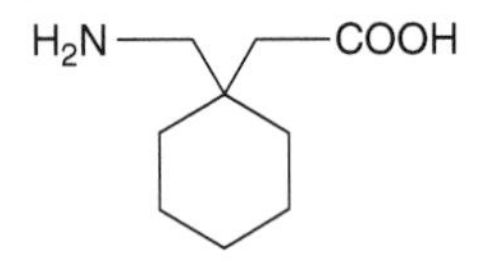

Empirical formula

$C_9H_{17}NO_2$ (Ananda *et al.*, 2003)

Molecular weight

171.24 (Ananda *et al.*, 2003).

Solubility

'Freely soluble' in water (Woods, 2001).

pK_a

3.7 and 10.7 (Zour *et al.*, 1992).

Optimum pH stability

Approximately pH 6 (Zour *et al.*, 1992).

Stability profile

Physical and chemical stability

Gabapentin is a white to off-white crystalline solid. It is freely soluble in water and both acidic and basic solutions (Woods, 2001; Summary of Product Characteristics, 2006).

Zour *et al.* (1992) have investigated the stability of gabapentin in buffered aqueous systems. The degradation of the drug was followed as a function of pH, buffer concentration, ionic strength and temperature. The results indicated that the degradation was proportional to the buffer concentration and temperature, with minimal degradation at around pH 6. Arrhenius plots suggested that a shelf-life of 2 years or more at room temperature may be obtained in an aqueous solution at pH 6. The energy of activation in the pH range of 4.5–7 was calculated to be between 34 and 40 kcal/mol, implying a relatively stable molecule.

In aqueous solution, gabapentin undergoes degradation via intramolecular cyclisation to 3,3-pentamethylene-4-butyrolactam ('lactam'). The specification for the maximum allowable limit for the lactam is 0.5% in any dosage form (Zour *et al.*, 1992).

Three major forms of gabapentin can exist in solution: the cation, the zwitterion and the anion. Zour *et al.* (1992) suggest a slower solvent-catalysed degradation rate for the zwitterionic species compared to the cationic or anionic species in the pH range 4.5–7.0. In this pH range, the majority of drug should be in the zwitterionic form (see pK_a values of 3.7 and 10.7).

Degradation products

Include 3,3-pentamethylene-4-butyrolactam (Zour *et al.*, 1992).

Stability in practice

Gabapentin can be formulated as a stable oral liquid, and is available for import as a licensed product. There is also one study that demonstrates the stability of possible extemporaneous preparations of gabapentin.

Nahata (1999) has investigated the stability of two oral 100 mg/mL gabapentin suspensions in plastic bottles at room temperature and under refrigeration. The contents of Neurontin capsules were ground in a pestle and mortar before the addition of one of the two following suspension vehicles:

- simple syrup 1:1 methylcellulose or
- Ora-Plus 1:1 Ora-Sweet.

Each suspension was stored in 10 amber plastic bottles and equally divided into storage at 4°C and 25°C. Triplicate samples were taken on days 0, 7, 14,

28, 42, 56, 70 and 91, and analysed by a stability-indicating HPLC method. pH and physical appearance were also monitored.

No substantial changes were noted in pH, appearance or odour over the three-month test period. The pH values of the Ora-based formulations were around 5.5–5.6, and the pH values of the methylcellulose and syrup formulations were approximately 7.1 throughout the study. All samples maintained over 90% of the starting concentration by day 56, regardless of formulation or storage temperature. Both Ora-based preparations retained stability for the entire 90-day period.

Microbiological studies were not carried out and the authors suggest that extended storage beyond 56 days would not be appropriate at room temperature. There may be an argument to restrict the shelf-life of the syrup and methylcellulose preparation further. This is because dilution of the syrup will result in the loss of its inherent antimicrobial effect, as well as dilution of the preservatives therein. Woods (2001) has suggested the addition of hydroxybenzoate preservatives at a concentration of 0.1% to allow the provision of a 30-day shelf-life at room temperature.

Bioavailability data

No information is available for the extemporaneous products. The licensed formulation in the USA is supported by bioavailability data as part of the licence application to the FDA.

References

Ananda K, Aravinda S, Vasudev PG, Maruga Poopathi Raja K, Sivaramakrishnan H, Nagarajan K *et al.* (2003). Stereochemistry of gabapentin and several derivatives: solid state conformations and solution equilibria. *Curr Sci* 85: 1002–1011.

CHIP3 Regulations (2002). Chemicals (Hazard Information and Packaging for Supply) Regulations. SI 2002/1689.

Lowey AR, Jackson MN (2008). A survey of extemporaneous dispensing activity in NHS trusts in Yorkshire, the North-East and London. *Hosp Pharmacist* 15: 217–219.

Medicines for Children (2003). *Medicines For Children*, 2nd edn. RCPCH and NPPG Publications.

Nahata MC (1999). Development of two stable oral suspensions for gabapentin. *Pediatr Neurol* 20: 195–197.

NHS Pharmaceutical Quality Assurance Committee (2004). Guidance for the purchase and supply of unlicensed medicinal products. Unpublished document available from regional quality assurance specialists in the UK or after registration from the NHS Pharmaceutical Quality Assurance Committee website: www.portal.nelm.nhs.uk/QA/default.aspx.

Summary of Product Characteristics (2006). Gabapentin Oral Solution (Neurontin) 50 mg/ml. Parke-Davis Pharmaceuticals, a division of Pfizer Pharmaceuticals Ltd.

Woods DJ (2001). *Formulation in Pharmacy Practice. eMixt, Pharminfotech*, 2nd edn. Dunedin, New Zealand. www.pharminfotech.co.nz/emixt (accessed 18 April 2006).

Zour E, Lodhi SA, Nesbitt RU, Silbering SB, Chaturvedi PR (1992). Stability studies of gabapentin in aqueous solutions. *Pharm Res* 9: 595–600.

Gliclazide oral liquid

Risk assessment of parent compound

A risk assessment survey completed by clinical pharmacists suggested that gliclazide has a medium therapeutic index. Inaccurate doses were thought to be associated with significant morbidity (Lowey and Jackson, 2008). Current formulations used at NHS trusts were generally of a low to moderate technical risk, based on the number and/or complexity of manipulations and calculations. Gliclazide is therefore regarded as a medium-risk extemporaneous preparation (Box 14.20).

Summary

There are no published stability reports for gliclazide suspension. The insoluble nature of the drug should minimise drug degradation and there are no obvious stability problems reported in the literature.

***Box 14.20** Preferred formula*

Using a typical strength of 8 mg/mL as an example:

- gliclazide 80 mg tablets × 10
- 'Keltrol' 0.4% or similar to 100 mL.

Method guidance

Tablets can be ground to a fine, uniform powder in a pestle and mortar. A small amount of 'Keltrol' may be added to form a paste, before adding further portions of suspending agent to approximately 75% of the final volume. Transfer to a measuring cylinder and use remaining 'Keltrol' to wash out the pestle and mortar of any remaining drug before making the suspension up to volume. Transfer to an amber medicine bottle.

Shelf-life

7 days refrigerated in amber glass. **Shake the bottle.**

Note: 'Keltrol' xanthan gum formulations vary slightly between manufacturers, and may contain chloroform and ethanol. The individual specification should be checked by the appropriate pharmacist.

It would appear logical to limit the expiry to a cautious 7 days, until further data are available.

Clinical pharmaceutics

Points to consider

- At typical paediatric strengths (8–16 mg/mL), almost all of the drug will be in suspension. **Shake the bottle to ensure uniformity of drug throughout the suspension.**
- There are no published stability reports for oral liquid formulations of gliclazide.
- 'Keltrol' is made to slightly different specifications by several manufacturers, but typically contains ethanol and chloroform. The exact specification should be assessed before use by a suitably competent pharmacist. Ethanol is a CNS depressant. Chloroform has been classified as a class 3 carcinogen (i.e. substances that cause concern owing to possible carcinogenic effects but for which available information is not adequate to make satisfactory assessments) (CHIP3 Regulations, 2002).
- The oral bioavailability of licensed formulations of gliclazide is excellent (Hong *et al.*, 1998) but there are no data for extemporaneous formulations.

Technical information

Structure

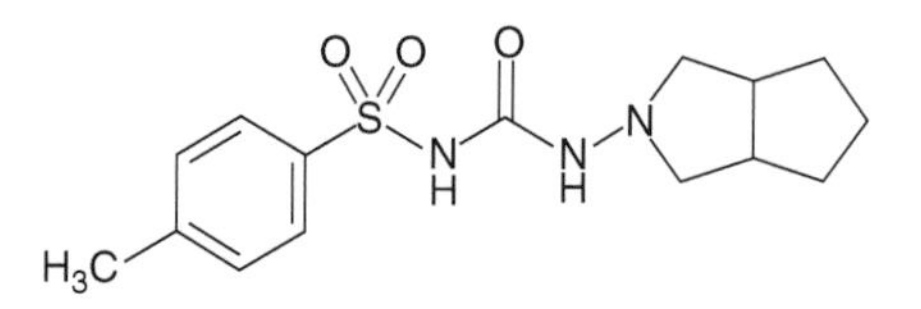

Empirical formula

$C_{15}H_{21}N_3O_3S$ (British Pharmacopoeia, 2007).

Molecular weight

323.4 (British Pharmacopoeia, 2007).

Solubility

Practically insoluble in water (British Pharmacopoeia, 2007).

pK_a

5.98 (Klimt *et al.*, 1970; Campbell *et al.*, 1991)

Optimum pH stability

Not known.

Stability profile

Physical and chemical stability

Gliclazide, the second generation sulfonylurea, is a white or almost white powder, characterised by poor solubility (Hong *et al.*, 1998; British Pharmacopoeia, 2007). Poor solubility may help to limit degradation of the drug when formulated as an oral suspension. The pH–solubility profile of gliclazide at 37°C shows much higher solubility at pH 1.2 and 7.2 than at pH 2.5–5.6.

Degradation products

No information available.

Stability in practice

There are no known published stability reports for gliclazide oral liquid.

Unpublished research conducted by Grassby (1991) investigated the stability of a 16 mg/mL suspension containing Diamicron tablets. The suspending agent has been described as:

- chloroform water concentrate 2.5 mL
- 'Keltrol' 400 mg
- methylhydroxybenzoate 150 mg
- propylhydroxybenzoate 15 mg
- filtered water to 100 mL.

Samples were stored in amber glass medical bottles in the refrigerator at 4°C, at room temperature (both exposed to and protected from light), and in an incubation oven at 40°C. After 150 days, there was less than 5% degradation observed in the samples stored at room temperature or in the fridge. However, a significant level of degradation occurred after 150 days for the samples stored in the oven. A cautious expiry of 14 days for refrigerated storage was suggested, as Grassby (1991) found a temporary increase in the drug concentration (up to 10% of the initial concentration over around 100 days). Although the assay was initially thought to be stability-indicating, Grassby (1991) suggests that a co-eluting degradation product may be accountable for the apparent increase in drug concentration.

The data suggest that no increase was seen until approximately day 40. Given this trend, and the relative insolubility of gliclazide, Grassby (1995) suggests that a 14-day expiry is justifiable, and that a longer one may well be possible.

Davison (1996) acknowledges the problems regarding the potential for a co-eluting degradation product during HPLC analysis and appears to endorse a restricted expiry of no more than 14 days.

In the development of of a HPLC method for gliclazide using solid-phase extraction, Noguchi *et al.* (1992) suggest that a stock solution of gliclazide 2 mg per 20 mL in methanol/water (1:1, v/v) was 'stable' for three months at 4°C, but data are not provided.

Bioavailability data

Overall bioavailability of gliclazide tablets is approximately 100%, although the rate at which absorption occurs may vary (Hong *et al.*, 1998). Absorption of gliclazide may be altered by its formulation as an extemporaneously prepared oral liquid. Patients starting or switching to an oral liquid form may warrant closer monitoring in the initial period.

References

British Pharmacopoeia (2007). *British Pharmacopoeia*, Vol. 1. London: The Stationery Office, pp. 955–956.

Campbell DB, Lavielle R, Nathan C (1991). The mode of action and clinical pharmacology of gliclazide: a review. *Diabetes Res Clin Pract* S21–S36.

CHIP3 Regulations (2002). Chemicals (Hazard Information and Packaging for Supply) Regulations. SI 2002/1689.

Davison H (1996). Unpublished memorandum, 28th October 1996, Belfast Regional Pharmaceutical Laboratory Service.

Grassby PF (1991). Stability testing report: stability of gliclazide suspension 80 mg/5ml. Welsh Pharmaceutical Research and Development Laboratories.

Grassby PF (1995). *UK Formulary of Extemporaneous Preparations*. Penarth: Paul F Grassby.

Hong SS, Lee SH, Lee YJ, Chung SJ, Lee MH, Shim CK (1998). Accelerated oral absorption of glicalzide in human subject from a soft gelatine capsule containing a PEG suspension of gliclazide. *J Control Release* 51: 185–192.

Klimt CF, Knatterud G, Meinert CL *et al.* (1970). A study of the effects of hypoglycemic agents on vascular complications in patients with adult-onset diabetes. I. Design, methods, and baseline results, II. Mortality results. *Diabetes* 19(Suppl2): 747–810.

Lowey AR, Jackson MN (2008). A survey of extemporaneous dispensing activity in NHS trusts in Yorkshire, the North-East and London. *Hosp Pharmacist* 15: 217–219.

Noguchi H, Tomita N, Naruto S, Nakano S (1992). Determination of glicalzide in serum by high-performance liquid chromatography using solid-phase extraction. *J Chromatogr* 583: 266–269.

Hydrocortisone oral liquid

Risk assessment of parent compound

A recent survey of clinical pharmacists suggested that hydrocortisone is regarded as a drug with a relatively wide therapeutic index. However, it has the potential to cause severe morbidity if administered in inaccurate dosages (Lowey and Jackson, 2008). Current formulations used at NHS trusts are generally of a low to moderate technical risk, based on the number and/or complexity of manipulations and calculations. Hydrocortisone is therefore regarded as a medium-risk extemporaneous preparation (Box 14.21).

Box 14.21 *Preferred formula*

Using a typical strength of 1 mg/mL as an example:

- hydrocortisone 20 mg tablets × 5
- Ora-Plus 50 mL
- Ora-Sweet to 100 mL.

Method guidance

Tablets can be ground to a fine, uniform powder in a pestle and mortar. A small amount of Ora-Plus may be added to form a paste, before adding further portions of Ora-Plus up to 50% of the final volume. Transfer to a measuring cylinder. The Ora-Sweet can be used to wash out the pestle and mortar of any remaining drug before making the suspension up to 100% volume. Transfer to an amber medicine bottle.

Shelf-life

28 days at room temperature in amber glass. **Shake the bottle.**

Summary

There is a wide body of evidence to suggest that hydrocortisone can be extemporaneously prepared in a range of formulations. Appropriate, simple suspending agents may be used to give a typical 4-week expiry for an extemporaneous product, providing the formulation is effectively preserved. Xanthan gum-based vehicles are known to be physically compatible with hydrocortisone (Nova Laboratories Ltd, 2003; Chong *et al.*, 2003), and have been used widely in the NHS (Lowey and Jackson, 2008).

The preparation based on Ora-Plus and Ora-Sweet is recommended for simplicity. The acidic pH of Ora-Sweet/Ora-Plus (approximately pH 4.2; Paddock Laboratories, 2006) is also likely to limit degradation. Note that the pH and individual formulation of 'Keltrol' can vary with supplier from around pH 4.5 to pH 6. The cherry flavour of Ora-Sweet may help to mask the drug's bitter taste.

Clinical pharmaceutics

Points to consider

At typical paediatric strengths (1–2 mg/mL), most of the drug will be in suspension. **Shake the bottle to ensure uniformity of drug throughout the suspension.**

- Ora-Plus and Ora-Sweet SF are free from chloroform and ethanol.
- Ora-Plus and Ora-Sweet SF contain hydroxybenzoate preservative systems.
- Ora-Sweet SF has a cherry flavour which may help to mask the bitter taste of the drug.
- The use of the free alcohol (from commercial tablets) may provide a better bioavailability than the sodium succinate or sodium phosphate salts.

Other notes

A licensed preparation of hydrocortisone cypionate was withdrawn by Pharmacia and Upjohn in 2001 after published data and clinical use suggested that the suspension was not bioequivalent to the tablet formulation (Merke *et al.*, 2001). The company had previously changed the suspending agent from tragacanth to xanthan gum. This followed a warning issued over the need for vigorous shaking of the preparation (for up to 10 minutes) in order to re-suspend a white precipitate of hydrocortisone cypionate.

Hydrocortisone displays polymorphism and also has a bitter taste (Florey, 1983).

Technical information

Structure

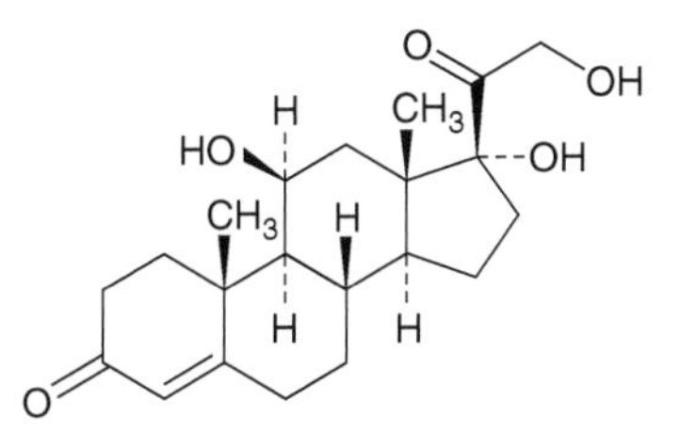

Empirical formula

$C_{21}H_{30}O_5$ (Lund, 1994).

Molecular weight

362.46 (Lund, 1994).

Solubility

1 in 3500 (approximately 0.29 mg/mL at 25°C) (Fleisher, 1986; Lund, 1994).

pK_a

11.05 (Trissel, 2005).

Optimum pH stability

pH 3.5–4.5 (Lund, 1994; Woods, 2001).

Stability profile

Physical and chemical stability

Hydrocortisone is acknowledged as being very stable as a solid (Florey, 1983). It is practically insoluble in water (Lund, 1994; British Pharmacopoeia, 2007). This poor solubility should help to limit any degradation that may occur via oxidative and non-oxidative reactions of the C-17 side-chain (Florey, 1983; Lund, 1994). Typical strengths of paediatric suspension are 1–2 mg/mL (Lowey and Jackson, 2008).

The rate of degradation is dependent on pH, and hydrocortisone is prone to degradation at very acidic (pH<2) and particularly at alkaline pH (Florey, 1983; Connors *et al.*, 1986; Lund, 1994). The rate of minimum degradation occurs at around pH 3.5–4.5 (Connors *et al.*, 1986; Fawcett *et al.*, 1995).

The oxidation of the C-17 side-chain is strongly catalysed by trace metal impurities (especially Cu^{2+}, Ni^{2+} and Fe^{3+} ions). Acetate and citrate buffer solutions also exert a catalytic effect. Enhanced degradation in phosphate and borate buffers may be due trace metals; carbonate buffers are not known to have an effect on degradation (Connors *et al.*, 1986). In phosphate or carbonate buffers at pH 7–9, up to 10% degradation occurs in 4 hours at 26°C.

The degradation rate is directly proportional to hydrogen ion and steroid concentrations, and the mechanism involves an acid-catalysed water elimination through an initial enolisation of the C-20 keto group (Hansen and Bundergaard, 1980).

The effect of temperature on the degradation rate has been determined for hydrocortisone in 1.0 M HCl over the range 60–80°C (Hansen and Bundergaard, 1980). The resulting Arrhenius plot suggested an activation energy of 23.3 kcal/mol. In 0.2 M borate buffer (pH 9.1) containing 0.2 ppm of Cu^{2+} ions over the range 37–60°C, the activation energy was determined as 14.3 kcal/mol.

Photolytic degradation has also been reported when hydrocortisone is solubilised in alcoholic solution (95% ethanol) and exposed to fluorescent lighting. The degradation appears to involve the A-ring rather than the side-chain (Hamlin *et al.*, 1960). However, prednisolone and methylprednisolone appeared to be much more susceptible to photolysis, containing two double bonds rather than just the single double bond found in hydrocortisone. The authors suggest that standard alcoholic solutions of these steroids should be stored in amber bottles in the refrigerator to ensure maintenance of the potency of the solutions (Hamlin *et al.*, 1960).

Degradation products

In common with some other corticosteroids, the predominant degradation products of hydrocortisone in methanolic or aqueous solution are 21-hydro derivatives (steroid glyoxals). In acidic conditions (pH < 2), the 17-deoxy-21-dehydro compound is the major product resulting from specific acid-catalysed water elimination (Hansen and Bundergaard, 1980). Other degradation products may include glycolic acids, the etienic acid and the 17-oxo-steroid.

It is possible that steroid glyoxals may be involved as pro-antigens in allergic reactions caused by corticosteroids.

Stability in practice

There are several studies to demonstrate the relative stability of hydrocortisone when formulated as a suspension.

Chong *et al.* (2003) describe the stability testing of a hydrocortisone suspension formulated to respond to the withdrawal of the Cortef suspension from the Canadian market. Formulations of 1 mg/mL and 2 mg/mL were achieved by crushing and suspending Cortef tablets in equal parts of Ora-Sweet and Ora-Plus. The suspensions were stored at 4 and 25°C for 91 days in amber plastic prescription bottles. After manual shaking of the suspension for 10 seconds, aliquots of 1 mL were removed from each of the six bottles (three at either temperature) each week and stored at –85°C until analysis by a validated stability-indicating HPLC method. Physical characteristics were also tested (pH, colour, odour, viscosity, precipitation and ease of resuspension).

Examples of both suspensions maintained at least 90% of initial concentration at both temperatures throughout the 91-day test period. No significant changes in pH were seen (less than ±0.3 pH units) and changes in colour and odour are described as 'slight, if any'.

Sources of error include the small shaking time and the use of a 1 mL aliquot for the removal of drug in a thick suspending agent. The inter-day variation shows up to ±19% variation of the initial concentration, and activation energies are not shown. Due to the relative stability of hydrocortisone in suspension, it would appear that any losses due to degradation are masked by the experimental error inherent in the testing of a suspension. Indeed, two of the test groups show a higher than 100% drug concentration on day 91.

Alternative 2.5 mg/mL formulations using tablets and powder have been described by Fawcett *et al.* (1995). These formulations contained sodium carboxymethylcellulose as a suspending agent, polysorbate 80 as a wetting agent, citric acid as a buffer, and 0.1% hydroxybenzoate preservatives. The suspensions were stored at 5, 25 and 40°C in high-density amber polyethylene bottles with polypropylene screw-top lids. One millilitre samples were removed after vigorous manual shaking on days 0, 15, 30, 60 and 91, and frozen at –70°C until analysis could occur. Aliquots (20 microlitres) were removed and diluted 500-fold with water before testing via a stability-indicating HPLC method. Due to the viscous nature of the formulations, the assays showed a high coefficient of variation (CV) of 4.67%; an internal reference of methylhydroxybenzoate was used to improve the precision (new CV 0.28%; $n = 10$).

Degradation of hydrocortisone was not significant in the formulations stored at 5 and 25°C for 90 days. Of the suspensions stored at 40°C, only the suspension prepared from tablets showed significant degradation (approximately 20% by day 90), with a rate constant of 6.28×10^{-3} mg/mL per day. The apparent pH of all the suspensions was 3.4 ± 0.1 and did not change on storage, regardless of whether decomposition occurred or not.

Dosage uniformity studies were also carried out. The intra-day CVs for repeated 1 mL sampling (using the same syringe) of the formulation prepared from tablets were 2.31% and 1.84% at 5 and 25°C, respectively. Inter-day CVs for sampling with different syringes over 10 days were 4.53% and 4.34%

at 5 and 25°C, respectively. Neither of these differences were statistically significant. This suggests that any increases in viscosity due to storage in a refrigerator may not significantly compromise dosage uniformity.

The preparation has been described as very bitter to taste, and it has been suggested that this may be improved by adding a flavour concentrate before giving the dose, and/or having a glass of water after the dose (Woods, 2001). Microbiological evaluation was not performed; therefore the recommended expiry was limited to 30 days.

Woods (2001) suggests that the use of hydrocortisone powder is preferred, and that the use of more soluble esters of hydrocortisone is not recommended, as they may have reduced bioavailability compared with the free alcohol (Fleisher *et al.*, 1986). However, the use of powder is not practical in most dispensaries in the UK.

Das Gupta (1985) has investigated the effect of some formulation adjuncts on the stability of hydrocortisone. The stability of hydrocortisone was tested in aqueous solutions containing either 20% ethyl alcohol, 50% glycerin or 50% propylene glycol with and without the presence of antioxidants, surfactants and a thickening agent. None of the antioxidants improved the drug stability, and cysteine hydrochloride and sodium lauryl sulfate had an adverse effect on the stability. Both at 60°C and at room temperature, propylene glycol 50% proved to be a better vehicle than glycerin (29.5% versus 44.5% decomposition over 21 days at 60°C).

Das Gupta (1985) concludes that hydrocortisone is very stable at room temperature in an aqueous buffered solution (pH 3.5) containing 50% by volume of propylene glycol without the presence of antioxidants. Although these results may be of more use to topical products, they add further evidence with regard to the relative stability of hydrocortisone.

Kristmundsdottir *et al.* (1996) have described the formulation and clinical use of a 0.3% w/v oral mouthwash solution of hydrocortisone whose solubility was improved by complex formation with 2-hydroxypropyl-β-cyclodextrin. Stability testing revealed little or no degradation over five weeks at 2.5 and 20°C but some decline (less than 10%) at 53°C. Maximum stability was shown at pH 4.5, and the stability was not improved by purging the solution with nitrogen.

Bioavailability data

Various prodrugs of hydrocortisone and other related steroids have been used in an attempt to increase the solubility of the parent drug. For example, cases of treatment failure for prednisolone tablets have been linked to poor dissolution rates (Fleisher *et al.*, 1986).

However, it has been suggested that the use of hydrocortisone free alcohol (as in the commercial tablets) provides a higher level of bioavailability for oral products when compared to sodium phosphate or sodium succinate salts (Fleisher *et al.*, 1986). Although hydrocortisone prodrugs may increase the solubility of the drug, they tend to be less stable than the parent compound, yielding lower and more variable plasma levels. The prodrugs may also be more vulnerable to intestinal esterases and hepatic metabolism (Fleisher *et al.*, 1986).

Another prodrug, hydrocortisone lysinate, has been found to be relatively unstable to hydrolysis, particularly if the pH is increased to around 7. This has implications given the likely pH seen in the gut lumen. Johnson *et al.* (1983) suggest that the lysinate may show an increased dissolution rate compared with the parent compound, but Fleisher *et al.* (1986) have found the overall absorption and permeability profile of the lysinate salt in animal models to be comparable to that of the parent compound.

A commercial formulation of hydrocortisone cypionate was withdrawn after findings suggested non-equivalence with the tablet formulation (Merke *et al.*, 2001).

Patel *et al.* (1984) have compared the absorption of four different tablet formulations and one suspension. However, no details are available as to the suspension's formulation.

References

British Pharmacopoeia (2007). *British Pharmacopoeia*, Vol. 1. London: The Stationery Office, pp. 1037–1039.

Chong G, Decarie D, Ensom MHH (2003). Stability of hydrocortisone in extemporaneously compounded suspensions. *J Inform Pharmacother* 13: 100–110.

Connors KA, Amidon GL, Stella ZJ (1986). *Chemical Stability of Pharmaceuticals: A Handbook for Pharmacists*, 2nd edn. New York: Wiley Interscience.

Das Gupta V (1985). The effect of some formulation adjuncts on the stability of hydrocortisone. *Drug Dev Ind Pharm* 11: 2083–2097.

Fawcett JP, Bouton DW, Jiang R, Woods DJ (1995). Stability of hydrocortisone oral suspensions made from tablets and powder. *Ann Pharmacother* 29: 928–990.

Fleisher D, Johnson LC, Stewart BH *et al.* (1986). Oral absorption of 21-corticosteroid esters: a function of aqueous solubility and enzyme activity and distribution. *J Pharm Sci* 75: 934–939.

Florey K, ed. (1983). *Analytical Profiles of Drug Substances*, Vol. 12. London: Academic Press, pp. 277–324.

Hamlin WE Chulski T, Johnson RH, Wagner JG (1960). A note on the photolytic degradation of anti-inflaminatory steroids. *J Am Pharm Assoc Sci Ed* 49: 253.

Hansen J, Bundgaard H (1980). Studies on the stability of corticosteroids V. The degradation pattern of hydrocortisone in aqueous solution. *Int J Pharm* 6: 307–319.

Johnson K, Amidon GL, Pogany S (1985). Solution kinetics of a water-soluble hydrocortisone prodrug: hydrocortisone-21-lysinate. *J Pharm Sci* 74: 87–89.

Kristmundsdottir T, Loftsson T, Holbrook WP (1996). Formulation and clinical evaluation of a hydrocortisone solution for the treatment of oral disease. *Int J Pharm* 139: 63–68.

Lowey AR, Jackson MN (2008). A survey of extemporaneous dispensing activity in NHS trusts in Yorkshire, the North-East and London. *Hosp Pharmacist* 15: 217–219.

Lund W, ed. (1994). *The Pharmaceutical Codex: Principles and Practice of Pharmaceutics*, 12th edn. London: Pharmaceutical Press.

Merke DP, Cho D, Calis KA, Keil MF, Chrousos GP (2001). Hydrocortisone suspension and hydrocortisone tablets are not bioequivalent in the treatment of children with congenital adrenal hyperplasia. *J Clin Endocrinol Metab* 86: 441–445.

Paddock Laboratories (2006). Ora-Plus and Ora-Sweet SF Information Sheets. www.paddocklaboratories.com (accessed 12 April 2006).

Patel RB, Rogge MC, Selen A, Goehl TJ, Shah VP, Prasad VK, Welling PG (1984). Bioavailability of hydrocortisone from commercial 20 tablets. *J Pharm Sci* 73: 964–966.

Trissel LA, ed. (2000). *Stability of Compounded Formulations*, 2nd edn. Washington DC: American Pharmaceutical Association.

Trissel LA (2005). *Handbook on Injectable Drugs*, 13th edn. Bethesda, MD: American Society of Health System Pharmacists.

Indometacin (indomethacin) oral liquid

Risk assessment of parent compound

A risk assessment survey completed by clinical pharmacists suggested that indometacin has a medium therapeutic index, with the potential to cause significant morbidity if inaccurate doses were to be administered (Lowey and Jackson, 2008). Current formulations used at NHS trusts generally show a low technical risk, based on the number and/or complexity of manipulations and calculations. Indometacin is therefore regarded as a medium-risk extemporaneous preparation (Box 14.22).

Box 14.22 *Preferred formula*

Using a typical strength of 5 mg/mL as an example:

- indometacin 50 mg capsules × 10
- xanthan gum 0.5% suspending agent ('Keltrol' or similar) to 100 mL.

Note: 'Keltrol' xanthan gum formulations vary slightly between manufacturers, and may contain chloroform and ethanol. The individual specification should be checked by the appropriate pharmacist.

Method guidance

The contents of the capsules can be ground to a fine, uniform powder in a pestle and mortar. A small amount of suspending agent may be added to form a paste, before adding further portions of suspending agent to approximately 75% of the final volume. Transfer to a measuring cylinder. Use further suspending agent to wash out the pestle and mortar of any remaining drug before making the suspension up to 100% volume. Transfer to an amber medicine bottle.

Shelf-life

28 days at room temperature in amber glass. **Shake the bottle.**

Summary

There is a reasonable body of evidence to support the relative stability of indometacin in oral liquid formulations. Consideration should be given to the products that are licensed abroad and available for importation. By using the product licensed in another mutually recognised country, there may be a greater level of quality assurance, as there will have been both stability and bioavailability data presented to the licensing authority as part of the licensing process. Some of these preparations contain small amounts of ethanol. The exact formulation should assessed by a suitably competent pharmacist before procurement for its fitness-for-purpose.

Pharmacists are referred to the document 'Guidance for the purchase and supply of unlicensed medicinal products – notes for prescribers and pharmacists' for further information (NHS Pharmaceutical Quality Assurance Committee, 2004).

The published data available appear to provide scope for extrapolation of a 28-day shelf-life for simple extemporaneous formulations as seen above.

Clinical pharmaceutics

Points to consider

- At typical strengths (e.g. 5 mg/mL), almost all of the drug will be in suspension. **Shake the bottle to ensure uniformity of drug throughout the suspension.**
- Indometacin exhibits polymorphism. In theory, changes in polymorphic form could alter particle size (Han *et al.*, 2006). Whether this occurs with indometacin is not known.
- 'Keltrol' is made to slightly different specifications by several manufacturers, but typically contains ethanol and chloroform. The exact specification should be assessed before use by a suitably competent pharmacist. Ethanol is a CNS depressant. Chloroform has been classified as a class 3 carcinogen (i.e. substances that cause concern owing to possible carcinogenic effects but for which available information is not adequate to make satisfactory assessments) (CHIP3 Regulations, 2002).
- There is a paucity of bioavailability data.

Other notes

- Suspensions are available for import. The formulation of these suspensions should be assessed before use by a suitably competent pharmacist.
- Other NSAIDs may be considered as alternatives, but this depends on the indication.

Technical information

Structure

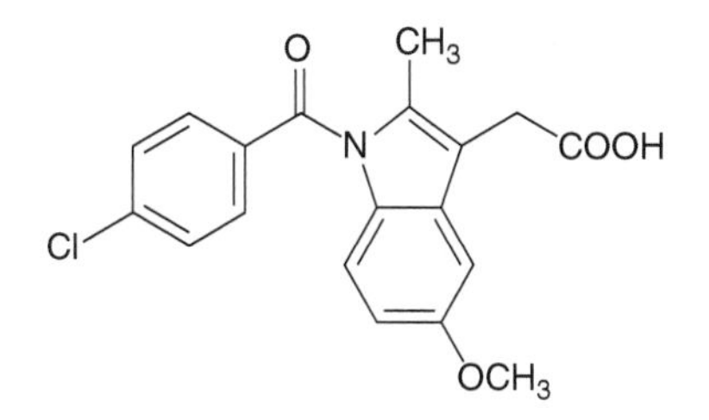

Empirical formula

$C_{19}H_{16}ClNO_4$ (British Pharmacopoeia, 2007).

Molecular weight

357.8 (British Pharmacopoeia, 2007).

Solubility

Practically insoluble in water (British Pharmacopoeia, 2007).

pK_a

4.5 (Carboxyl group) (Moffat, 1986).

Optimum pH stability

pH 3–5 (Connors *et al.*, 1986).

Stability profile

Physical and chemical stability

A white or yellow, crystalline powder, practically insoluble in water, sparingly soluble in alcohol (British Pharmacopoeia, 2007). Indometacin in aqueous solution and in solid form is light sensitive (Connors *et al.*, 1986) and should therefore be protected from light (British Pharmacopoeia, 2007). Indometacin is physically compatible with diluent A, a xanthan gum-based suspending agent (Nova Laboratories Ltd, 2003).

Aqueous degradation of indometacin has been described as first-order overall and specific base catalysed at 25°C (Connors *et al.*, 1986). Over a temperature range of 20.1–40.7°C in the presence of alkali, the rate constant–hydroxyl ion concentration profile was linear with a positive slope

(Hajratwala and Dawson, 1977; Singla *et al.*, 1991). Hydrolysis of the *N*-acyl group yields the corresponding acid and amine (Connors *et al.*, 1986).

The half-life at room temperature is reported to be around 200 hours in a pH 8.0 buffer and around 90 minutes in pH 10 solutions (Merck Sharp and Dohme, 1984).

Below pH 3, acid catalysis dominates, between pH 3 and 5 water attack is predominant, and specific base catalysis occurs above pH 7. Maximum stability is seen is the pH range 3–5. Indometacin degradation is also buffer-dependent (Connors *et al.*, 1986).

Alkaline hydrolysis shows a positive ionic strength effect. Indometacin degrades in the presence of non-ionic surfactants, exoylated lanolin and polysorbate 80 to a similar extent seen in solutions without surfactant (Dawson *et al.*, 1977; Connors *et al.*, 1986). However, cetrimonium bromide is reported to increase degradation considerably (Dawson *et al.*, 1977).

A 3% w/v indometacin aqueous solution with 30% w/v ethylcarbonate and 30% ethylurea has been reported to have a shelf-life of 25 months at 25°C (Pawelczyk and Knitter, 1978; Connors *et al.*, 1986). Ghanem *et al.* (1979) have also shown indometacin to display a 27-month shelf-life at room temperature, when solubilised with polysorbate 80 in the presence of 20% ethanol and 40% glycerol.

Photolytic degradation of aqueous indometacin exposed to UV light (254 nm wavelength) is a series of zero-order steps, the first of which is independent of oxygen. The remaining steps, however, require the presence of oxygen (Pawelcyk *et al.*, 1977; Connors *et al.*, 1986).

The degradation of indometacin can be minimised by the use of non-aqueous solvents to solubilise and increase the stability of the drug (Connors *et al.*, 1986). Amber-coloured containers and antioxidants have also been reported to reduce photolytic degradation (Connors *et al.*, 1986).

The solubility of indometacin increases with pH. Drug solubility in pH 1.2 and 7.2 buffered aqueous medium at 25°C was found to be 3.882 micrograms/mL and 767.5 micrograms/mL, respectively (Nodkodchi *et al.*, 2005).

Crystalline indometacin exhibits polymorphism (O'Brien *et al.*, 1984; Connors *et al.*, 1986). The number of different polymorphs is not certain, with suggestions ranging from two to four different forms (O'Brien *et al.*, 1984). Indometacin powder should be stored protected from light (British Phar- macopoeia, 2007).

Degradation products

Following hydrolysis of the amide moiety in both acidic and basic media, the corresponding carboxylic acid and amine are formed (Connors *et al.*, 1986).

In alkaline solutions, indometacin is reported to break down to 5-methoxy-2-methylindole-3-acetic acid and *p*-chlorobenzoic acid (O'Brien *et al.*, 1984; Tomida *et al.*, 1989).

Stability in practice

There is a fair body of evidence to describe the stability of indometacin in suspension and solution. However, some of the data relate to formulations that are flawed or unsuitable for young children. Products licensed outside the UK are also known to be available for import.

Indocid suspension, previously available from Thomas Morson Pharmaceuticals, was reported to be unstable in alkaline media and therefore it was recommended that it should not be mixed with an antacid. The Summary of Product Characteristics also stated that the suspension should not be diluted (Lund, 1994; ABPI, 1996). Formulation details are no longer available.

An indometacin preparation licensed in Germany is available for import (Indo-paed). It is reported to be stable for 3 years, but should be used within 30 days of opening and stored away from light (Summary of Product Characteristics [translation], 2006). The formulation contains 1% ethanol, along with simethicon emulsion, sorbic acid, sorbitol, tragacanth, purified water and flavourings. It is contraindicated in children under 2 years of age. The exact formula should be investigated before procurement by a suitably competent pharmacist. Each 5 mL of suspension contains up to 2.26 g of sorbitol. Sorbitol is recognised to have an osmotic laxative effect (Nash, 2003).

Further suspensions are (or have been) available in other countries, such as in the USA. Indocin suspension (Merck) has a pH of 4–5, while the Roxane oral suspension was reported to have a pH of 2.9 (Trissel, 2000). The Indocin suspension is reported to contain 25 mg indometacin, alcohol 1%, sorbic acid 0.1%, antifoam emulsions, flavours, purified water, sodium hydroxide or hydrochloric acid to adjust pH, sorbitol solution and tragacanth (Oral Suspension Drug Information, 2006). The safety and effectiveness has not been established in patients 14 years of age and younger.

Indometacin oral suspension has a United States Pharmacopeia monograph which states the pH should be between 2.5 and 5.0 (United States Pharmacopeia, 2003).

Das Gupta *et al.* (1978) have investigated the stability of a pediatric liquid dosage form of indometacin (2 mg/mL). The vehicle was simple syrup with 10% alcohol. Pharmaceutical grade powder or powder extracted from capsules were used as the drug source. The powder was wetted with alcohol using a pestle and mortar before mixing with simple syrup. 0.005% of methylhydroxybenzoate and 0.002% of propylhydroxybenzoate were added to the syrup as preservatives.

The suspensions were stored in amber bottles at 24°C and analysed by UV spectrophotometry. The sample prepared from capsule powder appeared to be more stable than did the one from pure powder. However, both formulations retained over 94% of the starting concentration after 224 days. The final pH values were 5.1 for the pure powder preparation and 5.5 for the capsule powder preparation.

Microbiological stability was not studied. Although the formulations appear to be stable, the use of a formulation with 10% alcohol may be inappropriate in some patients, particularly young children. As a comparator, the maximum quantity of alcohol allowed in American 'over-the-counter' medicines is 10% v/v for products labelled for use by people aged 12 years and older, and 5% for products intended for use by children aged 6–12 years. For children aged under 6 years, the limit is 0.5% v/v (Jass, 1995; Owen, 2003). Long-term use of syrup may also carry dental health concerns.

Stewart *et al.* (1985) reviewed the stability of six extemporaneously prepared paediatric formulations of indometacin in use at the time of the study. They were:

Formulation 1 (250 micrograms/mL)

- indometacin 25 mg
- disodium hydrogen orthophosphate (anhydrate) 904 mg
- potassium dihydrogen orthophosphate 46 mg
- concentrated chloroform water 2.5 mL
- purified water to 100 mL.

Formulation 2 (250 micrograms/mL)

- indometacin 25 mg
- alcohol 10 mL
- methyl hydroxybenzoate 5 mg
- propyl hydroxybenzoate 2 mg
- syrup to 100 mL.

Formulation 3 (250 micrograms/mL)

- indometacin 25 mg
- compound tragacanth powder 1.75 mg
- methyl hydroxybenzoate 5 mg
- propyl hydroxybenzoate 2 mg
- purified water to 100 mL.

Formulation 4 (4 mg/mL)

- indometacin 400 mg
- compound tragacanth powder 3 g

- orange syrup 10 mL
- sorbitol solution USP 10 mL
- methyl hydroxybenzoate 150 mg
- propyl hydroxybenzoate 50 mg
- absolute alcohol 4 mL
- distilled water to 100 mL.

Formulation 5 (2 mg/mL)

- indometacin 200 mg
- raspberry syrup 20 mL
- glycerin to 100 mL.

Formulation 6 (5 mg/mL)

- indometacin 500 mg
- imitation vanilla flavour 0.1 mL
- concentrated chloroform water 2.5 mL
- syrup to 100 mL.

The formulations containing 250 micrograms/mL were designed for closure of the patent ductus arteriosus and the formulations containing 200–500 mg were designed for the treatment of juvenile rheumatoid arthritis.

A large amount (10 L) of each formulation was prepared and dispersed using an automated mixer. The formulations were stored in clear and amber dispensing flat bottles (100 mL) and 500 mL screw-cap jars for viscosity measurements using a Brookfield viscometer (model LVT).

Samples were stored at the following temperatures at 2, 20, 35, 55°C and 'room temperature'. Stability-indicating HPLC analysis of samples was carried out after 1, 2, 4, and 12 weeks.

Stewart *et al.* (1985) present degradation profiles for each formulation. All formulations showed some evidence of instability over 100 days, particularly at higher temperatures. Of the lower strength formulations, formulation 1 showed the fastest degradation. This is unsurprising since formulation 1 is a buffered solution of indometacin (pH 8.15), while the other liquids were all suspensions. The pH values of formulations 2 and 3 were 5.80 and 5.05, respectively.

The higher strength formulations showed a greater degree of stability, since the relative proportion of drug in solution (and therefore open to hydrolysis) is lower.

Some of the suspensions stored in clear bottles showed slight discoloration, but there was not a large difference in drug content over 15 weeks. Stewart *et al.* (1985) suggest that the major degradation pathway is probably a hydroxyl ion catalysed hydrolysis rather than photolytic.

The lowest rates of degradation were seen with formulations 5 and 6, with over 90% of drug remaining after 100 days at 35°C. The authors state that the usefulness of formulation 5, however, is limited by the extent of caking and difficulty of re-dispersing. Formulations 2 and 3 showed relatively high coefficients of variation for dosage uniformity (10.3% and 10.7%) under patient use conditions.

However, most of the formulations studied by Stewart *et al.* (1985) are flawed in some way. Formulations 1 and 6 contain chloroform, which is recognised as a class 3 carcinogen. These are substances that cause concern owing to possible carcinogenic effects but for which available information is not adequate to make satisfactory assessments (CHIP3 Regulations, 2002). Formulations 2 and 4 contain alcohol, and formulation 5 has re-dispersibility problems.

Carter and Lewis (1982) have investigated the stability of indometacin solution at a concentration of 50 micrograms/mL. Using UV absorption methodology from the 1980 BP, the indometacin was made in Sorensen buffer at pH 7.2 and stored in either glass containers or plastic syringes. The authors suggest that the solution was stable for three months. The drug concentration was reported to be above 90% of the starting concentration, with no significant changes to the absorption spectrum.

Scanlon (1982) suggested that aqueous suspensions of indometacin for oral administration to close the patent ductus arteriosus should be abandoned. He suggests that the practice of dispersing a single 25 mg indometacin capsule in 10 mL distilled water produces uneven and inaccurate dosing, due to the poor water solubility of the drug. Scanlon suggests that dissolving the dose in alcohol produces more reproducible and accurate doses, and that some of the therapeutic failures seen may result from poor dosage accuracy.

Related products

Indometacin is available as an Indocid PDA lyophilised powder for reconstitution (as sodium trihydrate). Reconsitution at pH levels below 6 has been reported to cause precipitation of insoluble indometacin (Lund, 1994). The reconstituted injection has a pH between 6.0 and 7.5 (United States Pharmacopeia, 2003). It is reported to be stable for 16 days at room temperature, although the manufacturer advises that any unused solution is discarded as there is no preservative present (McEvoy, 1999; Trissel, 2000).

The injection should be stored below 30°C and protected from light (McEvoy, 1999).

Bioavailability data

No information is available for the extemporaneous preparations.

Bhat *et al.* (1980) have investigated the pharmacokinetics of oral and intravenous indometacin in preterm infants. For 'oral' administration, indometacin was suspended in saline and administered via the nasogastric tube and then flushed through with water. The effect of nasogastric administration on bioavailability is now known. The authors suggest a 13% oral bioavailability based on are under curve (AUC) methods.

The appropriate manufacturer or importer should be contacted with regard to imported products that are licensed outside the UK.

References

ABPI (1996). *ABPI Compendium of Data Sheets and Summaries of Product Characteristics 1996–1997*. London: Datapharm Publications.

Bhat R, Vidyasagar D, Fisher E, Hastreiter A, Ramirez JL, Burns L, Evans M (1980). The pharmacokinetics of oral and intravenous indomethacin in preterm infants. *Dev Pharmacol Ther* 1: 101–110.

British Pharmacopoeia (2007). *British Pharmacopoeia*, Vol. 1. London: The Stationery Office, pp. 783–785 and Vol. 2, pp. 21080–21081.

Carter MA, Lewis CW (1982). Stability of indomethacin solutions (letter). *Pharm J* 228: 569.

CHIP3 Regulations (2002). Chemicals (Hazard Information and Packaging for Supply) Regulations. SI 2002/1689.

Connors KA, Amidon GL, Stella ZJ (1986). *Chemical Stability of Pharmaceuticals: A Handbook for Pharmacists*, 2nd edn. New York: Wiley Interscience.

Das Gupta V, Gibbs CW Jr, Ghanekar AG (1978). Stability of pediatric liquid dosage forms of ethacrynic acid, indomethacin, methdopate hydrochloride and spironolactone. *Am J Hosp Pharm* 35: 1382–1385.

Dawson JE, Hajratwala BR, Taylor H (1977). Kinetics of indomethacin degradation II: presence of alkali plus surfactant. *J Pharm Sci* 66: 1259–1263.

Ghanem AH, Hassan ES, Hamdi AA (1979). Stability of indomethacin solubilized system. *Pharmazie* 34: 406–407.

Hajratwala BR, Dawson JE (1977). Kinetics of indomethacin degradation I: presence of alkali. *J Pharm Sci* 66: 27–29.

Han J, Beeton A, Long PF, Wong I, Tuleu C (2006). Physical and microbiological stability of an extemporaneous tacrolimus suspension for paediatric use. *J Clin Pharm Ther* 31: 167–172.

Jass HE (1995). Regulatory review. *Cosmet Toilet* 110(5): 21–22.

Lowey AR, Jackson MN (2008). A survey of extemporaneous dispensing activity in NHS trusts in Yorkshire, the North-East and London. *Hosp Pharmacist* 15: 217–219.

Lund W, ed. (1994). *The Pharmaceutical Codex: Principles and Practice of Pharmaceutics*, 12th edn. London: Pharmaceutical Press.

McEvoy J, ed. (1999). *AHFS Drug Information 1999*. Bethesda, MD: American Society of Health System Pharmacists.

Merck, Sharp and Dohme (1984). Indomethacin resume of essential information, June 1965. In: Florey K, ed. *Analytical Profiles of Drug Substances*, Vol. 13. London: Academic Press, pp. 211–238.

Moffat AC, Osselton MD, Widdop B (2003). *Clarke's Analysis of Drugs and Poison*, 3rd edn. London: Pharmaceutical Press. ISBN 978 0 85369 473 1.

Nash RA (2003). Sorbitol. In: Rowe RC, Sheskey PJ, Weller PJ, eds. *Handbook of Pharmaceutical Excipients*, 4th edn. London: Pharmaceutical Press and Washington DC: American Pharmaceutical Association.

NHS Pharmaceutical Quality Assurance Committee (2004). Guidance for the purchase and supply of unlicensed medicinal products. Unpublished document available from regional quality assurance specialists in the UK or after registration from the NHS Pharmaceutical Quality Assurance Committee website: www.portal.nelm.nhs.uk/QA/default.aspx.

Nodkodchi A, Javadzadeh Y, Siah Shadbad MR, Barzegar-Jalali M (2005). The effect of type and concentration of vehicles on the dissolution rate of a poorly soluble drug (indomethacin) from liquisolid compacts. *J Pharm Pharm Sci* 8: 18–25.

Nova Laboratories Ltd (2003). Information Bulletin – Suspension Diluent Used at Nova Laboratories.

O'Brien, M, McCAuley J, Cohen E (1984). Indomethacin. In: Florey K, ed. *Analytical Profiles of Drug Substances*; Vol. 13. London: Academic Press, pp. 211–238.

Oral Suspension Drug Information (2006). Indocin. www.drugs.com/pdr/indocin_oral_suspension.html (accessed 22 September 2006).

Pawelczyk E, Knitter B (1978). Kinetics of drug degradation. Part 58: A method of preparation and the stability of 3% aqueous indometacin solution. *Pharmazie* 33: 586–588.

Rowe RC, Sheskey PJ, Quinn ME, eds. (2009). *Handbook of Pharmaceutical Excipients*, 6th edn. London: Pharmaceutical Press.

Scanlon JW (1982). Oral aqueous suspension of indomethacin should be abandoned. *Pediatrics* 69: 507.

Singla AK, Babber P, Pathak K (1991). Effect of zinc ion on indomethacin degradation in alkaline aqueous solutions. *Drug Dev Ind Pharm* 17: 1411–1418.

Stewart PJ, Doherty G, Bostock JM, Petrie AF (1985). The stability of extemporaneously prepared paediatric formulations of indomethacin. *Aust J Hosp Pharm* 15: 55–60.

Summary of Product Characteristics [Translation] (2006). Indo-paed. Supplied by: H Pretorius, IDIS Technical Information Advisor, Pharmaceutical Services Department.

Tomida H, Kuwada N, Tsuruta Y, Kohashi K, Kiryu S (1989). Nucleophilic aminoalcohol-catalyzed degradation of indomethacin in aqueous solution. *Pharma Acta Helv* 64: 312–315.

Trissel LA, ed. (2000). *Stability of Compounded Formulations*, 2nd edn. Washington DC: American Pharmaceutical Association.

United States Pharmacopeia (2003). *United States Pharmacopeia*, XXVI. Rockville, MD: The United States Pharmacopeial Convention.

Isosorbide mononitrate oral liquid

Risk assessment of parent compound

A risk assessment survey completed by clinical pharmacists suggested that isosorbide mononitrate has a moderate therapeutic index, with the potential to cause generally mild side-effects if inaccurate doses were to be administered (Lowey and Jackson, 2008). Current formulations used at NHS trusts generally show a low technical risk, based on the number and/or complexity of manipulations and calculations. Isosorbide mononitrate is therefore regarded as a moderate-risk extemporaneous preparation (Box 14.23).

Summary

The use of oral liquid forms of isosorbide mononitrate is not widespread, and there is a paucity of data to support their stability or efficacy. Consideration

Box 14.23 *Preferred formula*

Using a typical strength of 4 mg/mL as an example:

- isosorbide mononitrate 40 mg **normal release** tablets × 10
- xanthan gum suspending agent 0.5% ('Keltrol' or similar) to 100 mL.

Method guidance

Tablets can be ground to a fine, uniform powder in a pestle and mortar. A small amount of suspending agent may be added to form a paste, before adding further portions up to 75% of the final volume. Transfer to a measuring cylinder. The remaining suspending agent can be used to wash out the pestle and mortar before making the suspension up to 100% volume. Transfer to an amber medicine bottle.

Shelf-life

7 days at room temperature in amber glass. **Shake the bottle.**

Note: 'Keltrol' xanthan gum formulations vary slightly between manufacturers, and may contain chloroform and ethanol. The individual specification should be checked by the appropriate pharmacist.

should be given to alternative forms of treatment (e.g. sublingual isosorbide dinitrate forms, nitrate patches).

If the use of an oral liquid is deemed necessary, a simple formula and a limited shelf-life should be used in order to minimise overall risk.

Clinical pharmaceutics

Points to consider

- At typical concentrations (4 mg/mL), the drug should be in solution. Dosage uniformity should not be an issue. A suspending agent is recommended in order to suspend any insoluble tablet excipients.
- Standard release tablets only should be used to make an extemporaneous oral liquid.
- There are no known data with regard to bioavailability of isosorbide mononitrate oral liquids.
- Isosorbide mononitrate is a major metabolite of isosorbide dinitrate.
- 'Keltrol' is made to slightly different specifications by several manufacturers, but typically contains ethanol and chloroform. The exact specification should be assessed before use by a suitably competent pharmacist. Ethanol is a CNS depressant. Chloroform has been classified as a class 3 carcinogen (i.e. substances that cause concern owing to possible carcinogenic effects but for which available information is not adequate to make satisfactory assessments) (CHIP3 Regulations, 2002).

Technical information

Structure

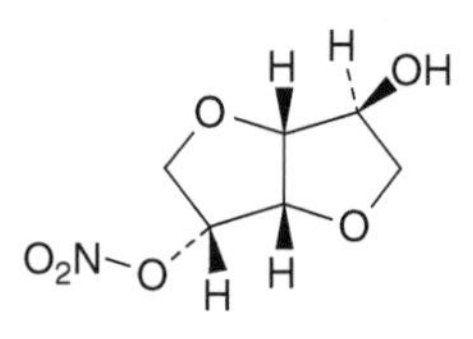

Empirical formula

$C_6H_9NO_6$ (British Pharmacopoeia, 2007).

Molecular weight

191.1 (British Pharmacopoeia, 2007).

Solubility

Freely soluble in water (100 mg/mL to 1 g/mL) (British Pharmacopoeia, 2007).

pK_a

Not known.

Optimum pH stability

Not known.

Stability profile

Physical and chemical stability

Undiluted isosorbide mononitrate is a white, crystalline powder, freely soluble in water and in alcohol (British Pharmacopoeia, 2007). The solubility of the diluted product depends on the diluent and its concentration (British Pharmacopoeia, 2007). Diluted isosorbide mononitrate is a mixture of isosorbide mononitrate with lactose or mannitol (Martindale, 2005).

Isosorbide mononitrate was stabilised by the formation of a 1:1 inclusion complex with β-cyclodextrin. Samples of powder stored at 75% relative humidity at both 60°C and 45°C for 30 days were shown to have only 8% and 63% of their initial concentration remaining, respectively (Lund, 1994). Complexed isosorbide mononitrate showed no decomposition over 30 days at 60°C and 75% relative humidity.

Isosorbide mononitrate should be stored protected from light (British Pharmacopoeia, 2007).

Isosorbide oral solution USP ($C_6H_{10}O_4$) has a pH range of 3.2–3.8. It should be preserved in tight containers (United States Pharmacopeia, 2007).

Isosorbide mononitrate is a degradation product of the hydrolysis of isosorbide dinitrate, which degrades more rapidly in base than acid. Isosobide dinitrate is stable as a solid at a temperature of 45°C for 12 months and at room temperature for 5 years (Sivieri and DeAngelis, 1975).

Degradation products

No data.

Stability in practice

There are few data to show the chemical, physical or microbiological stability of isosorbide mononitrate oral liquids.

Isosorbide mononitrate is known to be physically compatible with Diluent A, a xanthan gum-based suspending agent (Nova Laboratories Ltd, 2003).

An oral mixture containing isosorbide 50% w/v and sorbitol 5% w/v has been described (Brown *et al.*, 1983; Lund, 1994. The authors suggested a shelf-life of 12 months following an evaluation of chemical and microbiological stability. The excipients used included saccharin sodium, sodium citrate, potassium citrate, malic acid, potassium sorbate, ethyl alcohol, vanillin, peppermint oil, burnt sugar solution and distilled water.

Bioavailability data

No data.

References

British Pharmacopoeia (2007). *British Pharmacopoeia*, Vol. 1. London: The Stationery Office, pp. 1135–1137.

Brown AF, Fisher B, Harvey DA, Hoddinott DJ (1983). Analysis and microbiological stability of an isosorbide 50% mixture. *J Clin Hosp Pharm* 8: 339–344.

CHIP3 Regulations (2002). Chemicals (Hazard Information and Packaging for Supply) Regulations. SI 2002/1689.

Lowey AR, Jackson MN (2008). A survey of extemporaneous dispensing activity in NHS trusts in Yorkshire, the North-East and London. *Hosp Pharmacist* 15: 217–219.

Lund W, ed. (1994). *The Pharmaceutical Codex: Principles and Practice of Pharmaceutics*, 12th edn. London: Pharmaceutical Press.

Nova Laboratories Ltd (2003). Information Bulletin – Suspension Diluent Used at Nova Laboratories.

Sivieri LA, DeAngelis NJ (1975). Isosorbide dinitrate. In: Florey K, ed. *Analytical Profiles of Drug Substances*, Vol. 4. London: Academic Press, pp. 225–244 (erratum 1976, Vol 5: 556).

Sweetman SC (2009). *Martindale: The Complete Drug Reference*, 36th edn. London: Pharmaceutical Press.

United States Pharmacopeia (2003). *United States Pharmacopeia*, XXVI. Rockville, MD: The United States Pharmacopeial Convention.

Joulie's solution (oral phosphate supplement)

Joulie's solution typically contains approximately 1 mmol/mL of phosphate. However, an exact formulation is not widely agreed.

Risk assessment of parent compound

A risk assessment survey completed by clinical pharmacists suggested that Joulie's solution has a low therapeutic index. Inaccurate doses were thought to be associated with minor morbidity (Lowey and Jackson, 2008). Current formulations used at NHS trusts were of a high technical risk, based on the number and/or complexity of manipulations and calculations. Due mainly to its technical risks, Joulie's solution is regarded as a high-risk extemporaneous preparation (Box 14.24).

Summary

While Joulie's solution is used widely in the NHS, an exact formulation has yet to be agreed. The supervising pharmacist should take steps to validate and check the formulation to be prepared or procured.

***Box 14.24** Preferred formula*

Note: 'Specials' are now available from reputable manufacturers in the UK. Extemporaneous preparation should only occur where there is a demonstrable clinical need.

The following formula would provide 0.91 mmol/mL of phosphate and 0.76 mmol/mL of sodium:

- disodium hydrogen phosphate anhydrous BP 5.4 g
- phosphoric acid (87% m/m, specific gravity 1.71) 3.4 mL
- water for injections BP to 100 mL.

Shelf-life

7 days refrigerated in amber glass.

Given the lack of stability issues with this preparation, Joulie's solution would seem an ideal candidate to procure from a 'Specials' manufacturer (either NHS or non-NHS). Once again, the exact formula should be agreed before purchase. Terminally sterilised specials are available with no preservative, if the preservatives are deemed to be inappropriate for the target population.

Clinical pharmaceutics

Points to consider

- Joulie's formula is a solution rather than a suspension. Dosage uniformity should not be an issue. If crystallisation occurs, gentle warming is reported to redissolve the crystals (Mandeville Medicines, 2006).
- 'Specials' are available from several manufacturers. These items include those with preservatives, and those which are terminally sterilised. Preservatives such as the hydroxybenzoates, sodium benzoate and chloroform have been used.
- Chloroform is has been classified as a class 3 carcinogen (substances that cause concern owing to possible carcinogenic effects but for which available information is not adequate to make satisfactory assessments) (CHIP3 Regulations, 2002).
- The World Health Organization (WHO) acceptable daily intake of total benzoates, calculated as benzoic acid, has been estimated at up to 5 mg/kg of body weight (Jacobs, 2003). Parenteral use of sodium benzoate is not recommended in premature infants or infants of low birth weight. Sodium benzoate may displace bile from albumin binding sites.
- Hydroxybenzoates are widely used as antimicrobial preservatives in cosmetics, and oral and topical pharmaceutical formulations. Concern has been expressed over their use in infant parenteral products as bilirubin binding may be affected (Rieger, 2003). This is potentially hazardous in hyperbilirubinaemic neonates (Loria *et al.*, 1976). However, there is no proven link to similar effects when used in oral products. WHO/FAO (1974) suggested an estimated total acceptable daily intake for methyl-, ethyl- and propyl-hydroxybenzoates at up to 10 mg/kg body weight.
- There are no known bioavailability data available for this preparation.

Technical information

Table 14.3

Table 14.3 Technical information for Joulie's solution

	Disodium hydrogen phosphate	Phosphoric acid
Empirical formula	Na_2HPO_4 (United States Pharmacopoeia, 2003)	H_3PO_4 (Allen, 2005)
Molecular weight	141.96 (Anhydrous) 177.98 (Dihydrate) 268.03 (Heptahydrate) 358.08 (Dodecahydrate) (Allen, 2005)	98.00 (British Pharmacopoeia, 2007)
Solubility or miscibility	Very soluble in water (anhydrous 1 in 8, heptahydrate 1 in 4, dodecahydrate 1 in 3) (Allen, 2005)	Miscible with alcohol and water with the evolution of heat (Chambliss, 2003)
pK_a	2.15 and 7.2 (Allen, 2005)	Not known
Optimum pH stability	Not known	Not known

Stability profile

Physical and chemical stability

Joulie's solution is an oral phosphate supplement, containing sodium phosphate (also known as dibasic sodium phosphate or disodium hydrogen phosphate) and phosphoric acid (Allen, 2005).

Dihydrogen sodium phosphate (also known as sodium phosphate and dibasic sodium phosphate) is available in an anhydrous form, dihydrate, heptahydrate and as a dodecahydrate. The anhydrous form is a hygroscopic white powder, and the dihydrate exists as a white or almost white odourless crystal form (Allen, 2005). The heptahydrate has a colourless crystal, white granular or caked salt form; the dodecahydrate exists as colourless or transparent crystals (Allen, 2005).

Sodium phosphate is very soluble in water but practically insoluble in ethanol (Kearney, 2003). Each gram of dibasic sodium phosphate (anhydrous) represents approximately 14.1 mmol of sodium and 7.0 mmol of phosphate. Each gram of the dihydrate represents approximately 11.2 mmol of sodium and 5.6 mmol of phosphate. Each gram of the heptahydrate represents 7.5 mmol of the sodium and 3.7 mmol of the phosphate and each gram of the dodecahydrate represents approximately 5.6 mmol of sodium and 2.8 mmol of phosphate (Martindale, 2005).

Phosphoric acid (also known as hydrogen phosphate or orthophosphoric acid) occurs as a clear, colourless, syrupy liquid (British Pharmacopoeia, 2007). It is miscible with water and alcohol (British Pharmacopoeia, 2007).

The pH of a 1% w/w aqueous solution is 1.6. It should be stored in a glass container (British Pharmacopoeia, 2007). Phosphoric acid may solidify into a mass of colourless crystals when stored at low temperatures. The crystals do not melt under 28°C (British Pharmacopoeia, 2007).

Phosphoric acid should be stored in an airtight container in a cool, dry place. Stainless steel containers may be used (Chambliss, 2003).

Degradation products

No data.

Stability in practice

Several 'Specials' manufacturers are reported to produce Joulie's solution, with known shelf-lives ranging from 28 days to 2 years. Some manufacturers produce terminally sterilised solutions which therefore do not contain a preservative. The advantage is that potentially toxic preservatives are avoided. However, the unpreserved contents should be used within 7 days of opening the bottle, and stored at 2–8°C.

NHS 'Specials' include the following formulation (W. Goddard, Laboratory Manager, Quality Control West Midlands, University Hospitals Birmingham NHS Foundation Trust, personal communication, 2007):

- disodium hydrogen phosphate anhydrous BP 5.4 g
- phosphoric acid BP (87% m/m) 3.4 mL
- water for injections BP to 100 mL.

This is packaged in a type 1 glass container and following autoclaving is assigned an expiry of 12 months. Once opened, it should be discarded after 7 days (W. Goddard, personal communication, 2007). It provides 0.91 mmol/mL of phosphate and 0.76 mmol/mL of sodium.

Allen (2005) has proposed the following formula:

- dibasic sodium phosphate, heptahydrate 13.6 g
- phosphoric acid (85–88% w/v, specific gravity 1.71) 3.5 mL
- methyl hydroxybenzoate 100 mg
- propyl hydroxybenzoate 50 mg
- purified water to 100 mL.

Allen's formula provides approximately 1 mmol of both sodium and phosphate per millilitre.

The purified water is heated almost to boiling before the addition of the hydroxybenzoates. After cooling, the sodium phosphate is added and mixed well, before the phosphoric acid is added. Sufficient purified water is then added to make up to volume. Allen (2005) suggests an expiry date of 6 months for this preparation, though no supporting data are presented.

The hydroxybenzoates are widely used antimicrobial preservatives in cosmetics and oral and topical pharmaceutical formulations. Concern has been expressed over their use in infant parenteral products as bilirubin binding may be affected (Rieger, 2003). This is potentially hazardous in hyperbilirubinaemic neonates (Loria *et al.*, 1976). However, there is no proven link to similar effects when used in oral products. WHO/FAO (1974) suggested an estimated total acceptable daily intake for methyl-, ethyl- and propyl-hydroxybenzoates at up to 10 mg/kg body weight.

A formula historically used at St James' Hospital, Leeds (year of origin unknown) is as follows:

- sodium phosphate (disodium hydrogen phosphate [dodecahydrate]) 67.96 g
- chloroform water concentrated 12.5 mL
- phosphoric acid 34.6 g
- distilled water to 500 mL.

This formula provides approximately 1 mmol/mL of phosphate and 0.76 mmol/mL of sodium.

Chloroform has been classified as a class 3 carcinogen (substances that cause concern owing to possible carcinogenic effects but for which available information is not adequate to make satisfactory assessments) (CHIP3 Regulations, 2002). The Calderdale and Huddersfield NHS Foundation Trust Pharmacy Manufacturing Unit have now removed the chloroform water from the formula and restricted the shelf-life accordingly (D. Wallace, personal communication, 2006).

If crystallisation occurs, gentle warming is reported to re-dissolve the crystals (Mandeville Medicines, 2006).

Bioavailability data

No data.

References

Allen LV Jr (2005). Joulies solution. *Int J Pharm Compound* 9: 315.

British Pharmacopoeia (2007). *British Pharmacopoeia*, Vol. 2. London: The Stationery Office, p. 1641.

Chambliss WG (2003). Phosphoric acid. In: Rowe RC, Sheskey PJ, Weller PJ, eds. *Handbook of Pharmaceutical Excipients*, 4th edn. London: Pharmaceutical Press and Washington DC: American Pharmaceutical Association.

CHIP3 Regulations (2002). Chemicals (Hazard Information and Packaging for Supply) Regulations. SI 2002/1689.

Jacobs H (2003). Benzoic acid. In: Rowe RC, Sheskey PJ, Weller PJ, eds. *Handbook of Pharmaceutical Excipients*, 4th edn. London: Pharmaceutical Press and Washington DC: American Pharmaceutical Association, pp. 50–52.

Kearney AS (2003). Sodium phosphate, dibasic. In: Rowe RC, Sheskey PJ, Weller PJ, eds. *Handbook of Pharmaceutical Excipients*, 4th edn. London: Pharmaceutical Press and Washington DC: American Pharmaceutical Association, pp. 574–576.

Loria CJ, Excehverria P, Smith AL (1976). Effect of antibiotic formulations in serum protein: bilirubin interaction of newborn infants. *J Pediatr* 89: 479–482.

Lowey AR, Jackson MN (2008). A survey of extemporaneous dispensing activity in NHS trusts in Yorkshire, the North-East and London. *Hosp Pharmacist* 15: 217–219.

Mandeville Medicines (2009). Labelling directions. Joulie's Solution. Batch number: TN031411 Expiry 15/6/9.

Rieger MM (2003). Ethylparabens. In: Rowe RC, Sheskey PJ, Weller PJ, eds. *Handbook of Pharmaceutical Excipients*, 4th edn. London: Pharmaceutical Press and Washington DC: American Pharmaceutical Association, pp. 244–246.

Sweetman SC (2009). *Martindale: The Complete Drug Reference*, 36th edn. London: Pharmaceutical Press.

United States Pharmacopeia (2003). *United States Pharmacopeia*, XXVI. Rockville, MD: The United States Pharmacopeial Convention.

WHO/FAO (1974). Toxicological evaluation of certain food additives with a review of general principles and of specifications. Seventeenth report of the Joint FAO/WHO Expert Committee on Food Additives. World Health Organ. Tech. Rep. Ser. No. 539.

Knox mouthwash

Risk assessment of parent compound

A risk assessment survey completed by clinical pharmacists suggested that Knox mouthwash has a wide therapeutic index, with the potential to cause minor side-effects if inaccurate doses were to be administered (Lowey and Jackson, 2008). Current formulations used at NHS trusts for the mouthwash show a moderate technical risk, based on the number and complexity of manipulations and calculations. Knox mouthwash is therefore regarded as a moderate-risk extemporaneous product (Box 14.25).

Summary

There are no data available in the public domain to support any of the formulae used for Knox mouthwash. Although it is generally considered a low-risk product, its use should only be considered following a local risk assessment.

The use of licensed alternative products should be considered (e.g. doxycycline dispersible tablets and betamethasone soluble tablets).

Box 14.25 *Preferred formula*

Note: There are several formulae used for Knox mouthwash, using different strengths and different drugs. The following formula is offered only as an example.

A typical Knox mouthwash can be prepared as follows:

- triamcinolone acetonide 10 mg/mL injection × 5 mL
- erythromycin ethyl succinate suspension 50 mg/mL to 100 mL
- (final concentrations = triamcinolone acetonide 500 micrograms/mL and erythromcyin ethyl succinate 47.5 mg/mL).

A paediatric Knox mouthwash has also been described, using 25 mg/mL erythromycin ethyl succinate suspension (NHS Pharmaceutical Quality Assurance Committee Survey, 2006).

Shelf-life

7 days refrigerated in amber glass. **Shake the bottle.**

Clinical pharmaceutics

Points to consider

- The mouthwash should be shaken well before use, as settling may occur, leading to poor dosage uniformity.
- The issues surrounding excipients will depend on the constituent products used for the mouthwash. Triamcinolone acetonide injection may contain benzyl alcohol. Although serious adverse effects (diarrhoea, CNS depression, respiratory failure) are possible, the concentrations of benzyl alcohol normally employed as a preservative are not normally associated with such effects (Brunson, 2003).
- Erythromycin ethyl succinate is odourless and almost tasteless (Lund, 1994).
- Patients are advised to spit out the mouthwash to minimise systemic absorption.
- Triamcinolone acetonide is known to have a bitter taste (Ungphaiboon *et al.*, 2005) and to exist in two polymorphic forms (Sieh, 1982). In theory, changes in polymorphic form could alter particle size (Han *et al.*, 2006). Whether this occurs with triamcinolone in Knox mouthwash is not known.

Other notes

- Some NHS centres have assigned 14-day shelf-lives, where the expiry of the erythromycin suspension allows this (Lowey and Jackson, 2008).
- While most centres now use erythromycin, some may still use extemporaneously prepared oxytetracycline syrup based on historical Knox formulae (Lowey and Jackson, 2008).

Technical information

Table 14.4

Stability profile

Physical and chemical stability

Erythromycin ethyl succinate is a white, crystalline powder, hygroscopic in nature. It is practically insoluble in water but freely soluble in ethanol, acetone, chloroform, macrogol 400 and methanol. It should be kept in an airtight container, protected from light, and stored at a temperature not exceeding 30°C (Lund, 1994; British Pharmacopoeia, 2007). Erythromycin ethyl succinate Oral Suspension USP should be stored in a cold place (Lund, 1994).

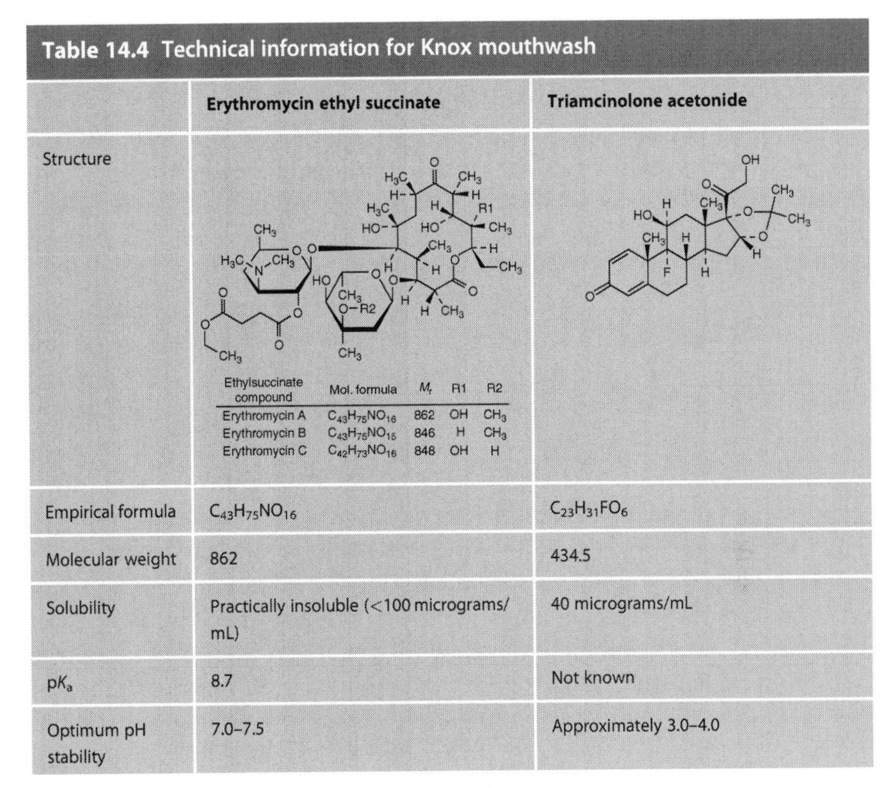

Table 14.4 Technical information for Knox mouthwash

	Erythromycin ethyl succinate	Triamcinolone acetonide
Structure	(structure)	(structure)
Empirical formula	$C_{43}H_{75}NO_{16}$	$C_{23}H_{31}FO_6$
Molecular weight	862	434.5
Solubility	Practically insoluble (<100 micrograms/mL)	40 micrograms/mL
pK_a	8.7	Not known
Optimum pH stability	7.0–7.5	Approximately 3.0–4.0

Ethylsuccinate compound	Mol. formula	M_r	R1	R2
Erythromycin A	$C_{43}H_{75}NO_{16}$	862	OH	CH_3
Erythromycin B	$C_{43}H_{75}NO_{15}$	846	H	CH_3
Erythromycin C	$C_{42}H_{73}NO_{16}$	848	OH	H

Adapted from Moffat (1986), Connors *et al.* (1986), Lund (1994), British Pharmacopoeia (2007).

Approximately 1.17 g of the ethyl succinate is equivalent to 1 g of erythromycin (Martindale, 2005). A 1% aqueous suspensions of erythromycin ethyl succinate has a pH of 6.0–8.5; the oral suspension has a pH of 6.5–8.5, and the reconstituted dry oral suspension has a pH of 7.0–9.0 (United States Pharmacopeia, 1999).

The decomposition of erythromycin base in aqueous solutions is affected by pH, with maximum stability occurring in the region pH 7.0–7.5. Decomposition in both acid and alkali follows first-order kinetics. The activation energy for the hydrolysis of erythromycin at pH 7 is stated as 77.8 kJ/mol. An Arrhenius plot of different salts of erythromycin suggest that the half-life of the ethyl succinate form is approximately 33 days at 25°C in sodium citrate buffer at pH 7 (Lund, 1994).

Koch (1979) reported that erythromycin base is unstable in both acidic and alkaline solutions, showing maximum stability between 6.0 and 9.5. Its aqueous, alcoholic solution buffered at pH 7.0–8.0 was stable for one week when refrigerated; no data or definitions of stability are given (Koch, 1979).

Erythromycin is not sensitive to photolytic degradation as a solid or as a solution (pH 4 and 8; Lund, 1994).

Triamcinolone acetonide is a white or almost white, crystalline powder, practically insoluble in water and sparingly soluble in alcohol. It should be stored protected from light (British Pharmacopoeia, 2007).

Triamcinolone acetonide is reported to be very stable as a solid. In common with other corticosteroids, the α-ketol side-chain becomes vulnerable to oxidative rearrangement in alkaline solutions. It is possible that the A-ring may be prone to photolytic degaradation, as is the case for hydrocortisone and prednisolone (Florey, 1972).

The pH rate profile for the pseudo-first order decomposition of triamcinolone is reported to show specific acid and base catalysis, with maximum stability reported at approximately 3.4 (Das Gupta, 1983; Connors *et al.*, 1986; Ungphaiboon *et al.*, 2005). Above pH 5.5, the rate of degradation increases rapidly (Das Gupta, 1983). An increase in the ionic strength leads to an increase in the rate of decomposition (Connors *et al.*, 1986). Das Gupta (1983) has suggested that that if pH is maintained as a constant, ethanol is a better solvent than glycerin/propylene glycol for improving stability.

When the pH of a triamcinolone acetonide solution is above 7, the rate of decomposition decreases with increasing ionic strength (Das Gupta, 1983).

Buffering agents and antioxidants should be considered to increase the stability of triamcinolone actenoide formulations (Ungphaiboon *et al.*, 2005).

Degradation products

No data.

Stability in practice

There are no known stability reports for Knox mouthwash.

Licensed suspensions of erythromycin ethyl succinate are assigned a shelf-life of 7–14 days at room temperature after reconstitution in line with its product licence. Triamcinolone acetonide is generally stable in solution. There have been no known reports of a physical incompatibility when these two agents are mixed together.

Ungphaiboon *et al.* (2005) have studied the formulation and efficacy of triamcinolone acetonide mouthwash for treating oral lichen planus. The mouthwash was prepared by dissolving 0.1 g triamcinolone acetonide and menthol 0.05 g in ethanol 10 mL. Propylene glycol 30 mL, glycerin 20 mL and 70% sorbitol 20 mL were then added. Sodium saccharin 0.1 g, sodium metabisulfite 0.02 g and disodium ethylene diamine tetraacetate 0.1 g were dissolved in 5 mL water. The above components were then combined and made up to 100 mL with 70% sorbitol.

The co-solvents glycerin and propylene glycol were used to limit the amount of ethanol in the preparation, as the authors suggest that high concentrations of ethanol may irritate the damaged mucosa. Relatively small concentrations of glycerine and propylene glycol were used, however, as the acrid tastes of these agents may also lead to burning sensations. Saccharin sorbitol and sorbitol were used to help mask the bitter drug taste, and the sodium metabisulfite and disodium ethylene diamine tetraacetate were used as antioxidants.

Accelerated stability testing was performed at 45, 60, 70 and 80°C and 75% relative humidity. HPLC methodology was combined with spectrophotometry and visual inspection for signs of formulation failure.

All samples remained transparent and colourless. The pH of the formulation was 5.11; the authors deemed that the use of a buffer was not necessary, and the pH remained stable over 1000 days at ambient temperature. The Arrhenius plot from the accelerated stability studies suggested an activation energy of 16.23 kcal/mol and a predicted shelf-life (time to lose 10% of the original drug concentration) as 6.17 years at 30°C. The authors also describe the results of longer term stability testing at 'ambient' temperature (described as approximately 30°C), with a calculated shelf-life of 4.25 years.

Ungphaiboon *et al.* (2005) went on to test the acceptability and efficacy of the mouthwash in a small cohort of patients. Some of the patients reportedly commented that the mouthwash was too sweet.

This study shows improved stability when compared to extemporaneous preparations of 0.1% formulations described by Vincent (1991) and Marek (1999). However, the complexity of the formulation limits its preparation in the UK to units with significant formulation expertise, notably 'Specials' manufacturing units.

Excipients that have been used in presentations of erythromycin as the ethyl succinate include: Allura Red AC (E129), aluminium magnesium silicate, amberlite, calcium hydrogen phosphate, carmellose sodium, citric acid, DandC Red No.30, iron oxide, macrogol, magnesium stearate, mannitol, methyl and propyl hydroxybenzoates, polysorbate 60, propylene glycol, Quinoline Yellow (E104), saccharin sodium, sodium citrate, sodium starch glycollate, sorbic acid, sorbitan mono-oleate, starch, sucrose, sugar, Sunset Yellow FCF (E110), titanium dioxide (E171), and vitamin E (Lund, 1994).

Bioavailability data

Laine *et al.* (1987) have compared the bioavailability of different crystalline forms of erythromycin. This information is of limited use as Knox mouthwash is not usually swallowed.

References

British Pharmacopoeia (2007). *British Pharmacopoeia*, Vol. 1. London: The Stationery Office, pp. 783–785 and Vol. 2, pp. 2090–2092.

Brunson EL (2003). Benzyl alcohol. In: Rowe RC, Sheskey PJ, Weller PJ, eds. *Handbook of Pharmaceutical Excipients*, 4th edn. London: Pharmaceutical Press and Washington DC: American Pharmaceutical Association.

Connors KA, Amidon GL, Stella ZJ (1986). *Chemical Stability of Pharmaceuticals: A Handbook for Pharmacists*, 2nd edn. New York: Wiley Interscience.

Das Gupta V (1983). Stability of triamcinolone acetonide solutions as determined by high-performance liquid chromatography. *J Pharm Sci* 72: 1453–1456.

Florey K (1972). Triamcinolone acetonide. In: Florey K, ed. *Analytical Profiles of Drug Substances*, Vol. 1. London: Academic Press, pp. 397–421.

Han J, Beeton A, Long PF, Wong I, Tuleu C (2006). Physical and microbiological stability of an extemporaneous tacrolimus suspension for paediatric use. *J Clin Pharm Ther* 31: 167–172.

Koch WL (1979). Erythromycin. In: Florey K, ed. *Analytical Profiles of Drug Substances*, Vol. 8. London: Academic Press, pp. 159–177.

Laine E, Kahela P, Rajala R, Haikkila T, Saarnivaara K, Piippo I (1987). Crystal forms and bioavailability of erythromycin. *Int J Pharm* 38: 33–38.

Lowey AR, Jackson MN (2008). A survey of extemporaneous dispensing activity in NHS trusts in Yorkshire, the North-East and London. *Hosp Pharmacist* 15: 217–219.

Lund W, ed. (1994). *The Pharmaceutical Codex: Principles and Practice of Pharmaceutics*, 12th edn. London: Pharmaceutical Press.

Marek CL (1999). Issues and opportunities: compounding for dentistry. *Int J Pharm Compound* 3: 4–7.

Moffat AC, Osselton MD, Widdop B (2003). *Clarke's Analysis of Drugs and Poison*, 3rd edn. London: Pharmaceutical Press. ISBN 978 0 85369 473 1.

Sieh DH (1982). Triamcinolone acetonide. In: Florey K, ed. *Analytical Profiles of Drug Substances*, Vol. 11. London: Academic Press, pp. 615–661.

Sweetman SC (2009). *Martindale: The Complete Drug Reference*, 36th edn. London: Pharmaceutical Press. IBSN 978 0 85369 840 1.

Ungphaiboon W, Nittayananta W, Vuddhakul V, Maneenuan Kietthubthew S, Wongpoowarak W, Phadoongsombat N (2005). Formulation and efficacy of triamcinolone acetonide for treating oral lichen planus. *Am J Health Syst Pharm* 62: 485–491.

United States Pharmacopeia (1999). *United States Pharmacopeia*, XXIV. Rockville, MD: The United States Pharmacopeial Convention.

Vincent SD (1991). Diagnosing and managing oral lichen planus. *Am J Dent Assoc* 122: 93–96.

Levothyroxine sodium oral liquid

Risk assessment of parent compound

A risk assessment survey completed by clinical pharmacists suggested that levothyroxine has a moderate therapeutic index. Inaccurate doses were thought to be associated with significant morbidity (Lowey and Jackson, 2008). Current formulations used at NHS trusts were of a low technical risk, based on the number and/or complexity of manipulations and calculations. Levothyroxine is therefore regarded as a medium-risk extemporaneous preparation. However, there are several specific technical issues with respect to levothyroxine. These issues require consideration on an individual patient basis (Box 14.26).

Box 14.26 *Preferred formula*

The preparation of an extemporaneous preparation is no longer recommended as a licensed preparation is now available in the UK.

Using a typical strength of 10 micrograms/mL as an example:

- levothyroxine sodium 100 micrograms tablet × 10
- preserved xanthan gum 0.4% (or similar) suspending agent to 100 mL.

Method guidance

Tablets can be ground to a fine, uniform powder in a pestle and mortar. A small amount of xanthan gum suspending agent may be added to form a paste, before adding a further portion up to 75% of the final volume. Transfer to a measuring cylinder. The remaining suspending agent can be used to wash out the pestle and mortar of any remaining drug before making the suspension up to 100% volume. Transfer to an amber medicine bottle.

Shelf-life

7 days refrigerated in amber glass. **Shake the bottle.**

Note: 'Keltrol' xanthan gum formulations vary slightly between manufacturers, and may contain chloroform and ethanol. The individual specification should be checked by the appropriate pharmacist.

Summary

The vast majority of NHS experience detected in the survey related to the use of suspensions based on xanthan gum ('Keltrol'). Despite some concerns expressed in the pharmaceutical literature, such formulations appear to have been used successfully in the NHS for several years.

Levothryoxine oral solution is now available as a licensed product in the UK. The use of tablets dispersed in water as an interim measure may be considered, depending on the dose required by the patient. However, the licensed product should be used as soon as it is available and practical to switch therapy.

Clinical pharmaceutics

Points to consider

- At typical concentrations (5–100 micrograms/mL), most or all of the drug should be in solution. Dosage uniformity is not likely to be an issue. However, given the technical issues surrounding the use of levothyroxine, it would seem logical to shake the bottle to maintain dosage uniformity, and to distribute any insoluble tablet excipients evenly.
- 'Keltrol' is made to slightly different specifications by several manufacturers, but typically contains ethanol and chloroform. The exact specification should be assessed before use by a suitably competent pharmacist. Ethanol is a CNS depressant. Chloroform has been classified as a class 3 carcinogen (i.e. substances that cause concern owing to possible carcinogenic effects but for which available information is not adequate to make satisfactory assessments) (CHIP3 Regulations, 2002).
- Levothyroxine sodium is the sodium salt of the levo isomer of the thyroid hormone thryoxine. While problems with racemisation are possible, there is no known evidence of this process occurring. Studies by Patel *et al.* (2003) found no problems with racemisation.
- Levothyroxine is a tasteless powder (Trissel, 2000).
- There have been doubts raised over the bioequivalence and efficacy of some unlicensed levothyroxine oral liquids.

Other notes

- Levothyroxine sodium is commercially available in some countries as a lyophilised powder for injection that is reconstituted immediately before use. The injection is reported to have been administered orally, but this method may be expensive and wasteful (Boulton *et al.*, 1996).
- Tablets may be crushed and dispersed in water immediately before administration (Boulton *et al.*, 1996). The usefulness of this strategy

depends on the individual patient circumstances, particularly the dose required.

- At least one oral drop (5 micrograms per drop) formulation is reported to be marketed and licensed in Europe. This formulation should be assessed before purchase by a suitably competent pharmacist.
- Levothyroxine is acknowledged as a typical candidate for tenfold prescribing errors, particularly given the nature of the dosage units (as micrograms). Levothyroxine was found to account for 19% of tenfold prescribing errors ($n = 200$) in a tertiary care teaching hospital (Lesar, 2003).

Technical information

Structure

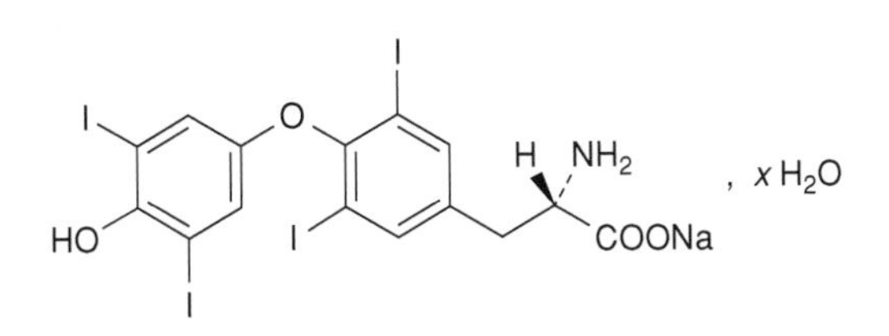

Empirical formula

$C_{15}H_{10}I_4NaO_4.xH_2O$ (British Pharmacopoeia, 2007).

Molecular weight

799 (British Pharmacopoeia, 2007).

Solubility

Approximately 150 micrograms/mL at 25°C (Budavari, 1996).

pK_a

2.40 (carboxyl), 6.87 (phenolic) and 9.35 (amino) (Won, 1992; Patel *et al.*, 2003).

Optimum pH stability

Basic pH (Won, 1992; Boulton *et al.*, 1996).

Stability profile

Physical and chemical stability

Levothyroxine sodium is an almost white or slightly brownish-yellow powder, or a fine crystalline powder, very slightly soluble in water and slightly soluble in alcohol (British Pharmacopoeia, 2007). It dissolves in dilute solutions of alkali hydroxides (British Pharmacopoeia, 2007). It is also reported to be a tasteless, hygroscopic powder (Budavari, 1996; McEvoy, 1999; Trissel, 2000; Martindale, 2005). However, Patel *et al.* (2003) found the powder to be non-hygroscopic under normal processing conditions (less than 30% relative humidity). The powder was stable in the solid state for six months at 40°C and 75% relative humidity whether stored in open or closed containers.

Levothyroxine has three ionisable moieties and can exist as the cation, zwitterion, anion and dianion, depending on the pH of the solution (Won, 1992). At typical strengths of oral liquid (10–100 micrograms/mL), most or all of the drug should be in solution, and therefore susceptible to aqueous degradation mechanisms.

Won (1992) studied the kinetics of the deiodination of levothyroxine in aqueous solution over the pH range 1–12, and found that the degradation followed first-order kinetics. As the pH of the solution increased, the degradation reduced. The log k–pH profile indicated that the kinetics of the deiodination included proton attack on the anion and dianion in alkaline solution.

The stability of the solid state was investigated at elevated temperatures. Levothyroxine sodium underwent deamination on heating, following biphasic first-order kinetics. The greater part of the degradation occurred at the beginning of the heating. Won (1992) suggests that a threshold temperature may exist somewhere between 50 and 60°C, above which levothyroxine sodium degrades rapidly.

In aqueous media, UV radiation may result in deiodination of levothyroxine sodium (Won, 1992). Such deiodination is proportional to the decrease in the pH of the aqueous solution (Won, 1992).

Saturated aqueous solutions have a pH of 8.35–9.35 (Budavari, 1996).

Levothyroxine sodium powder should be stored in an airtight container, protected from light, at 2–8°C (British Pharmacopoeia, 2007). It is sensitive to light and pH (Boulton *et al.*, 1996). American reference texts suggest that while storage is possible at a 'controlled room temperature' (McEvoy, 1999; United States Pharmacopeia, 2003), storage between 8 and 15°C has been required to maintain the potency for some commercial products (McEvoy, 1999; Trissel, 2000). Refrigerated storage has also been suggested, as well as protection from air and moisture (McEvoy, 1999; United States Pharmacopeia, 2003).

Degradation products

The deiodination of levothyroxine yields free iodine and liothyronine (T3). Liothyronine is known to exhibit pharmacological activity on the thyroid gland and may degrade further. When subjected to heat, its principal degradation products are triiodothyroacetic acid, triiodothyroacetic acid amide and triiodothyroethane (Andre *et al.*, 1996).

Stability in practice

There is a fair body of evidence to support the stability of levothyroxine in extemporaneously prepared oral liquids. The bioavailability of such liquids is less certain. Therefore, the licensed formulations should be used wherever possible.

Boulton *et al.* (1996) have investigated the stability of a 25 micrograms/mL extemporaneously compounded levothyroxine sodium liquid. A comparison of liquids made from pure powder and commercially available tablets was included in the study (with and without 0.1% methyl hydroxybenzoate 0.1%).

For the tablet-based formulations, the appropriate number of tablets were ground into fine powder in a pestle and mortar, before trituration with glycerol. Likewise, the powder-based formulations were triturated with glycerol. The preservative was added at this stage (if appropriate), before the liquid was brought to volume with distilled water.

The authors note that the powder-based preparations produced clear, colourless solutions, whereas the tablet-based preparations led to a suspension of excipients in a yellow liquid. One litre of each tablet-based formula was prepared and 300 mL of each powder-based preparation. Samples were transferred to 50 mL amber high-density polyethylene bottles with polypropylene screw-top lids.

Five bottles of each tablet-based preparations were stored in the dark at each of three temperatures: 2–8, 23–27 and 38–42°C. The powder-based samples were stored only at the elevated temperature range. A 1 mL sample was removed on days 0, 3, 8, 14, 22, 31 and 61 after inversion of the samples by hand. Analysis was via a stability-indicating HPLC method. Forcing the degradation of levothyroxine led to the levothyroxine peak decaying to baseline, while the liothyronine peak became larger.

Samples were also taken on days 0, 14 and 31 for microbiological testing (total count of aerobic organisms) of tablet-based formulations.

The authors found significant levels of degradation in all of the formulations studied. The tablet-based formulations showed poorer stability in the presence of preservative. The powder-based formulations showed better stability in the presence of preservative (at 40°C).

The tablet-based formulations without preservative showed around 5–6% loss in 8 days and 10–12% loss in 14 days when stored at either 2–8 or 23–27°C (Boulton *et al.*, 1996; Trissel, 2000).

The authors suggest t_{90} (shelf-life) values of 15.7 days, 11.2 days and 8.1 days at 4°C, 25°C and 40°C, respectively. Storage at 40°C led to 11% drug loss in 8 days, and the addition of methyl hydroxybenzoate lowered pH and stability, but did not reduce the number of viable organisms when compared with the unpreserved suspension.

In conclusion, Boulton *et al.* (1996) suggest that the following formulation is stable for 8 days when stored in amber high-density polyethylene bottles at 4°C:

- levothyroxine sodium tablets 100 micrograms × 25
- glycerol 40 mL
- water to 100 mL.

Woods (2001), one of the original authors, states that the stability is maintained at room temperature, but that temperatures above 27°C should be avoided.

Alexander *et al.* (1997) have also studied the stability of an extemporaneously formulated levothyroxine sodium syrup (40 micrograms/mL) compounded from commercial tablets. The authors used sorbitol rather than sucrose, as sorbitol has been reported to increase the stability of drugs sensitive to the presence of sucrose (Alexander *et al.*, 1997). The formulation was buffered to pH 7.5 with a phosphate buffer, as levothyroxine is reported to be more stable and more soluble at a basic pH. Glycerin was added to alleviate any undesirable tastes, and ethanol was used to improve drug solubility and prevent precipitation. The final formula was as follows:

- levothyroxine tablets 200 mg × 200
- sorbitol USP 70% solution 500 mL
- ethanol 50 mL
- glycerin 99%, USP 250 mL
- parabens (hydroxybenzoates) 10 mL
- sodium bisulfite 1 g
- sodium EDTA 1 g
- sodium dihydrogen phosphate 4.768 g
- disodium phosphate 17.719 g
- banana flavour (details not stated) 10 mL
- water to 1,000 mL.

The suspension was analysed using a stability-indicating HPLC method. Prior to the withdrawal of samples, an inspection was made for signs of turbidity and ease of pouring. The inter- and intra-day precision of the assay procedure showed a relative standard deviation of less than 2% percent based on four injections of identical concentration over 4 days.

Samples of the liquid were stored at 40, 50, 60 and 70°C for up to 124 hours (in excess of 5 days). Due to the viscosity of the preparation, the authors state that the volumetric pipettes had to be completely drained with the diluent, occasionally leading to drug recovery rates of over 100%. However, this error should remain consistent throughout the results.

There were no colour changes at any temperature and all samples remained homogeneous with no apparent changes in viscosity. However, Alexander *et al.* (1997) suggest that the samples stored at 60 and 70°C may have maintained a lower viscosity than the samples stored at 30°C.

The first-order rate constant was calculated for each temperature. The apparent activation energy was calculated to be 39.5027 kJ/mol. The authors suggest a shelf-life (t_{90}) of approximately 15 days at 25°C and approximately 47 days at 5°C (Alexander *et al.*, 1997).

While this work provides useful data, particularly for 'Specials' manufacturers who are considering making a levothyroxine sodium preparation, the complexity of the formula limits its use in a dispensary situation. The ethanol content of 5% may also be undesirable, particularly in younger patients.

C. Hiller (personal communication, 2006) describes the stability of levothyroxine sodium powder in xanthan gum 0.4% ('Keltrol') suspending agent. Several 'Specials' manufacturers produce similar preparations. Hiller describes several years' experience with such a preparation, using a shelf-life of three months. The preparation has been tested in-house over much longer periods than this, however, using a stability-indicating assay, with no problems reported. The product has been used in the NHS for several years without any major problems reported with regard to bioavailability.

Related preparations – tablets

Das Gupta *et al.* (1990) have investigated the effect of excipients on the stability of levothyroxine sodium tablets, using a stability-indicating HPLC procedure. The researchers found that one of the brands of tablets contained excipients which appeared to act as a catalyst to hasten the decomposition of the drug after extraction. This has implications with regard to which brand of tablets is chosen as a starting material for an oral liquid. Furthermore, the manufacturer of this brand had changed the types of lids/bottles used for the drug three times within one year.

The degradation due to exposure to light decreased significantly when the samples were filtered in amber-coloured vials, suggesting that light may be necessary for the excipient(s) to act as a catalyst (Das Gupta *et al.*, 1990).

Patel *et al.* (2003) devised a series of experiments to test the effect of excipients on the stability of levothyroxine sodium pentahydrate tablets. The stability of the drug in slurries was testing the presence of HPMC, povidone, croscarmellose sodium, sodium starch glycolate, crospovidone, stearic acid, magnesium stearate and fumed silica. Various pHs were used for testing.

In all the excipient slurries, the drug degraded more at pH 3 than at pH 11. For each excipient, stability improved as the pH increased from acidic to basic, inferring that the addition of a basic pH modifier improves the stability of levothyroxine sodium pentahydrate in the presence of various excipients. The primary degradation pathways observed with mass spectroscopy were deiodination and deamination. No racemisation was observed for any of the tablet batches.

Patel *et al.* (2003) conclude by suggesting that the use of basic pH modifiers may be a potential technique for improving the stability of levothyroxine sodium pentahydrate tablets.

Related preparations – intravenous

Kato *et al.* (1986) found that protection from light and storage under refrigeration improved the shelf-life of a levothyroxine sodium injection. Cano *et al.* (1981) found two IV formulations to be stable for 30 days, and samples of sodium levothyroxine in 0.01 N methanolic sodium hydroxide remained stable at 5°C for six months (Brower *et al.*, 1984).

Bioavailability data

Levothyroxine products have had a history of bioequivalence and bioavailability problems (Wertheimer and Santella, 2005). Given that there have been known concerns with levothyroxine sodium tablets, it is logical that any liquid preparations derived from crushed tablets may also be open to controversy.

For example, in the US market between 1990 and 1997, there were 10 recalls of levothyroxine tablets, equating to 100 million tablets (Johnson, 2003). Published and unpublished research has often shown conflicting data, and researchers have criticised the methodologies used to assess bioavailability and bioequivalence (Garnick *et al.*, 1984; Blakesley *et al.*, 2004; Blakesley, 2005).

The original extracts of animal thyroid hormone were replaced with L-thyroxine (T4) in the late 1950s and 1960s, after these extracts were found to lose their potency with time, particularly if not stored correctly (Koutras, 2003). The fact that L-thyroxine (T4) degrades to a compound that retains pharmacological activity (liothyronine, T3) complicates the situation. Moreover, as exogenous levothyroxine is biochemically and physiologically indistinguishable from endogenously produced thyroxine, this precludes an easy method of distinguishing the exogenous and endogenous sources of measured T4 in the blood (Blakesley, 2005).

Various authors have raised doubts about the clinical effectiveness of a range of levothyroxine preparations. Oliveira *et al.* (1997) and Peran *et al.* (1997) noticed in Spain that some batches of levothyroxine appeared to have a reduced effect, and Koutras (2003) reports similar problems in Greece.

Hennessey (2003) found considerable differences in potency in American batches of levothyroxine tablets. He concluded that levothyroxine products were 'unstable, influenced by factors such as light, temperature, air exposure, humidity and the use of some excipients which may actually accelerate the degradation of active ingredients'. The Food and Drug Administration (FDA) has acknowledged the existence of such batch-to-batch variation, noting that the 'active ingredient degrades quickly with exposure to light, moisture, oxygen and carbohydrate excipients' (Nightingale, 1998). As a result, many products are manufactured using an overage to allow for such degradation, resulting in inconsistent tablet-to-tablet equivalence (Wertheimer and Santella, 2005).

As the drug itself is known to have such issues with regard to potency, it follows that problems may be evident with oral liquid formulations. In 2004, a series of letters were exchanged in the UK pharmaceutical literature after Bird (2004) noted the therapeutic failure of a xanthan gum ('Keltrol 0.4%') based levothyroxine suspension, used in a hypothyroid infant (Bird, 2004; Perrin, 2004; Morgan and Huma, 2004; Wells, 2004; Nunn, 2004). The suspension was remade freshly but the patient still did not respond, until the suspension was replaced with tablets dispersed in water.

Perrin (2004) suggests that levothyroxine should not be suspended in xanthan gum in this way, due to the low dose and the known susceptibility of the drug to oxidation. Perrin (2004) notes the potential effects of shaking the suspension and its storage conditions, in addition to the possibility of the drug becoming bound to the gum.

However, levothyroxine suspensions in xanthan gum vehicles have been used with anecdotal supporting evidence for many years in the NHS. Morgan and Huma (2004) present a case study of the successful use of levothyroxine (10 micrograms/mL) in a 0.5% xanthan gum base. An 800 g premature infant responded to treatment and thyroid function tests normalised. They argue against the use of approximate doses derived from tablets dispersed or dissolved in water, stating that a rapid correction of thyroid function test or inexact doing would be clinically unacceptable in such a scenario. Whether the authors used pure drug powder or crushed tablets is not stated.

Von Heppe *et al.* (2004) investigated the claim that inadequate control of thyroid function tests in some infants with congenital hypothyroidism may be due to the lack of absorption of L-thyroxine in a tablet form. The authors researched the use of an L-thyroxine solution (licensed in Germany; 1 drop = 50 micrograms) in 28 newborn patients, and went on to conclude that the use of an L-thyroxine solution improves the practicability and individualised dosage in newborns and infants. Data from the study suggested that the high L-thyroxine dosage which is required in newborns and young infants with congenital hypothyroidism reflects an increased requirement for thyroid hormones and is not due to reduced absorption from tablets.

Wells (2004) advocates the use of pure drug powder rather than tablets and states that xanthan gum is not necessary as the resulting liquid will be a solution rather than a suspension. However, Wells suggests the use of filtering systems, and this approach has been criticised for its lack of practicality in standard dispensaries (Nunn, 2004). Although the filtration may prevent the use of some unnecessary excipients in neonates, any drug bound to these excipients will also be removed by filtration. Nunn (2004) also warns of the dangers of weighing out such small amounts of pure drug.

Koytchev and Lauschner (2004) studied the bioequivalence of levothyroxine tablets in healthy volunteers, compared to reference tablets and an oral solution. Although the three formulations appeared to be approximately bioequivalent, details of the oral solution were not supplied.

Yamamoto (2003) studied the use of crushed or chewed levothryoxine tablets in three patients with persistently elevated levels of TSH despite taking 200, 150 and 125 micrograms of levothyroxine sodium tablets per day. The patients were not known to have any gastrointestinal conditions that may have interfered with the absorption of the drug. Serum TSH levels were documented to normalise after the patients took the tablets after they had been pulverised. The author suggests that the results may indicate that the drug is absorbed less well from tablets than it is from powder, due to the slow dissolution of the tablets. The study is obviously limited by the small number of patients involved.

Ogawa *et al.* (2000) also report a possible case report describing malabsorption of levothyroxine from tablets. Initial testing on a 51-year-old female patient revealed no problems with absorption or metabolism of levothyroxine following administration of oral liquid, pulverised tablets via nasogastric tube, or intravenous administration.

Bioavailability studies have also been published for the licensed levothyroxine drops in France and Germany (Chassard *et al.*, 1991; Grussendorf *et al.*, 2004).

References

Alexander KS, Kothapalli MR, Dollimor D (1997). Stability of an extemporaneously formulation levothryoxine sodium syrup compounded from commercial tablets. *Int J Pharm Compound* 1: 60–64.

Andre M, Domanig R, Riemer E, Moser H, Groeppelin A (1996). Identification of the thermal degradation products of G-triiodothyronine sodium (liothyronine sodium) by reversed-phase high-performance liquid chromatography with photodiode-array UV and mass spectrometric detection. *J Chromatrogr A* 752: 287–294.

Bird K (2004). Is levothyroxine suspension effective? [letter] *Pharm J* 273: 680.

Blakesley VA (2005). Current methodology to assess bioequivalence of levothyroxine sodium products is inadequate. *AAPS J* 7: E42–E46.

Blakesley V, Awni W, Locke C, Ludden T, Granneman GR (2004). Are bioequivalence studies of levothyroxine sodium formulations in euthyroid volunteers reliable? *Thyroid* 14: 191–200.

Boulton DW, Fawcett JP, Woods DJ (1996). Stability of an extemporaneously compounded levothyroxine sodium oral liquid. *Am J Health Syst Pharm* 53: 1157–1161.

British Pharmacopoeia (2007). *British Pharmacopoeia*, Vol. 2. London: The Stationery Office, pp. 1242–1243.

Brower JF, Toler DY, Reepmeyer JC (1984). Determination of sodium levothyroxine in bulk, tablet and injection formulations by high-performance liquid chromatography. *J Pharm Sci* 73: 1315–1317.

Budavari S, ed. (1996). *The Merck Index*, 12th edn. Rahway, NJ: Merck and Co.

Cano SM, Mallart L, Martinez J *et al.* (1981). *Farm Clin* 4: 304.

Chassard D, Kerihuel JC, Caplain H, Tran Quang N, Thebault JJ (1991). Comparative bioequivalence study of a new reference l-thyroxine solution in normal healthy volunteers. *Eur J Drug Metab Pharmacokinet* Spec No. 3: 324–326.

CHIP3 Regulations (2002). Chemicals (Hazard Information and Packaging for Supply) Regulations. SI 2002/1689.

Das Gupta V, Odom C, Bethea C, Plattenburg J (1990). Effect of excipients on the stability of levothyroxine sodium tablets. *J Clin Pharm Ther* 15: 331.

Garnick RL, Burt GF, Long DA, Bastian JW, Aldred JP (1984). High-performance liquid chromatographic assay for sodium levothyroxine in tablet formulations: content uniformity applications. *J Pharm Sci* 73: 75–77.

Grussendorf M, Vaupel R, Wegsschelder K (2004). Bioequivalence of l-thyroxine tablets and a liquid l-thyroxine solution in the treatment of hypothyroid patients. *Med Klin* 99: 639–644.

Hennessey JV (2003). Levothyroxine a new drug? Since when? How could that be? *Thyroid* 13: 279–282.

Johnson SB (2003). Endogenous substance bioavailability and bioequivalence: levothyroxine sodium tablets. US Food and Drug Administration Pharmaceutical Science Advisory Committee; March 13, 2003. http://www.fda.gov/ohrms/dockets/ac/03/slides/3926s2.htm.

Kato Y, Saito M, Kolzumi H *et al.* (1986). *Jap J Hosp Pharm* 12: 253.

Koutras DA (2003). The treacherous use of thyroxine preparations. Stability of thyroxine preparations. *Hormones* 2: 159–160.

Koytchev R, Lauschner R (2004). Bioequivalence study of levothyroxine tablets compared to reference tablets and an oral solution. *Arzneimittelforschung* 54: 680–684.

Lesar TS (2003). Tenfold medication dose prescribing errors. *Am J Nurse Pract* 7: 31–32, 34–38, 43.

Lowey AR, Jackson MN (2008). A survey of extemporaneous dispensing activity in NHS trusts in Yorkshire, the North-East and London. *Hosp Pharmacist* 15: 217–219.

McEvoy J, ed. (1999). *AHFS Drug Information.* Bethesda, MD: American Society of Health System Pharmacists.

Morgan R, Huma Z (2004). Levothyroxine suspension effective [letter]. *Pharm J* 273: 785.

Nightingale SM (1998). Therapeutic equivalence of generic drugs. Letter to health practitioners. US Food and Drug Administration, January 28 1998. http://fda.gov/cder/news/nightgenlett.htm.

Nunn AJ (2004). Levothyroxine: liquid preparations should be a high priority [letter]. *Pharm J* 273: 880.

Ogawa D, Otsuka F, Mimura U, Ueno A, Hashimoto H, Kishida M, Ogura T, Makino H (2000). Pseudomalabsorption of levothyroxine: a case report. *Endocr J* 47: 45–50.

Oliveira G, Almaraz MC, Soriguer F, Garriga MJ, Gonzalez-Romero Tinahones F, Ruis de Adama MS (1997). Altered bioavailability due to changes in the formulation of a commercial preparation of levothyroxine in patients with differentiated thyroid carcinoma. *Clin Endocrinol* 46: 707–711.

Patel H, Stalcup A, Dansereau R, Sakr A (2003). The effect of excipients on the stability of levothyroxine sodium pentahydrate tablets. *Int J Pharm* 264: 35–43.

Peran S, Garriga MJ, Morreale de Escobar G, Asuncion M, Peran M (1997). Increase in plasma thyrotropin levels in hypothyroid patients during treatment due to a defect in the commercial preparation. *J Clin Endocrinol Metab* 82: 3192–3195.

Perrin JH (2004). Do not suspend levothyroxine [letter]. *Pharm J* 273: 748.

Sweetman SC (2009). *Martindale: The Complete Drug Reference*, 36th edn. London: Pharmaceutical Press.

Trissel LA, ed. (2000). *Stability of Compounded Formulations*, 2nd edn. Washington DC: American Pharmaceutical Association.

United States Pharmacopeia (2003). *United States Pharmacopeia*, XXVI. Rockville, MD: The United States Pharmacopeial Convention.

Von Heppe JH, Krude H, L'Allemand D, Schabel D, Gruters A (2004). The use of L-T4 as liquid solution improves the practicability and individualised dosage in newborns and infants with congenital hypothyroidism. *J Pediatr Endocrinol Metab* 17: 967–974.

Wells IJ (2004). Suspensions are not the solution [letter]. *Pharm J* 273: 852.

Wertheimer AI, Santella BS (2005). The levothyroxine spectrum: bioequivalence and cost considerations. *Formulary* 40: 258–271.

Won CM (1992). Kinetics of degradation of levothyroxine in aqueous solution and in solid state. *Pharm Res* 9: 131–137.

Yamamoto T (2003). Tablet formulation of levothyroxine is absorbed less well than powdered levothyroxine. *Thyroid* 13: 1177–1181.

Lorazepam oral liquid

Risk assessment of parent compound

A survey of clinical pharmacists suggested that lorazepam has a moderate therapeutic index, with the potential to cause serious morbidity if inaccurate doses are administered (Lowey and Jackson, 2008). Current formulations used at NHS trusts were generally of a low to moderate technical risk, based on the number and/or complexity of manipulations and calculations. Lorazepam is therefore regarded as a medium- to high-risk extemporaneous preparation (Box 14.27).

Summary

There would appear to be sufficient evidence to justify the preparation of extemporaneous suspensions of lorazepam. Given the relative stability of the drug, lorazepam may be a candidate for purchase from a 'Specials' manufacturer, providing an adequate specification can be agreed. However, the data from the 2 mg/mL and 1 mg/mL preparations may not be directly transferable to the lower strength preparations (e.g. 100 or 200 micrograms/mL) that are in common use in neonatal and paediatric wards in the UK (Lowey and Jackson, 2008). This is because a significant percentage of the drug will be in solution in the lower strength preparations, and thus susceptible to degradation. A more conservative shelf-life of 7 days is therefore recommended for these lower concentrations in the absence of any supporting stability data.

Preparations containing propylene glycol and polyethylene glycol, including imports and solutions prepared from the injection, may not be suitable for neonates and young infants due to toxicity concerns.

Clinical pharmaceutics

Points to consider

- The proportion of the drug in solution or suspension is heavily dependent on the strength of the oral liquid. As a precaution, **shake the bottle before use to improve dosage uniformity.**
- It has been suggested that the injection should not be used to make oral preparations due to propylene glycol content.
- Ora-Plus and Ora-Sweet SF are free from chloroform and ethanol.

Box 14.27 *Preferred formula*

Using a typical strength of 200 micrograms/mL as an example:

- lorazepam tablets 1 mg × 10
- Ora-Plus 25 mL
- Ora-Sweet to 50 mL.

Method guidance

Tablets can be ground to a fine, uniform powder in a pestle and mortar. A small amount of Ora-Plus may be added to form a paste, before adding further portions of Ora-Plus up to 50% of the final volume. Transfer to a measuring cylinder. The Ora-Sweet can be used to wash out the pestle and mortar before making the suspension up to 100% volume. Transfer to an amber medicine bottle.

Shelf-life

7 days refrigerated in amber glass.

Alternative

An alternative formula would be:

- lorazepam tablets 1 mg × 10
- xanthan gum 0.5% ('Keltrol' or similar) to 50 mL.

Shelf-life

7 days refrigerated in amber glass. **Shake the bottle.**

Note: 'Keltrol' xanthan gum formulations vary slightly between manufacturers, and may contain chloroform and ethanol. The individual specification should be checked by the appropriate pharmacist.

- Ora-Plus and Ora-Sweet SF contain hydroxybenzoate preservative systems.
- 'Keltrol' is made to slightly different specifications by several manufacturers, but typically contains ethanol and chloroform. The exact specification should be assessed before use by a suitably competent pharmacist. Ethanol is a CNS depressant. Chloroform has been classified as a class 3 carcinogen (i.e. substances that cause concern owing to possible carcinogenic effects but for which available information is not adequate to make satisfactory assessments) (CHIP3 Regulations, 2002).
- There are no bioavailability data available.

- It has been suggested that the injection can be used in an unlicensed manner via the sublingual and rectal routes (Guy's and St Thomas' and Lewisham Hospitals, 2005).
- The possible clinical alternatives to the use of lorazepam oral liquid will depend on the individual indication. They may include the use of other benzodiazepine preparations.

Other notes

- A 2 mg/mL lorazepam concentrated suspension is available in the USA (Intensol, Roxane Laboratories, Licence No. NDC 0054-3532-44). The strength may be too high for neonates, and the safety and effectiveness has not been demonstrated in children under 12 years. The solution (30 mL) has a graduated dropper capable of giving doses down to 0.25 mL (500 micrograms). It carries a shelf-life of 90 days when stored in the fridge (2–8°C) and protected from light. This concentrate should be assessed for its fitness-for-purpose by the appropriate pharmacist(s) before use in an individual patient. It is known to be formulated in a mixture of propylene glycol and polyethylene glycol (acting as solubilizing agents) (Lee *et al.*, 2004). Although propylene glycol is generally acknowledged to be safe in small quantities (Ruddick, 1972), larger amounts may cause side-effects such as hyperosmolarity, seizures, lactic acidosis and cardiac toxicity (in both children and adults) (Martin and Finberg, 1970; Arulananatham and Genel, 1978; Cate and Hedrick, 1980; Glasgow *et al.*, 1983; Lee *et al.*, 2004). Propylene glycol and polyethylene glycol may also act as osmotic laxatives. Marshall (1995) has documented diarrhoea as a result of enteral administration of the liquid formulations of benzodiazepines.
- During periods of shortage of this preparation, the American Society of Health System Pharmacists (ASHP) website suggested the use of extemporaneously prepared alternatives, using Ora-Plus and Ora-Sweet as the suspending agent (ASHP, 2005).

Technical information

Structure

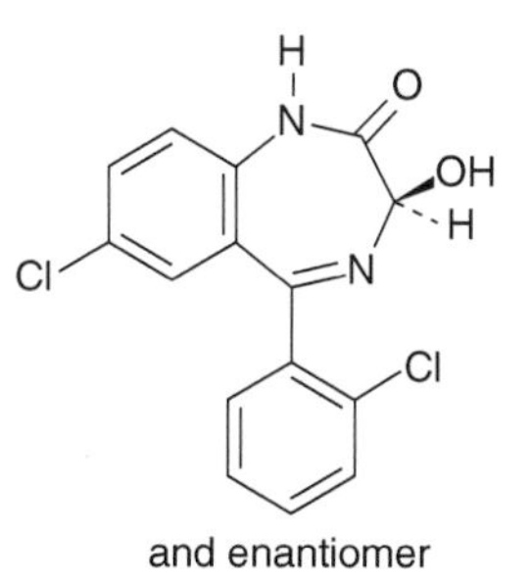

Empirical formula

$C_{15}H_{10}Cl_2N_2O_2$ (British Pharmacopoeia, 2007).

Molecular weight

321.2 (British Pharmacopoeia, 2007).

Solubility

80 micrograms/mL in water (Lund, 1994).

pK_a

1.3 and 11.5 (Lund, 1994).

Optimum pH stability

Not known.

Stability profile

Physical and chemical stability

Lorazepam is a white, or nearly white, practically odourless, crystalline powder, with a solubility of 80 micrograms/mL in water (Rutgers and Shearer, 1980; Lund, 1994). Therefore, the amount of drug in solution will depend on the choice of the strength of the formulation. A 100 micrograms/mL strength will have most of the drug in solution and be open to aqueous degradation kinetics, while a 500 micrograms/mL strength will have the majority of drug in suspension.

Lorazepam can degrade by dehydration, losing a molecule of water and rearranging to form 6-chloro-4-(*o*-chlorophenyl)-2-quinazolinecarboxaldehyde. This can further disproportionate and either oxidise to the corresponding quinazolinecarboxylic acid or reduce to the quinoazoline alcohol (Rutgers and Shearer, 1980).

Extreme pHs should avoided; acid hydrolysis ultimately yields 2-amino-2′,5-dichlorobenzophenone. In base, lorazepam is reported to rearrange to 7-chloro-5-(2-chloro-phenyl)-4,5-dihydro-2*H*-1,4–1,4-benzodiazepin-2,3 (1*H*)-dione (Rutgers and Shearer, 1980).

In common with most other benzodiazepines, lorazepam is hydrolysed in acidic solutions to a series of degradation products (Panderi *et al.*,

1998). Panderi *et al.* (1998) carried out a kinetic investigation of the degradation of lorazepam in acidic aqueous solutions using HPLC. The authors confirmed the formation of 6-chloro-4-(*o*-chlorophenyl)-2-quinazolinecarboxaldehyde as the main degradant in hydrochloric acid solutions of 0.01, 0.1 and 1.0 M. During accelerated stability studies, this product was found to break down further to an unknown compound. Under more acidic conditions (hydrochloric acid 4.0 M), lorazepam degraded to (2-amino-5-chlorophenyl)(2-chlorophenyl)methanone.

Enough data for lorazepam in 0.1 M HCl were produced by Panderi *et al.* (1998) to draw Arrhenius plots. The activation energy was calculated to be 23.78 kcal/mol, with a half-life of 6011 minutes (~100 hours) at 25°C.

Degradation products

May include 2-amino-2′,5-dichlorobenzene. Other products may include 6-chloro-4-(*o*-chlorophenyl)-2-quinazolinecarboxaldehyde, and the corresponding quinazolinecarboxylic acid or quinoazoline alcohol. Rearrangement in base may yield 7-chloro-5-(2-chlorophenyl)-4,5-dihydro-2*H*-1,4–1,4-benzodiazepin-2,3(1*H*)-dione (Rutgers and Shearer, 1980).

Stability in practice

A body of evidence from published and unpublished sources appears to suggest that lorazepam can be formulated as an oral liquid with a reasonable shelf-life.

During a shortage of the licensed 2 mg/mL preparation mentioned above, the ASHP recommended that 0.5 mg/mL suspensions could be prepared by crushing lorazepam tablets and mixing firstly with a little water, then with either cherry syrup or a 1:1 mixture of Ora-Sweet and Ora-Plus. The ASHP website recommended an expiry of 14 days when stored in the refrigerator (ASHP, 2005).

If a 1 mg/mL suspension is required, the ASHP have referenced the work carried out by Lee *et al.* (2004), who also used Ora-Plus and Ora-Sweet as their suspending agent of choice. Two different brands of lorazepam tablet were used to produce 1 mg/mL suspensions. A 2 mg/mL strength was reported to be inappropriate due to its viscosity. A pestle and mortar method was not used as the authors suggested that such a method resulted in significant amounts of drug loss during transfer to the final bottle. Instead, each suspension was prepared by dispersing 180 2 mg tablets in a 'minimum' volume of sterile water in an amber glass bottle. A slurry was formed by shaking the bottle, before sequential additions of Ora-Plus. Ora-Sweet was then added to make the volume up to 360 mL (1 mg/mL). The actual formulae used were as follows:

Formula 1 – lorazepam tablets from Mylan Pharmaceuticals

- sterile water 144 mL
- Ora-Plus 108 mL
- Ora-Sweet 83 mL.

Formula 2 – lorazepam tablets from Watson Laboratories

- sterile water 48 mL
- Ora-Plus 156 mL
- Ora-Sweet 83 mL.

Each suspension was then divided into 10 smaller amber bottles, split evenly between storage at 4 and 22°C. A stability-indicating HPLC assay was used after modification to allow analysis of a suspension (Gunawan and Treiman, 1988). Samples were removed from each bottle after manual shaking for 1 minute on days 2, 3, 7, 14, 21, 28, 42 63 and 91. A visual inspection was also carried out at this time.

Each 1 mL aliquot was vortexed for 30 seconds before two 100 microlitre samples were placed in separate microcentrifuge tubes before the addition of diazepam as in internal standard. A 25 microlitre injection volume was finally used after further preparation. Six calibration standards from 0.05 mg/mL to 1.5 mg/mL were freshly prepared on each day of the assay, and further quality assurance samples were prepared using the commercially available Intensol solution.

The initial actual concentrations ranged from 0.98 to 1.04 mg/mL. For the suspensions prepared with Mylan tablets, the percentage concentration remaining was greater than 90% regardless of the storage temperature, even when taking into account standard deviation. For the suspensions prepared with Watson tablets, the suspensions stored at 4°C showed little or no degradation over the 91-day period, with 99.4% of the initial concentration remaining on day 91 (standard deviation (SD) = 2.7). However, the suspensions based on Watson tablets that were stored at 22°C were found to drop slightly below the 90% barrier by day 91 (88.9%; SD = 1.4). Taking into account the standard deviation, the result on day 63 was also borderline (90.9%; SD = 1.1).

The authors conclude that the suspensions are stable for two months when stored at either 22°C or 4°C, and for three months when stored exclusively in the fridge (Lee *et al.*, 2004). Whether the differences seen are attributable to the different brands or the different formulae is not clear; the Mylan product required significantly more water for dispersal, resulting in further differences in the composition of the other agents in the vehicle.

The authors also acknowledge that this method of preparation dilutes the preservative systems in the Ora-Plus and Ora-Sweet components, and therefore suggest that refrigeration of the final suspension may be more appropriate to help limit microbial contamination.

An in-house report from a regional quality control laboratory in 1987 describes the formulation and stability of a lorazepam oral solution (T.J. Munton, Regional QA Pharmacist, North Bristol NHS Trust, personal communication, 2006). The report was commissioned after problems with precipitation were encountered with a formulation reportedly recommended by Wyeth. The formula is reported to have contained compound tragacanth powder, alcohol and chloroform water. A new, clear solution was therefore developed, according to the following formula:

- lorazepam powder 20 mg
- ethanol (absolute) 5 mL
- propylene glycol 10 mL
- glycerol 40 mL
- double strength chloroform water to 100 mL.

The product was packed into amber glass containers at 4 and 25°C, and into a clear glass container for storage on a south-facing window ledge. The results of HPLC analysis suggested that this formula is stable, when protected from light, for up to one month at 4°C and for up to two weeks at room temperature. The product is described as 'fairly bland and sweet but pleasant to take'. The product on the window ledge showed significantly more degradation, and developed a slight yellow colour after 27 days.

However, the usefulness of this formula is limited by the presence of ethanol, propylene glycol and chloroform. The latter has been classified as a class 3 carcinogen (substances that cause concern owing to possible carcinogenic effects but for which available information is not adequate to make satisfactory assessments) (CHIP3 Regulations, 2002).

Other related preparations

Nahata *et al.* (1993) investigated the use of an oral preparation based on the dilution of lorazepam injection with bacteriostatic water from 4 mg/mL to 1 mg/mL. After 7 days storage, the mean lorazepam concentration was 88% of the original concentration at 22°C and 90% of the original concentration at 4°C.

Other studies have considered the stability of injectable solutions (Stiles *et al.*, 1996; Share *et al.*, 1998; Norenberg *et al.*, 2004). The data are not directly applicable to tablet-based extemporaneous preparations, particularly as the stability of the injectable preparation can be affected by the solubilizing agents present in the admixture (Boullata *et al.*, 1996; Boullata and Gelone, 1996; Volles, 1996). As the injection is diluted, the concentration of polyethylene glycol decreases, leading to the potential for precipitation (Levanda, 1998). This may have implications if the injection is used to prepare an oral liquid. Some injectable preparations also contain benzyl alcohol.

Solutions of lorazepam injection that are diluted for use should be discarded if they are discoloured or contain a precipitate (McEvoy, 1999).

Bioavailability data

No data available.

References

Arulananatham K, Genel M (1978). Central nervous system toxicity associated with the ingestion of propylene glycol. *J Pediatr* 93: 515–516.

ASHP (American Society of Health System Pharmacists) (2005). Drug shortage resource centre. www.ashp.org (accessed 27 September 2005).

Boullata JI, Gelone SP (1996). More on usability or lorazepam admixtures for continuous infusion [letter]. *Am J Health Syst Pharm* 53: 2754.

Boullata JI, Gelone SP, Mancano MA *et al.* (1996). Precipitation of lorazepam infusion [letter]. *Ann Pharmacother* 30: 1037–1038.

British Pharmacopoeia (2007). *British Pharmacopoeia*, Vol. 2. London: The Stationery Office, pp. 1266–1267.

Cate J, Hedrick R (1980). Propylene glycol intoxication and lactic acidosis. *N Engl J Med* 303: 1237.

CHIP3 Regulations (2002). Chemicals (Hazard Information and Packaging for Supply) Regulations. SI 2002/1689.

Glasgow A, Boeckx R, Miller M, MacDonald M, August G, Goodman S (1983). Hyperosmolarity in small infants due to propylene glycol. *Pediatrics* 72: 353–355.

Gunawan S, Treiman D (1988). Determination of lorazepam in plasma of patients during status epilepticus in high-performance liquid chromatography. *Ther Drug Monit* 10: 172–176.

Guy's and St Thomas' and Lewisham Hospitals (2005). *Paediatric Formulary*. London: Guy's and St Thomas' and Lewisham NHS Trust.

Lee W-M, Lugo RA, Rusho WJ, Mackay M, Sweeley J (2004). Chemical stability of extemporaneously prepared lorazepam suspension at two temperatures. *J Pediatr Pharmacol Ther* 9: 254–258.

Levanda M (1998). Noticeable difference in admixtures prepared from lorazepam 2 and 4 mg/ml. *Am J Health Syst Pharm* 55: 2305.

Lowey AR, Jackson MN (2008). A survey of extemporaneous dispensing activity in NHS trusts in Yorkshire, the North-East and London. *Hosp Pharmacist* 15: 217–219.

Lund W, ed. (1994). *The Pharmaceutical Codex: Principles and Practice of Pharmaceutics*, 12th edn. London: Pharmaceutical Press.

Marshall JD, Farrah HC, Kearns GL (1995). Diarrhea asoociated with enteral benzodiazepine solutions. *J Pediatr* 126: 657–659.

Martin G, Finberg L (1970). Propylene glycol: a potentially toxic vehicle in liquid dosage form. *J Pediatr* 77: 877–878.

McEvoy J, ed. (1999). *AHFS Drug Information*. Bethesda, MD: American Society of Health System Pharmacists.

Nahata MC, Morosco RS, Hipple TF (1993). Stability of lorazepam diluted in bacteriostatic water for injection at two temperatures. *J Clin Pharm Ther* 18: 69–71.

Norenberg JP, Achuism LE, Steel TH, Anderson TL (2004). Stability of lorazepam in 0.9% sodium chloride stored in polyolefin bags. *Am J Health Syst Pharm* 61: 1039–1041.

Panderi IE, Archontaki HA, Gikas EE, Parrisi-Poulou M (1998). Kinetic investigation on the degradation of lorazepam in acidic aqueous solutions by high performance liquid chromatography. *J Liq Chromatogr Rel Technol* 21: 1783–1795.

Ruddick J (1972). Toxicology, metabolism, and biochemistry of 1,2-propanediol. *Toxicol Appl Pharmacol* 21: 102–111.

Rutgers JG and Shearer CM (1980). Lorazepam. In: Florey K, ed. *Analytical Profiles of Drug Substances*, Vol. 9. London: Academic Press, pp. 397–426.

Share MJ, Harrison RD, Folstad J, Fleming RA (1998). Stability of lorazepam 1 and 2 mg/ml in glass bottles and polypropylene syringes. *Am J Health Syst Pharm* 55: 2013–2015.

Stiles ML, Allen LV, Prince SJ (1996). Stability of desferioxamine mesylate, floxuridine, fluorouracil, hydromorphone hydrochloride, lorazepam, and midazolam hydrochloride in polypropylene infusion-pump syringes. *Am J Health Syst Pharm* 53: 1583–1588.

Volles DF (1996). More on usability of lorazepam admixtures for continuous infusion. *Am J Health Syst Pharm* 53: 2753–2754.

Magnesium glycerophosphate oral liquid

Risk assessment of parent compound

A risk assessment survey completed by clinical pharmacists suggested that magnesium glycerophosphate has a wide therapeutic index, but with the potential to cause significant side-effects if inaccurate doses were to be administered (Lowey and Jackson, 2008). Current formulations used at NHS trusts generally show a low technical risk, based on the number and/or complexity of manipulations and calculations. Magnesium glycerophosphate is therefore generally regarded as a moderate-risk extemporaneous preparation (Box 14.28).

Box 14.28 *Preferred formula*

Magnesium glycerophosphate oral liquid is available as a 'Special' from NHS and non-NHS 'Specials' manufacturers.

Using a typical strength of 1 mmol/mL as an example:

- magnesium glycerophosphate PhEur powder 194.4 g
- Nipasept 0.1% solution to 100 mL.

Shelf-life

14 days at room temperature in amber glass.

Alternative

An alternative formula would be:

- magnesium glycerophosphate PhEur powder 194.4 g
- preserved xanthan gum 0.5% ('Keltrol' or similar) to 100 mL.

Shelf-life

14 days at room temperature in amber glass.

Note: 'Keltrol' xanthan gum formulations vary slightly between manufacturers, and may contain chloroform and ethanol. The individual specification should be checked by the appropriate pharmacist.

Summary

Magnesium glycerophosphate oral liquid is known to be available from both NHS and non-NHS 'Specials' manufacturers. There is little stability information in the public domain, but these suppliers should be able to provide suitable data to verify the shelf-lives used. The specification of the product should be assessed by a suitably competent pharmacist before purchase. Extemporaneous preparation is no longer recommended under normal circumstances.

Extension of the stated expiry date may well be possible with further testing.

Clinical pharmaceutics

Points to consider

- Although the exact solubility of magnesium glycerophosphate is not known, it is available at typical strengths (e.g. 1 mmol/mL) as a solution. Dosage uniformity should not be an issue.
- There are no known data with respect to bioavailability of oral magnesium glycerophosphate liquid formulations.
- Nipasept contains a hydroxybenzoate preservative system. This is a widely used antimicrobial preservative in cosmetics, and oral and topical pharmaceutical formulations. Concern has been expressed over its use in infant parenteral products as bilirubin binding may be affected (Rieger, 2003). This is potentially hazardous in hyperbilirubinaemic neonates (Loria *et al.*, 1976). However, there is no proven link to similar effects when used in oral products. The World Health Organization has suggested an estimated total acceptable daily intake for methyl-, ethyl- and propyl-hydroxybenzoates at up to 10 mg/kg body weight (WHO/FAO, 1974).
- 'Keltrol' is made to slightly different specifications by several manufacturers, but typically contains ethanol and chloroform. The exact specification should be assessed before use by a suitably competent pharmacist. Ethanol is a CNS despressant. Chloroform has been classified as a class 3 carcinogen (i.e. substances that cause concern owing to possible carcinogenic effects but for which available information is not adequate to make satisfactory assessments) (CHIP3 Regulations, 2002).
- There have been reports of bacterial contamination of some raw material sources (A. Murphy, University College London Hospital, NHS Pharmaceutical Quality Assurance Committee, personal communication, 2006).
- Unlicensed oral sachets are also available as an alternative product (S. Keady, Lead Divisional Clinical Pharmacist, Women and Children's Services, UCLH NHS Foundation Trust, personal communication, 2007).

Other notes

- Each gram of magnesium glycerophosphate (anhydrous) represents approximately 5.1 mmol of magnesium. Magnesium glycerophosphate (anhydrous) 8 g is approximately equivalent to 1 g of magnesium (Martindale, 2005).

Technical information

Structure

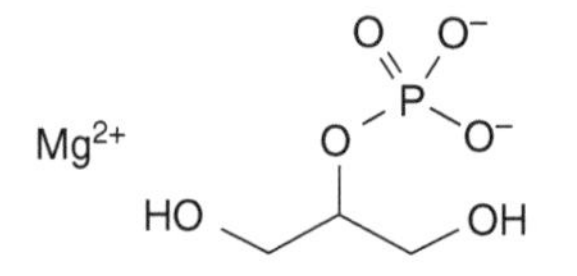

Empirical formula

$C_3H_7MgO_6P$ (British Pharmacopoeia, 2007).

Molecular weight

194.4 (British Pharmacopoeia, 2007).

Solubility

Not known.

pK_a

Not known.

Optimum pH stability

Not known.

Stability profile

Physical and chemical stability

Magnesium glycerophosphate is a mixture, in variable proportions, of magnesium(*RS*)-2,3-dihydroxypropylphosphate and magnesium 2-hydroxy-1-(hydroxymethyl)ethyl phosphate, which may be hydrated. Magnesium glycerophosphate contains not less than 11% and not more 12.5% of magnesium, calculated with reference to the dried substance (British Pharmacopoeia, 2007).

It is a white, hygroscopic powder, practically insoluble in alcohol. It dissolves in dilute solutions of acids (British Pharmacopoeia, 2007).

Magnesium glycerophosphate should be stored in an airtight container (British Pharmacopoeia, 2007).

Degradation products

No information.

Stability in practice

There is no known information available in the public domain to support the stability of magnesium glycerophosphate oral solutions.

Bioavailability data

No data.

References

British Pharmacopoeia (2007). *British Pharmacopoeia*, Vol. 2. London: The Stationery Office, pp. 1291–1292.

CHIP3 Regulations (2002). Chemicals (Hazard Information and Packaging for Supply) Regulations. SI 2002/1689.

Loria CJ, Excehverria P, Smith AL (1976). Effect of antibiotic formulations in serum protein: bilirubin interaction of newborn infants. *J Pediatr* 89: 479–482.

Rieger MM (2003). Ethylparabens. In: Rowe RC, Sheskey PJ, Weller PJ, eds. *Handbook of Pharmaceutical Excipients*, 4th edn. London: Pharmaceutical Press and Washington DC: American Pharmaceutical Association.

Sweetman SC (2009). *Martindale: The Complete Drug Reference*, 36th edn. London: Pharmaceutical Press.

WHO/FAO (1974). Toxicological evaluation of certain food additives with a review of general principles and of specifications. Seventeenth report of the Joint FAO/WHO Expert Committee on Food Additives. World Health Organ. Tech. Rep. Ser. No. 539.

Menadiol sodium phosphate oral liquid

Risk assessment of parent compound

A risk assessment survey completed by clinical pharmacists suggested that menadiol sodium phosphate has a wide therapeutic index. Inaccurate doses were thought to be potentially associated with significant morbidity (Lowey and Jackson, 2008). Formulations in use at the NHS trusts surveyed were of a low technical risk, based on the number and/or complexity of manipulations and calculations. Menadiol sodium phosphate is therefore regarded as a moderate-risk extemporaneous preparation (Box 14.29).

Summary

There are no data available in the public domain to support the formulation above. It is used in a limited number of NHS centres (Lowey and Jackson, 2008).

Box 14.29 *Preferred formula*

Using a typical strength of 1 mg/mL as an example:

- menadiol sodium phosphate tablets (equivalent to 10 mg menadiol) × 10
- methylcellulose solution 2% to 100 mL.

Method guidance

Place the tablets into a mortar and grind to a fine powder with a pestle. Add a small quantity of methylcellulose and mix to form a smooth paste. Add sufficient methylcellulose to make a suspension, then pour into a measure. Rinse the pestle and mortar with further portions of methylcellulose, adding the rinsings to the measure. Make up to volume with methylcellulose, mix well, and transfer to an amber glass bottle.

Shelf-life

7 days refrigerated in amber glass. **Shake well before use.**

The choice of methylcellulose as a suspending agent is based on successful historical use of the suspension. Settling of the suspending agent may occur; vigorous shaking is recommended before administration.

Given the lack of supporting data, a cautious shelf-life and storage conditions are recommended. The patient's response to treatment should be monitored closely. Alternative treatments should be considered where possible.

Clinical pharmaceutics

Points to consider

- At typical concentrations (1 mg/mL), almost all of the drug is likely to be in solution rather than suspension.
- There are no known bioavailability data for this preparation.

Other notes

- Fat-soluble phytomenadione (vitamin K1) is available for oral administration as a licensed Konakion MM Paediatric preparation (BNF, 2007).

Technical information

Structure

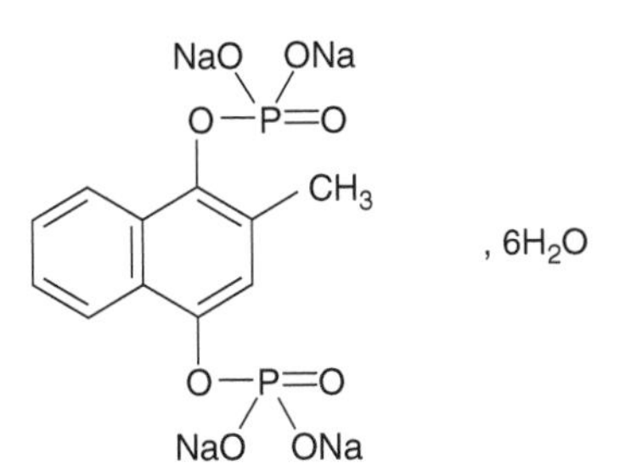

Empirical formula

$C_{11}H_8Na_4O_8P_2.6H_2O$ (British Pharmacopoeia, 2007).

Molecular weight

530.2 (British Pharmacopoeia, 2007).

Solubility

Very soluble in water (British Pharmacopoeia, 2007) (less than 1 mL of solvent required per gram of solute).

pK_a

Not known.

Optimum pH stability

Not known.

Stability profile

Physical and chemical stability

Menadiol sodium phosphate is a white to pink, crystalline powder with a characteristic odour (British Pharmacopoeia, 2007). It is hygroscopic in nature (British Pharmacopoeia, 2007). Menadiol sodium phosphate should be stored at a temperature not exceeding 8°C in airtight containers, and protected from light (Martindale, 2005).

Menadiol is very soluble in water. It is licensed for oral administration to prevent vitamin K deficiency in malabsorption syndromes (BNF, 2007), but is also used for other unlicensed indications.

Vitamin K is generally recognised as one of the more labile vitamins, and may present problems of instability in some dosage forms (de Ritter, 1982)

Menadiol phosphate injection BP is formulated at pH 7.0–8.0 (British Pharmacopoeia, 2007). Menadiol sodium diphosphate injection USP is formulated at pH 7.5–8.5 (United States Pharmacopeia, 2003).

Degradation products

Not known.

Stability in practice

No data.

Bioavailability data

No data.

References

British Pharmacopoeia (2007). *British Pharmacopoeia*. London: The Stationery Office, Vol. 2, pp. 1331–1332 and Vol. 3, p. 2374.

de Ritter E (1982). Vitamins in pharmaceutical formulations. *J Pharm Sci* 71: 1073–1096.

Joint Formulary Committee (2010). *British National Formulary,* No. 59. London: British Medical Association and Royal Pharmaceutical Society of Great Britain.

Sweetman (2009). *Martindale: The Complete Drug Reference*, 36th edn. London: Pharmaceutical Press.

United States Pharmacopeia (2003). *United States Pharmacopeia*, XXVI. Rockville, MD: The United States Pharmacopeial Convention.

Metformin hydrochloride oral liquid

Risk assessment of parent compound

A risk assessment survey completed by clinical pharmacists suggested that metformin hydrochloride has a moderate therapeutic index. Inaccurate doses were thought to be associated with serious morbidity (Lowey and Jackson, 2008). Current formulations used at NHS trusts were of a moderate technical risk, based on the number and/or complexity of manipulations and calculations. Metformin hydrochloride is therefore regarded as a high-risk extemporaneous preparation (Box 14.30).

Summary

There is a very limited amount of anecdotal evidence in the public domain to support the stability or the successful clinical use of xanthan gum-based metformin suspensions. However, a licensed metformin oral liquid is now available which should be used wherever possible.

Clinical pharmaceutics

Points to consider

- At typical concentrations (100 mg/mL), most or all of the drug will be in solution.
- Ora-Plus and Ora-Sweet SF are free from chloroform and ethanol.
- Ora-Plus and Ora-Sweet SF contain hydroxybenzoate preservative systems.
- 'Keltrol' is made to slightly different specifications by several manufacturers, but typically contains ethanol and chloroform. The exact specification should be assessed before use by a suitably competent pharmacist. Ethanol is a CNS depressant. Chloroform has been classified as a class 3 carcinogen (i.e. substances that cause concern owing to possible carcinogenic effects but for which available information is not adequate to make satisfactory assessments) (CHIP3 Regulations, 2002).
- No bioavailability data are known to be available for the extemporaneously prepared metformin oral liquids.

***Box 14.30** Preferred formula*

A licensed preparation is now available. Extemporaneous preparation should no longer be carried out except in extenuating circumstances.

Using a typical strength of 100 mg/mL as an example:

- metformin hydrochloride 500 mg tablets × 20
- Ora-Plus to 50 mL
- Ora-Sweet or Ora-Sweet SF to 100 mL.

Method guidance

Tablets can be ground to a fine, uniform powder in a pestle and mortar. A small amount of Ora-Plus may be added to form a paste, before adding further portions of Ora-Plus up to 50% of the final volume. Transfer to a measuring cylinder. The Ora-Sweet can be used to wash out the pestle and mortar before making the suspension up to 100% volume. Transfer to an amber medicine bottle.

Shelf-life

7 days at room temperature in amber glass.

Alternative

An alternative formula would be:

- metformin hydrochloride 500 mg tablets × 20
- xanthan gum 0.5% suspending agent ('Keltrol' or similar) to 100 mL.

Shelf-life

7 days at room temperature in amber glass

Note: 'Keltrol' xanthan gum formulations vary slightly between manufacturers, and may contain chloroform and ethanol. The individual specification should be checked by the appropriate pharmacist.

Technical information

Structure

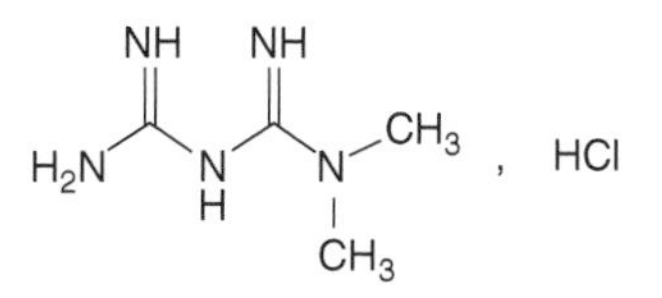

Empirical formula

$C_4H_{11}N_5$,HCl (British Pharmacopoeia, 2007).

Molecular weight

165.6 (British Pharmacopoeia, 2007).

Solubility

Freely soluble in water (1 g/mL to 100 mg/mL) (British Pharmacopoeia, 2007).

pK_a

2.8 and 11.5 (at 32°C) (Moffat, 1986; Summary of Product Characteristics, 2006).

Optimum pH stability

Not known.

Stability profile

Physical and chemical stability

Metformin is an amphoteric compound, with two pK_a values (2.8 and 11.5). It is a white, crystalline, hygroscopic, almost odourless powder (Pharmaceutical Codex, 1973).

The solubility of metformin means that most or all of the drug is in solution at a typical strength of 100 mg/mL, and thus open to aqueous degradation kinetics.

In-house testing at Leeds Teaching Hospitals NHS Trust suggests that metformin is physically compatible with Ora-Plus 1:1 Ora-Sweet SF (Lowey and Watkinson, 2006), as well as with xanthan gum (Nova Laboratories Ltd, 2003).

Metformin hydrochloride should be stored in an airtight container (Lund, 1973).

Degradation products

Not known.

Stability in practice

No data available. Metformin oral liquid is now available as a licensed product in the UK.

Bioavailability data

There are no bioavailability data available for the extemporaneous preparations mentioned above.

After oral administration, the absorption of metformin is slow and incomplete even from solution and rapidly dissolving tablets (Sirtoli *et al.*, 1978, Pentikainen *et al.*, 1979; Noel, 1980; Tucker *et al.*, 1981; Pentakainen, 1986).

References

British Pharmacopoeia (2007). *British Pharmacopoeia*, Vol. 2. London: The Stationery Office, pp. 1347–1349.

CHIP3 Regulations (2002). Chemicals (Hazard Information and Packaging for Supply) Regulations. SI 2002/1689.

Lowey AR, Jackson MN (2008). A survey of extemporaneous dispensing activity in NHS trusts in Yorkshire, the North-East and London. *Hosp Pharmacist* 15: 217–219.

Lowey AR, Watkinson A (2006). In-house testing, Leeds Teaching Hospitals, October 2006.

Moffat AC, Osselton MD, Widdop B (2003). *Analysis of Drugs and Poisons*, 3rd edn. London: Pharmaceutical Press. ISBN 978 0 85369 4731.

Noel M (1980). Kinetic study of normal and sustained release dosage forms of metformin in normal subjects. *J Int Biomed Data* 1: 9–20.

Nova Laboratories Ltd (2003). Information Bulletin – Suspension Diluent Used at Nova Laboratories.

Pentikainen PJ (1986). Bioavailability of metformin. Comparison of solution, rapidly dissolving tablet, and three sustained release products. *Int J Clin Pharmacol Ther Toxicol* 24: 213–220.

Pentikainen PJ, Neuvonen PJ, Penttila A (1979). Pharmacokinetics of metformin after intravenous and oral absorption in man. *Eur J Clin Pharmacol* 16: 195–202.

Pharmaceutical Codex (1973). *The Pharmaceutical Codex*, 10th edn. London: Pharmaceutical Press, p. 296.

Sirtoli CR, Franceschini G, Galli-Kienle M, Cighetti C, Galli G, Bondioli A (1978). Disposition of metformin (*N*,*N*-dimethhyl-biguanide) in man. *Clin Pharmacol Ther* 24: 683–693.

Summary of Product Characteristics (2006). Riomet. www.riomet.com.

Tucker GT, Casey C, Phillips PJ, Connor H, Ward JD, Woods HF (1981). Metformin kinetics in health in healthy subjects and in patients with diabetes mellitus. *Br J Clin Pharmacol* 12: 235–246.

Midazolam hydrochloride oral liquid

Risk assessment of parent compound

A survey of clinical pharmacists has suggested that midazolam has a moderate therapeutic index with the potential to cause severe morbidity if inaccurate doses are given (Lowey and Jackson, 2008). Current formulations used at NHS trusts were generally of a low to moderate technical risk, based on the number and/or complexity of manipulations and calculations. Midazolam hydrochloride is therefore regarded as a medium-risk extemporaneous preparation (Box 14.31).

Summary

Midazolam oral solution is widely used as an extemporaneous preparation. The experience in the NHS largely consists of using the injection orally, diluted in a syrup-based vehicle to aid palatability. The adverse effects of

Box 14.31 *Preferred formula – oral use only*

Note: Imported products and 'Specials' are now available. Extemporaneous preparation should only be undertaken where there is a demonstrable clinical need.

Using a typical strength of 1 mg/mL as an example:

- midazolam injection 10 mg/2 mL 10 mL
- syrup BP (with preservative) to 50 mL.

Method guidance

To the injection, add a small amount of syrup BP in order to form a paste. Then add further portions of syrup BP up to approximately 75% of the final volume. Transfer to a measuring cylinder. The remaining syrup BP can be used to wash out the pestle and mortar before making the suspension up to 100% volume. Transfer to an amber medicine bottle.

Shelf-life

28 days refrigerated in amber glass. **Shake the bottle.**

syrup, including implications for diabetics and dental health, should be noted. However, use of midazolam oral liquid is typically limited to the short term. Note that the formulation above is for oral use only. Other formulations may be available for other routes (e.g. intranasal, buccal).

Given the relative stability of the drug, midazolam may be a suitable candidate to purchase from a 'Specials' manufacturer (commercial or NHS production unit). The purchasing pharmacist must take adequate steps to satisfy themself as to the quality and formulation of the item being purchased. A 'Special' should take into account the issues of choosing appropriate preservatives and the possible need for pH buffering.

A licensed product is available for import from the USA. This product should be assessed by a suitably competent person before purchase.

Clinical pharmaceutics

Points to consider

- The solubility of midazolam hydrochloride is pH dependent. Although most preparations seen are solutions, patients and carers should be advised to **shake the bottle** as a precaution in case midazolam precipitates from the solution.
- Midazolam has a very bitter taste which can affect patient compliance. This may be partly due to the benzyl alcohol in the injection.
- Approaches to improve the palatability of parenteral forms of midazolam that are administered orally have included the use of fruit syrups, apple juice, honey, chocolate syrup, flavoured gelatin, cola and carbonated beverages, and grape-flavoured soft drinks (Peterson, 1990; Rosen and Rosen, 1991; Mishra *et al.*, 2005)
- A licensed pre-mixed cherry-flavoured solution is available in the USA (Versed syrup, Roche). This formulation is compounded with sorbitol, glycerin, citric acid anhydrous, sodium citrate, sodium benzoate, sodium saccharin, artificial cough syrup flavour, artificial bitterness modifier and water (Roche Laboratories, 1998; Cote *et al.*, 2002). The pH is adjusted to around 3 with hydrochloric acid.
- There is a wide body of published literature regarding bioavailability of midazolam. Issues have been raised with regard to the relative bioavailabilities of the commercial and extemporaneous preparations.

Other notes

- Midazolam preparations have been administered by various methods, including the oral, buccal, rectal and intranasal routes. This may carry further pharmaceutical considerations. For example, solutions with a pH of around 3 may cause some transient irritation to mucosa. 'Specials' that target specific routes are known be available for purchase. The formulation of any unlicensed preparations should be assessed for their appropriateness before purchase and/or use. For example, the addition of some ethanol may be used in order to facilitate the development of a formulation with a higher (less irritant) pH. A local risk–benefit analysis should be carried out, on an individual patient basis if deemed necessary.

- The tablet preparations are typically midazolam maleate; the injections are midazolam hydrochloride.

Technical information

Structure

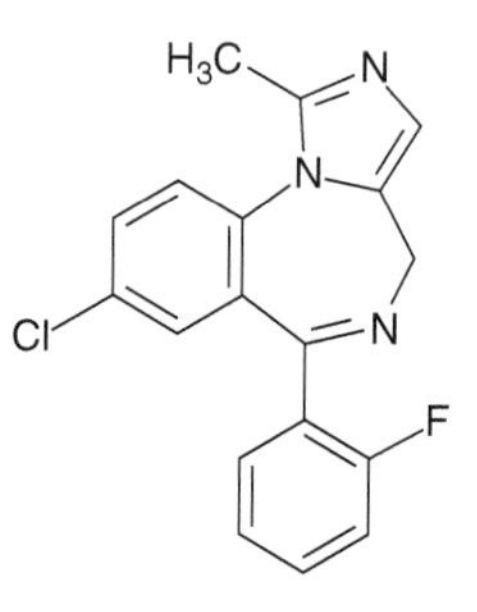

Empirical formula (base)

$C_{18}H_{13}ClFN_3$ (British Pharmacopoeia, 2007).

Molecular weight (base)

325.8 (British Pharmacopoeia, 2007).

Solubility

Midazolam base is practically insoluble in water (British Pharmacopoeia, 2007). Solubility of midazolam hydrochloride is pH dependent.

Midazolam hydrochloride solubility–pH profile at 25°C is shown in Table 14.5 (Trissel, 2000).

pK_a

6.15 (Allen, 1991).

Optimum pH stability

Midazolam hydrochloride is reported to be stable at pH 3–3.6 (McEvoy, 1999; Trissel, 2000).

Table 14.5 Midazolam hydrochloride solubility–pH profile at 25°C

pH	Solubility (mg/mL)
6.2	0.24
5.1	1.09
3.8	3.67
3.4	10.3
2.8	>22

From Trissel (2000).

Stability profile

Physical and chemical stability

Midazolam is a white or yellowish, crystalline powder, practically insoluble in water but freely soluble in alcohol (British Pharmacopoeia, 2007). The solubility of the hydrochloride salt is reported to increase in acidic pH (Trissel, 2000). Allen (1991) suggests that a vehicle with an approximate pH of 4 is needed to dissolve midazolam to concentrations of 2 mg/mL. A higher pH may lead to precipitation and formation of a suspension rather than a solution. Given the relatively small amount of drug in standard midazolam preparations, this precipitate may not be easily apparent. Acidic suspending agents are therefore recommended (e.g. cherry syrup, orange syrup, Ora-Plus, Ora-Sweet) (Allen, 1991).

Midazolam hydrochloride is reported to be stable at pH 3–3.6 (McEvoy, 1999; Trissel, 2000). The drug is known to exhibit its best water solubility at pH values less than 4, and to exhibit greater lipid solubility at high pH values (Forman and Souney, 1987; Trissel, 2000). This has considerable implications with regard to bioavailability (see below).

The fused imidazole ring modifies the properties inherent in the classic benzodiazepines, including with regard to stability in aqueous solutions (Gerecke, 1983). The ring is relatively basic, with the nitrogen at position 2 displaying a pK_a of 6.15 ± 0.1 (Walser *et al.*, 1978). This basicity allows the preparation of salts, including the hydrochloride.

Midazolam normally exists in an equilibrium of both an open- and a closed-ring structure; the proportion is pH dependent (Walser *et al.*, 1978; Cote *et al.*, 2002). At a pH of less than 3.3, an open-ring structure (benzophenone) may be formed (Han *et al.*, 1976; Andersin, 1991; Hagan *et al.*, 1993; Walker *et al.*, 1997): this occurs when the azomethine bond is cleaved (Selkämaa and Tammilehto, 1989). Gerecke (1983) suggests that an aqueous solution of the hydrochloride (5 mg/mL midazolam) at pH 3.3 consists of

80–85% of the closed-ring form and 15–20% of the open-ring form. The open-ring structure has been reported to interfere with assays of midazolam and cannot be completely separated from midazolam (Walker *et al.*, 1997). However, Walker *et al.* (1997) suggest that the benzophenone is not a true degradant, as it can readily convert to midazolam at pH 7.4 and 37°C.

The fused imidazole ring confers on the molecule an increased stability to hydrolysis (Gerecke, 1983). It is reported that a 5 mg/mL aqueous solution of midazolam (as the hydrochloride) at pH 3.3 showed no detectable amount of degradation products after being heated for four weeks at 80°C (Aldamy, 1983).

Preliminary experiments by Selkämaa and Tammilehto (1989) have reportedly shown midazolam to be relatively thermostable. However, photochemical decomposition was reported to occur readily, with the drug totally degrading after exposure to radiation from a high-pressure mercury lamp for around 5 hours, and after exposure to ordinary daylight for three months.

Selkämaa and Tammilehto (1989) carried out further work to isolate and identify the breakdown products of photochemical degradation. Midazolam was dissolved in ethanol (0.5%) and stored in 10 mL aliquots in clear glass ampoules before exposure to a high-pressure mercury lamp or to daylight on a windowsill. Thin-layer chromatography revealed several degradation products. The three main degradants were isolated by flash chromatography and identified as 6-chloro-2-methyl-4-(2′-fluorophenyl)-quinazoline, *N*-desalkylflurazepam and 7-chloro-2-[(1-ethoxyethylimino) ethoxymethyl]-5-(2′-fluorophenyl)-3*H*-1,4-benzopdiazepine. Several other minor products were identified by GC-MS.

Midazolam injection should be protected from light (British Pharmacopoeia, 2007). The injection (as a hydrochloride) has a pH adjusted to around 3 (Trissel, 2000) and is reported to have a shelf-life of 2 years after preparation (Allen, 1991). The free base of midazolam is lipophilic (Gerecke, 1983).

Degradation products

Photodegradation products include 6-chloro-2-methyl-4-(2′-fluorophenyl)-quinazoline, *N*-desalkylflurazepam and 7-chloro-2-[(1-ethoxyethylimino) ethoxymethyl]-5-(2′-fluorophenyl)-3*H*-1,4-benzopdiazepine (Selkämaa and Tammilehto, 1989). Other minor products have also been reported (Selkämaa and Tammilehto, 1989).

Stability in practice

Several studies demonstrate the relative stability of midazolam in oral liquid formulations. The palatability of some of the formulations may be open to question.

Walker *et al.* (1997) investigated the stability of parenteral midazolam hydrochloride in a sweetened, flavoured and coloured oral solution. Concentrations of 0.35 mg/mL, 0.64 mg/mL and 1.03 mg/mL were produced by diluting 1, 2 or 3 mL of a commercially available parenteral Versed 5 mg/mL preparation in 13 mL of the following vehicle:

- simple syrup 50 mL
- pure orange extract 0.12 mL
- red food colour 1 drop
- yellow food colour 1 drop
- distilled water to 100 mL.

Samples of each concentration were prepared in triplicate and stored in high-density polyethylene containers at room temperature (23 ± 2°C). A 5 microlitre sample was withdrawn from each of the nine bottles on days 0, 1, 2, 6, 7, 9, 13, 21 and 102. No details are given about from which portion of the liquid the samples were taken and whether prior shaking was carried out. However, as solutions, these issues may not be significant. The HPLC method described by Hagan *et al.* (1993) was used to assay drug content, after proving that the assay was stability-indicating by degrading midazolam in acid and alkali at elevated temperatures.

On each study day, the solutions were also checked visually for colour, clarity and particulates against a black-and-white background. Each of the solutions are reported to appear as clear orange-coloured liquids. However, the authors later contradict this statement by suggesting that all the solutions remained clear and colourless throughout the duration of the study (Walker *et al.*, 1997).

All of the solutions retained more than 94% of the starting concentration after 102 days. The statistical methods could not detect any significant degradation, but the lack of data points between days 21 and 102 will limit the ability of the statistical test to detect a definite trend. Although the concentrations studied would appear to be unusual choices, the results do give an indication of the stability of midazolam hydrochloride over a useful concentration range. Microbiological stability was not investigated, and the use of food dyes may not be deemed appropriate by some healthcare practitioners.

Mehta *et al.* (1993) have studied the chemical stability of 1 mg/mL midazolam syrup. The syrup was prepared by adding midazolam injection (Hypnovel 5 mg/mL; as hydrochloride) to syrup BP (containing methyl hydroxybenzoate as an antimicrobial preservative). Four batches of 50 mL were prepared and then split further into three portions for storage in 30 mL tightly closed bottles at 'room temperature' (no details stated) and 4°C, in the presence and absence of light. Samples were removed and assayed by HPLC every week for up to six weeks. Details of the sampling method and stability-indicating HPLC method are provided.

Assay results indicate that no significant degradation occurred at either storage temperature over the test period; drug content was maintained within 5% of the starting value. The standard deviation values range up to 3.4%; this may reflect the use of a 1 mL sample volume. Moreover, the consistency and viscosity of syrup may also have contributed to experimental error. There were no significant changes reported in the pH of the syrups over the study period of six weeks (initial pH approximately 3.5). The presence of light had no effect on the stability of the drug, and the liquids remained clear and colourless.

Mehta *et al.* (1993) conclude that such a formulation of midazolam syrup is stable at 4°C for at least four weeks. The lack of a room temperature recommendation is not explained. Microbiological investigations were not carried out. Methyl-hydroxybenzoate is used as the preservative. Although hydroxybenzoates (also known as 'parabens') are widely used as antimicrobial preservative in cosmetics, and oral and topical pharmaceutical formulations, concern has been expressed over their use in infant parenteral products as bilirubin binding may be affected (Rieger, 2002). This is potentially hazardous in hyperbilirubinaemic neonates (Loria *et al.*, 1976). However, there is no proven link to similar effects when used in oral products. The World Health Organization has suggested an estimated total acceptable daily intake for methyl-, ethyl- and propyl-hydroxybenzoates at up to 10 mg/kg body weight (WHO/FAO, 1974).

Soy *et al.* (1994) studied the stability of a 1 mg/mL oral midazolam solution and its efficacy when used by the oral route in pre-surgical paediatric patients. The formula of the oral solution was as follows:

- midazolam hydrochloride (from Dormicum ampoules 5 mg/mL) 100 mg
- sodium saccharin 240 mg
- lemon or strawberry favour 1 drop
- purified water 80 mL.

A duplicate solution was prepared without the active drug for comparison. Stability was studied for two months using a stability-indicating HPLC method based on the method used by Steedman *et al.* (1992). Assays were undertaken on days 0, 4, 21, 35 and 73. The authors state that HPLC analysis confirmed that there was no loss of midazolam throughout the study. Sample chromatagrams are shown, but only two data points are presented for consideration. The spectrophotometry results suggest a concentration of 0.92 micrograms/mL on day 4 and 0.89 micrograms/mL on day 73. This suggests that even on day 4, the preparation is close to the standard threshold of 90% of the initial concentration remaining. Given the relative stability of midazolam, the initial preparation and analysis stages may be open to considerable error.

Although the authors state that 'there were no significant differences in the control and test solutions' pH value during the study', this statement is misleading. Although the inter-day variation seen in each of the two solutions pH values appears to be minimal and the difference between the two pH values appears relatively consistent, the mean pH values were very different (midazolam solution mean pH 3.79; non-active solution mean pH 6.56).

Gregory *et al.* (1993) comment particularly on the problem of benzyl alcohol in parenteral midazolam products leaving a bitter taste when used orally. Two solutions were formulated:

Solution A (2.5 mg/mL)

- midazolam 5 mg/mL injection 15 mL
- simple syrup NF* 14.5 mL
- peppermint oil 0.5 mL.

Solution B (3 mg/mL)

- midazolam 5 mg/mL injection 18 mL
- simple syrup*, NF 11.4 mL
- peppermint oil 0.6 mL.

*Simple syrup NF contains 85 g of sucrose per 100 mL of purified water.

The solutions were stored in amber glass bottles and assayed during day 1 (12–16 hours after preparation), on day 14 and on day 38. All assays were carried out in triplicate using a reversed-phase stability-indicating HPLC method.

Solution A showed a concentration of only 2.28 mg/mL on day 1 (91% of the expected 2.5 mg/mL). The results at day 14 was similar at 2.26 mg/mL (90% of the initial concentration), but had dropped to 1.82 mg/mL (72%) by day 38.

Solution B showed a concentration of 2.82 mg/mL (94% of the expected 3 mg/mL) on day 1. A similar result of 2.91 mg/mL (97%) was found on day 14, but this had reduced to 2.24 mg/mL (75%) by day 38. The correlations of variation for replicate varied from 0.7% to a significant 7.0%. The authors suggest that the differences from day 1 to day 14 are likely to be due to experimental error, and suggest that the solutions are stable for 14 days at room temperature.

Full details of the sampling methodology are not stated and microbiological stability was not investigated. The investigation is severely limited by the lack of an adequate number of data points, making accurate interpretation of the results impossible. Extrapolation of a shelf-life from these results is not recommended. The storage temperature is also not stated.

Steedman *et al.* (1992) have investigated the stability of midazolam hydrochloride in a flavoured, dye-free oral solution. The authors chose a concentration of 2.5 mg/mL as a balance between maximizing palatability and minimizing the administration volume. The formula consisted of:

- midazolam hydrochloride injection 5 mg/mL (Versed) 180 mL
- flavoured, dye-free syrup 180 mL.

The resulting 360 mL were divided into 20 mL aliquots in amber glass bottles with child-resistant closures. Six bottles were stored at each of the following temperatures: 7, 20 and 40°C. The pH of a separate 10 mL sample was also taken at 21°C and compared with the pH of syrup alone.

A stability-indicating HPLC method was used to assay drug content on days 0, 1, 3, 5, 14, 21, 35 and 56. Samples of 1 g were removed from each of the 18 bottles. Details are not provided with regard to shaking or the portion of the liquid sampled, but these issues are of less importance given that the drug should be in aqueous solution. The samples were also inspected for signs of microbial growth, and changes in colour, turbidity and odour.

The pH of the syrup vehicle alone was found to be 4.6, and the pH of the product 4.2. The timing of these results is not stated, and the authors do not show repeat pH measurement results over the course of the study. The pH of the vehicle is known to be crucial with respect to the solubility of midazolam. It has been suggested that a 2.5 mg/mL solution should have a pH of no more than 4.2 to ensure solubility (McEvoy, 1999). Although no signs of any physical changes were seen, the borderline pH result would suggest that solubility problems may be possible.

Throughout the 56 days, the mean drug content on the three groups of solutions remained above 90% of the starting concentration. However, it is difficult to quantify the amount of drug degradation due to considerable inter-day variation. Considerable standard deviations are seen of up to 4.6% ($n = 6$). The background 'experimental noise' therefore may be concealing small amounts of degradation. This precludes the prediction of a longer shelf-life from the data.

Steedman *et al.* (1992) conclude by suggesting that injectable midazolam (as the hydrochloride salt) in syrup at a concentration of 2.5 mg/mL remained stable in amber glass bottles for 56 days at 7, 20 or 40°C. No gross microbiological growths were witnessed, but full microbiological validation was not carried out. Given the use of syrup as a vehicle, concerns over microbiological growth may limit the practical shelf-life to one month or less.

Related preparations

Bhatt-Mehta *et al.* (1993) have investigated the stability of midazolam hydrochloride in extemporaneously prepared flavoured gelatin, in an effort to alleviate the bitter taste associated with midazolam. Gelatin cups of 5 mg and 15 mg were prepared and stored at 4°C for 14 days or at –30°C for 28 days.

All samples were reported to have maintained at least 96% of the initial midazolam concentration throughout the respective test periods. A stability-indicating HPLC method was used. The authors suggest that a midazolam concentration of 1 mg/mL of gelatin should not be exceeded due to problems with a very bitter aftertaste.

Hagan *et al.* (1993) have investigated the stability of midazolam hydrochloride 0.5 mg/mL in 5% dextrose injection or 0.9% sodium chloride over 30 days (in polyolefin bags). Three containers of midazolam hydrochloride in 5% dextrose injection and three containers of midazolam in 0.9% sodium chloride injection were stored at room temperature (22–24°C) in the dark. An identical set of six solutions were stored at 2–6°C in a glass-doored refrigerator. A stability-indicating HPLC methodology was used.

The authors found little or no degradation over the 30 days. The effect of light on degradation was not investigated; Hagan *et al.* (1993) suggest containers should be protected from the light to prevent possible degradation due to normal room lighting.

Bioavailability data

There is a wide body of evidence describing the oral, buccal, rectal and intranasal administration of commercially available and extemporaneously prepared midazolam in various vehicles. This evidence includes bioavailability data, and is too exhaustive to be discussed fully here.

However, some specific issues can be raised. For example, the onset of action and the bioavailability of the commercial solution have been questioned, when compared to extemporaneous formulations (Brosius and Bannister, 2003, Khalil *et al.*, 2003).

Brosius and Bannister (2003) have compared the sedation scores and plasma levels of midazolam following the administration of a commercially available and extemporaneously prepared oral formulation of midazolam. The Versed syrup (Roche Laboratories, Nutley, NJ, USA) was compared with a standardised mixture of the injection in a thick grape syrup (Syrpalta; Humco Lab, Texarkana, TX, USA). The authors suggest that IV midazolam in Syrpalta syrup yields more reliable sedation and correspondingly higher plasma drug levels than an equivalent dose of the commercially available preparation.

Khalil *et al.* (2003) compared the same two formulations and suggest that children who received the midazolam/Syrpalta mixture had less anxiety at 15 minutes ($P = 0.046$) and at parental separation ($P < 0.001$) than those who received the pre-mixed midazolam solution. The authors conclude that the extemporaneous formulation has a faster onset of action than the commercial solution. It is suggested that the higher pH and increased viscosity may have contributed to the faster onset of action.

The absorption of oral drugs depends on the length of time the drug is in contact with the mucosa, the local pH, the quantity and flow of saliva, and the physico-chemical features of the drug and site of absorption (Pimlott and Addy, 1985; Karl *et al.*, 1993; Khalil *et al.*, 2003). For example, Zhang Jie *et al.* (2002) have shown that the absorption of midazolam in dogs is strongly pH dependent. This suggests that the absorption from blends and commercial syrups may be improved by slightly increasing the pH; the typical pH of the commercially available midazolam solution is 2.8–3. Lammers *et al.* (2002) have postulated that at pH 4.0–4.5, the imidazole ring of midazolam closes, making it more lipophilic, hence improving lipid solubility and accelerating absorption of the drug across mucosal membranes. However, the solubility of midazolam is known to be pH dependent. Increasing the pH too far will cause precipitation of the drug.

References

Aldamy L (1983). Personal communication cited in: Gerecke M. Chemical structure and properties of midazolam compared with other benzopdiazepines. *Br J Clin Pharmacol* 16(Suppl1): 11S–16S.

Allen LV (1991). Midazolam oral solution. *US Pharmacist* 16: 66–67.

Andersin R (1991). Solubility and acid-base behaviour of midazolam in media of different pH, studied by ultraviolet spectrophotometry with multicomponent software. *J Pharm Biomed Anal* 9: 451–455.

Bhatt-Mehta V, Johnson CE, Kostoff L, Rosen DA (1993). Stability of midazolam hydrochloride in extemporaneously prepared flavored gelatin. *Am J Hosp Pharm* 50: 472–474.

British Pharmacopoeia (2007). *British Pharmacopoeia*, Vol. 2. London: The Stationery Office, pp. 1398–1401.

Brosius KK, Bannister CF (2003). Midazolam premedication in children: a comparison of two oral dosage formulations on sedation score and plasma midazolam levels. *Anesth Analg* 96: 392–395.

Cote CJ, Cohen IT, Suresh S, Rabb M, Rose JB, Weldon BC, *et al.* (2002). A comparison of three doses of a commercially prepared oral midazolam syrup in children. *Anesth Analg* 94: 37–43.

Forman JK, Souney PF (1987). Visual compatibility of midazolam hydrochloride with common preoperative injectable medications. *Am J Hosp Pharm* 44: 2298–2299.

Gerecke M (1983). Chemical structure and properties of midazolam compared with other benzopdiazepines. *Br J Clin Pharmacol* 16(Suppl1): 11S–16S.

Gregory DF, Koestner JA, Tobias JD (1993). Stability of midazolam prepared for oral administration. *South Med J* 86: 771–772.

Hagan RL, Jacobs LF, Pimsler M, Merritt GJ (1993). Stability of midazolam hydrochloride in 5% dextrose injection or 0.9% sodium chloride injection over 30 days. *Am J Hosp Pharm* 50: 2379–2381.

Han WW, Yakatan GJ, Maness DD (1976). Kinetics and mechanism of hydrolysis of 1,4 benzodiazepines. I: Chlordiazepoxide and demoxepam. *J Pharm Sci* 65: 1198–1204.

Karl HW, Rosenberger JL, Larach MG, Ruffle JM (1993). Transmucosal administration of midazolam for premedication of pediatric patients. *Anaesthesiology* 78: 885–891.

Khalil SN, Vije HN, Kee SS, Farag A, Hanna E, Chuang AZ (2003). A paediatric trial comparing midazolam/Syrpalta mixture with premixed midazolam syrup (Roche). *Paediatr Anaesth* 13: 205–209.

Lammers CR, Rosner JL, Crockett DE, Chhokra R, Brock-Utne JG (2002). Oral midazolam with antacid increases the speed of onset of sedation in children prior to general anaesthesia. *Pediatr Anaesth* 12: 26.

Loria CJ, Excehverria P, Smith AL (1976). Effect of antibiotic formulations in serum protein: bilirubin interaction of newborn infants. *J Pediatr* 89: 479–482.

Lowey AR, Jackson MN (2008). A survey of extemporaneous dispensing activity in NHS trusts in Yorkshire, the North-East and London. *Hosp Pharmacist* 15: 217–219.

McEvoy J, ed. (1999). *AHFS Drug Information.* Bethesda, MD: American Society of Health System Pharmacists.

Mehta AC, Hart-Davies S, Bedford C (1993). Chemical stability of midazolam syrup. *Hosp Pharm Pract* 3: 224226.

Mishra LD, Sinha GK, Bhaskar Rao P, Sharma V, Staya K, Gairola R (2005). Injectable midazolam as oral premedicant in pediatric neurosurgery. *J Neurosurg Anesthiol* 17: 193–198.

Peterson MD (1990). Making midazolam palatable for children [letter]. *Anaesthesiology* 73: 1053.

Pimlott SJ, Addy M (1985). Evaluation of a method to study the uptake of prednisolone sodium phosphate from different oral mucosal sites. *Oral Surg Oral Med Oral Pathol* 60: 35–37.

Rieger MM (2002). Methylparabens. In: Rowe RC, Sheskey PJ, Weller PJ, eds. *Handbook of Pharmaceutical Excipients*, 4th edn. London: Pharmaceutical Press and Washington DC: American Pharmaceutical Association.

Roche Laboratories (1998). Versed (midazolam HCl) syrup [package insert]. Nutley, NJ: Roche Laboratories, Inc.

Rosen DA, Rosen KR (1991). A platable gelatine vehicle for midazolam and ketamine. *Anesthesiology* 75: 914–915.

Selkämaa R, Tammilehto S (1989). Photochemical decomposition of midazolam I. Isolation and indentification of products. *Int J Pharm* 49: 83–89.

Soy D, Lopez L, Salvador L, Parra L, Roca M, Chabas E, Codina C, Modamio P, Marino EL, Ribas J (1994). Stability of an oral midazolam solution for premedication in paediatric patients. *Pharm World Sci* 16: 26–264.

Steedman SL, Koonce JR, Wynn JE, Brahen NH (1992). Stability of midazolam hydrochloride in a flavoured, dye-free oral solution. *Am J Hosp Pharm* 49: 615–618.

Trissel LA, ed. (2000). *Stability of Compounded Formulations*, 2nd edn. Washington DC: American Pharmaceutical Association.

Walker SE, Grad HA, Haas DA, Mayer A (1997). Stability of parenteral midazolam in an oral formulation. *Anesth Prog* 44: 17–22.

Walser A, Benjamin LE, Flynn T, Mason C, Schwarz R, Fryer RI (1978). Quinazolines and 1,4-benzopdiazepines. Synthesis and reactions of imidazo[1,5-a][1,4]benzopdiazepines. *J Org Chem* 43: 936–944.

WHO/FAO (1974). Toxicological evaluation of certain food additives with a review of general principles and of specifications. Seventeenth report of the Joint FAO/WHO Expert Committee on Food Additives. World Health Organ. Tech. Rep. Ser. No. 539.

Zhang Jie, Niu Suyi, Zhang Hao *et al.* (2002). Oral mucosal absorption of midazolam in dogs is strongly pH dependent. *J Pharm Sci* 91: 980–982.

Morphine sulfate oral liquid

Risk assessment of parent compound

A risk assessment survey completed by clinical pharmacists suggested that morphine sulfate has a moderate therapeutic index. Inaccurate doses were thought likely to lead to a significant risk of death (Lowey and Jackson, 2008). Current formulations used at NHS trusts included those with a high technical risk, based on the number and/or complexity of manipulations and calculations. Morphine sulfate is therefore regarded as a very high-risk extemporaneous product (Box 14.32).

Summary

The majority of oral morphine liquids prepared in a recent survey were based on dilution of licensed products with water or, occasionally, syrup (Lowey and Jackson, 2008). The licensed products used as starting materials are morphine sulfate injections or Oramorph oral solution, with a typical shelf-life of 7 days (Lowey and Jackson, 2008).

Oramorph oral solution is known to contain 10% alcohol (96% v/v), sucrose and hydroxybenzoate preservatives. Until their recent withdrawal by the manufactures, Oramorph unit dosage vials (UDVs) were also used as a starting material. (The UDVs were free from alcohol, sucrose and preservatives.)

The evidence for the stability of morphine in aqueous solution would appear to justify these shelf-lives for the formulae above. Extension of the shelf-life may well be possible when further data are available.

Traditional use of preserved syrup BP as a suspending agent is not recommended as the typical pH of syrup BP (approximately 7.5–9.5) does not promote morphine stability. Furthermore, the hydroxybenzoates preservatives contained in preserved syrup BP are thought to be effective at a pH range of approximately 4–8 (Rieger, 2003).

Further work is required to investigate the use of a suitable suspending agent with a low pH and effective preservative system (e.g. Ora-Plus 1:1 Ora-Sweet SF).

***Box 14.32** Preferred formulae*

Note: 'Specials' are now available. Extemporaneous preparation should only be undertaken where there is a demonstrable clinical need.

Using a typical strength of 100 micrograms/mL as an example:

- morphine sulfate (Oramorph oral solution) 10 mg/5 mL 2.5 mL

water for irrigation/injection to 50 mL.

Shelf-life

7 days refrigerated in amber glass.

Note: Oramorph oral solution contains 10% alcohol (96% v/v).

Most dispensaries and wards in the UK currently store dilute morphine solutions in a locked controlled drug cupboard at room temperature. Unpreserved systems should be stored in a fridge where possible to limit microbial contamination. Chemical stability of morphine should not be problematic over this time frame.

Clinical pharmaceutics

Points to consider

- At typical concentrations (100–500 micrograms/mL) used in the extemporaneous preparations, all of the drug should be in solution. Therefore, dosage uniformity should not be an issue.
- Oramorph oral solution contains 10% alcohol (96% v/v), sucrose and hydroxybenzoate.
- Morphine has a bitter taste (Preechagoon *et al.*, 2005)
- There is a lack of bioavailability data for extemporaneously prepared morphine sulfate oral liquids.
- The headspace in bottles should be minimised to limit oxidation.

Technical information

Structure

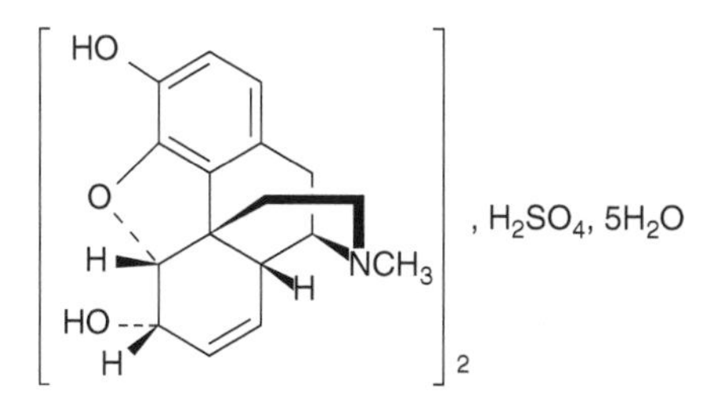

Empirical formula

$C_{34}H_{40}N_2O_{10}S,5H_2O$ (British Pharmacopoeia, 2007).

Molecular weight

759 (British Pharmacopoeia, 2007).

Solubility

1 in 16 of water (or approximately 62.5 mg/mL) (Martindale, 2005).

pK_a

8.31 (amino group) and 9.51 (phenolic group) at 25°C (Connors *et al.*, 1986).

Optimum pH stability

Quoted as 3.2 (Trissel, 2000). Avoid alkaline pH (degradation and precipitation risks) (Martindale, 2005).

Stability profile

Physical and chemical stability

Morphine sulfate is an odourless white or almost white, crystalline powder, soluble in water, insoluble in chloroform and very slightly soluble in alcohol (British Pharmacopoeia, 2007). Solubility increases to 1 in 1 of water at 80°C. When exposed to air, morphine sulfate gradually loses its water of hydration. When exposed to light for prolonged periods, the powder will darken slightly (Martindale, 2005).

Ionescu-Matui *et al.* (1948) suggested that morphine degrades by oxidation and subsequent dimerisation. A complete mechanism was given by Yeh and Lach (1960). No degradation occurs in the absence of oxygen, leading to the suggestion that a free radical reaction was involved. The possible reaction is described in more detail by Vermeire and Remon (1999).

Morphine sulfate solution stability is dependent on pH and the presence of oxygen, and the primary mechanism for degradation is oxidation of the phenolic group (Trissel, 2000). Trissel suggests that the solution is relatively stable at acidic pH, especially below pH 4. Optimum pH has been stated as 3.2, and degradation increases with increasing pH.

Morphine is reported to be very susceptible to degradation in aqueous alkaline media in the presence of oxygen and heavy metal ions (Helliwell and

Game, 1981). Vermiere and Remon (1999) suggest that oxygen, sunlight, UV irradiation, iron and organic impurities catalyse the degradation of morphine. Yeh and Lach (1961) studied the influence of oxygen, temperature, molarity of buffer, ionic strength and morphine concentration on the degradation kinetics of morphine. The data suggested that in the presence of excess oxygen, the degradation rate and extent increased with increasing pH. In closed ampoules, the reaction stopped after a certain time, probably due to a lack of oxygen. The effect of temperature was deemed less important than that of pH. Degradation rate was found to be independent of the molarity and the ionic strength of the buffer. Yeh and Lach (1961) suggest the degradation of morphine in excess oxygen to be pseudo-first order.

Morphine salts are sensitive to changes in pH; morphine is liable to be precipitated out of solution at alkaline pH. Aqueous solutions of morphine sulfate have a pH of approximately 4.8 (Budavari, 1996). Morphine sulfate injection has a pH between 2.5 and 6.5 (United States Pharmacopeia, 1999). The reason for the yellow to brown discoloration of morphine is not clear. The discoloration may be due to the further degradation of the initial degradation products (Vermeire and Remon, 1999). However, Vermeire and Remon (1999) have found that even in dark brown morphine solutions, degradation of the parent drug remains below 5%. Therefore, if the colour is due to the further breakdown of pseudomorphine, this may only occur to a small extent with the formation of strongly coloured products.

Morphine sulfate powder should be protected from light and stored in airtight containers (Martindale, 2005). Several investigators have studied the effect of light on the degradation of morphine, as light is known to catalyse some oxidation reactions (Vermeire and Remon, 1999). However, it would appear from the work carried out by Hung *et al.* (1988) and Oustric-Mendes *et al.* (1997) that light appears to accelerate only slightly the degradation of morphine. Although Strong *et al.* (1994) found much greater levels of degradation, the exact details of the light source used are not provided, and the authors failed to detect the presence of pseudomorphine in the degraded samples, even where more than 10% loss had occurred (Strong *et al.*, 1994). Furthermore, unusually high levels of drug loss were found in the samples that were protected from light (Strong *et al.*, 1994; Vermerie and Remon, 1999).

The oral solution should be protected from freezing (McEvoy, 1999). Storage in the fridge may not be considered necessary; Vermeire and Remon (1999) and Altman *et al.* (1990) suggest that concentrated morphine solutions should be stored in the fridge to prevent reported problems with precipitation. Storage of morphine solutions at elevated temperatures in some pump reservoirs has led to increased morphine concentrations due to water evaporation (Vermeire and Remon, 1999).

Degradation products

Pseudomorphine (2,2′-bimorphine) and morphine-*N*-oxide are reportedly formed in an approximate ratio of 9:1. A small amount of methylamine may also be produced (Deeks *et al.*, 1983; Hor *et al.*, 1997). Orr *et al.* (1982) have suggested that apomorphine may be a degradation product of morphine, although whether this occurs under ambient conditions is not certain.

Pseudomorphine is inactive orally or subcutaneously but is reported to show many properties of morphine following intravenous administration in dogs and cats (Travell, 1932; Schmidt and Livingstone, 1933; Deeks *et al.*, 1983). Unlike morphine, however, vomiting frequently occurred following administration. Pseudomorphine cannot cross the blood–brain barrier (Misra and Mule, 1972).

Morphine-*N*-oxide has been found to have weak analgesic activity (Fennessey, 1968). It has intravenous and subcutaneous toxicities, respectively, 3.2 and 8 times less than those of morphine in mice (Fennessey and Fearn, 1969). Chronic administration of morphine-*N*-oxide has no teratogenic effects in mice (Fennessey and Fearn, 1969).

Stability in practice

Several studies suggest that morphine can be prepared as a relatively stable oral liquid, when formulated correctly.

Preechagoon *et al.* (2005) have studied the formulation development and stability testing of oral morphine solution using a preformulation approach. The factors expected to affect the stability of morphine were studied (i.e. vehicle, antioxidant, chelating agent and pH). The master formula consisted of:

- morphine sulfate 0.2% w/v or 2 mg/mL
- syrup USP 50% w/v
- paraben concentrate 1% w/v
- purified water qs.

A formal fractional-ordered randomised blocked design was used to test the importance of further formulation factors, namely sodium metabisulfite, EDTA, glycerin and pH. Eight different formulations were devised to test these factors. Morphine was dissolved in the water; then the other ingredients were added and mixed with a magnetic stirrer. The total volume and pH were then adjusted. Three bottles of each formulation were prepared and kept in amber glass bottles at room temperature for 35 days.

Physical and chemical analysis occurred on days 0, 7, 14, 28 and 35. The resulting data were used to refine the formulations. A pH of 4 was preferred to pH 6 as lower levels of degradation occurred. Unexpectedly, preparations containing sodium metabisulfite presented stability problems. Further results suggested that the presence of EDTA and glycerin also stabilised the formulation.

Further work involved the addition of sorbitol, sodium chloride, sodium citrate buffer, tartaric acid and four different flavours for taste-masking of the morphine and for adjusting the pH of the preparations. Despite this, a bitter taste and a faint yellow colour were evident. The four final formulations are shown in Table 14.6.

The stability of these formulations was tested in the refrigerator and at 25°C and 75% relative humidity. No precipitation and colour changes were observed over 13 months at either storage temperature. A slight increase in pH occurred (<0.3 units). Viscosity was not measured. After 13 months, all four samples retained over 97% of morphine content. No significant differences were seen either between formulations or between storage conditions. The authors suggest that the low pH was able to minimise any degradation.

While the formulations are too complex for use at dispensary level, they provide useful detail with regard to the effectiveness of each element of the formula. Interestingly, the preformulation formula consisting of morphine

Table 14.6 Four formulations used to test oral morphine solution

Component	Formula 1 (F1)	Formula 2 (F2)	Formula 3 (F3)	Formula 4 (F4)
Morphine sulfate	0.36 g	0.36 g	0.36 g	0.36 g
Glycerin	67.5 mL	81 mL	27 mL	90 mL
Syrup USP	67.5 mL	54 mL	90 mL	40.5 mL
Sorbitol	22.5 mL	36 mL	13.5 mL	22.5 mL
EDTA (w/v)	0.1%	0.1%	0.1%	0.1%
Paraben concentrate (v/v)	1%	1%	1%	1%
Sodium chloride solution[a]	2.4 mL	2.2 mL	2.2 mL	2.7 mL
Sodium citrate buffer (0.05 M)	9 mL	13.5 mL	9 mL	9 mL
Tartaric acid solution[b]	2.5 mL	2.0 mL	2.2 mL	1.4 mL
Purified water to	180 mL	180 mL	180 mL	180 mL
pH adjusted to	3.82	3.98	3.87	4.0

[a] Sodium chloride stock solution is 20 g in 50 mL purified water.
[b] Tartaric acid stock solution is 10 g in 35 mL purified water.

sulfate 200 mg, syrup USP 50 mL, parabens concentrate 1 mL and purified water (to 100 mL, pH adjusted to 4) also demonstrated stability throughout the study period. The authors conclude that pH is one of the major factors influencing the stability of morphine in solution.

EDTA and glycerin were shown to stabilise morphine; sodium metabisulfite decreased drug stability. This finding contradicts the previous work by Yeh and Lach (1971) and Oustric-Mendes *et al.* (1997) who suggested that the addition of sodium metabisulfite may inhibit the transformation of morphine to pseudomorphine. Other ingredients such as tartaric acid and sodium citrate did not appear to affect stability. Obvious degradation was documented in the presence of light. No microbial contamination was found over 13 months, measured using total viable count test.

Information from Boehringer Ingelheim Ltd in 2006 suggested that Oramorph 10 mg in 5 mL can be diluted with either syrup BP or water (L. Houghton, Boehringer Ingelheim Ltd Medicines Information Department, personal communication, 2006). An informal 'in-house' study of the dilution of Oramorph oral solution with preserved syrup 1:1 is reported to provide scope for a 30-day shelf-life, when stored at 2–8°C. However, no data are available. If water or unpreserved syrup is used as a diluent, Boehringer Ingelheim suggest a shelf-life of 7 days when stored at 2–8°C (L. Houghton, personal communication, 2006).

It has been suggested that the solubility and stability of morphine are not considered problematic, and that problems are more likely to result from dilution of the preservative system (B. Wetherill, Deputy Chief Pharmacist, County Durham and Darlington Acute Hospitals NHS Trust, personal communication, 2006). Therefore, less concentrated solutions may necessitate shorter shelf-lives.

Oramorph UDVs also contained disodium edetate, citric acid anhydrous and purified water. Oramorph oral solution 10 mg in 5 mL also contains alcohol, corn syrup, sucrose, methyl hydroxybenzoate, propyl hydroxybenzoate and purified water (Summary of Product Characteristics, 2006). The content of alcohol (96% v/v) is approximately 10% v/v (L. Houghton, personal communication, 2006).

Beaumont (1982) has studied the stability of morphine and diamorphine in chloroform water and alternative solutions, in line with the drug's use in 'Brompton's cocktails' (formerly using cocaine as the opiate). The analysis used an in-house assay based on an ion-pairing agent in conjunction with RP-HPLC.

In order to study the effect of pH and buffer strength on stability, a range of buffer solutions at pH 2.2–8.0 was prepared using McIlvaine's citric acid–sodium phosphate buffer system. These were added to solutions of morphine and diamorphine in double strength chloroform water (final concentrations 1 mg/mL).

Each solution was then divided into two parts for separate storage in the fridge at room temperature. Analysis was then performed daily in three replicate runs and compared against the degradation of the drugs in unbuffered chloroform water (pH approximately 6.5).

The morphine solutions were found to be very stable; no breakdown could be detected in any solution after three months, even when stored at room temperature. Beaumont (1982) therefore endorses the use of the simplest formulation – 1 mg/mL morphine in unbuffered chloroform water – in preference to the more unstable diamorphine solutions where possible. For such a preparation, Beaumont (1982) suggests a shelf-life of three months (one month from opening).

Helliwell and Game (1981) studied the stability of morphine in Kaolin and Morphine Mixture BP. The authors suggest that morphine may well be open to degradation, given the alkaline pH of kaolin and morphine mixture (~pH 8).

Two litres of kaolin and morphine mixture BP were prepared from a commercially available powder by adding 80 mL Chloroform and Morphine Tincture BP, and making up to volume with water. The volume was packed down into 12 × 300 mL clear glass bottles, with a varying amount of liquid, in order to test the effect of the volume of headspace. The bottles were stored undisturbed in the dark and assayed periodically over six months.

The bottles with the least headspace displayed the best stability. Indeed, there was no noticeable change in morphine content for the 300 mL bottles that contained 300 mL of mixture. The losses increased with a decreasing volume in the same size bottle.

Helliwell and Jennings (1982) studied the stability of morphine in Ammonium Chloride and Morphine Mixture BP, using the data from the kaolin and morphine experiment above as a benchmark. Similar results were found. The anhydrous content of 0.0052% w/v was maintained for 25 weeks when stored as a 300 mL volume in a 300 mL bottle, but the drug content declined to 0.0016% when stored as a 50 mL aliquot in a 300 mL bottle.

Excipients that have been used in oral liquid preparations of morphine sulfate include Allura Red AC (E129), disodium edetate, glycerol, invert sugar, sodium benzoate, sodium chloride and sucrose (Lund, 1994).

Related preparations – parenteral use

Morphine sulfate solutions for intravenous use have been described as relatively stable (Martindale, 2005). Vecchio *et al.* (1988) found that concentrations of 0.04 mg/mL and 0.4 mg/mL (in sodium chloride 0.9% or glucose 5%) retained more than 90% of the original concentration when stored at either 4 or 23°C for 7 days, whether exposed to or protected from light.

Walker *et al.* (1988) have suggested that 5 or 10 mg/mL solutions of morphine sulfate in glucose or sodium chloride retained more than 95% of the original concentration when stored at 23°C for 30 days in portable infusion pump cassettes.

Grassby (1991) found that morphine sulfate 2 mg/mL in sodium chloride 0.9% was stable for six weeks when stored in polypropylene syringes at ambient temperatures in the light or dark. A similar solution containing 0.1% sodium metabisulfite lost 15% of its potency over the same time period. Stability of these solutions, with or without metabisulfite, was found to be unacceptable when stored in glass syringes in the dark (Grassby and Hutchings, 1993).

Gove *et al.* (1985) has reported that morphine sulfate injection is stable for at least 69 days when stored at room temperature in plastic syringes with plastic syringe caps. Hung *et al.* (1988) found less than 3% loss of morphine sulfate in 12 weeks at 22°C in the presence of light for injection repackaged into plastics syringes. Lower losses were seen when protected from light and stored in the fridge.

Duafala *et al.* (1990) studied the stability of morphine sulfate in one brand of PVC container, one brand of glass syringe and two brands of disposable infusion devices. Solutions of 15 and 2 mg/mL were used to fill the containers before storage at 23–25 and 4°C. Stability was also determined at 31°C in the disposable infusion devices to simulate near-patient use. The solutions remained stable for at least 12 days in all containers and devices at each temperature tested. No substantial changes were seen with regard to pH or physical characteristics.

Deeks *et al.* (1983) studied the stability of morphine sulfate 0.2 mg/mL in sodium chloride 0.9% for intrathecal injection, using HPLC analysis to assay the breakdown product pseudomorphine. Ampoules from each batch were autoclaved at 115°C for 30 minutes zero, one, two or three times and then stored protected from light at ambient room temperature (14–22°C) or 32°C for 48 weeks.

No precipitation or colour changes were seen. The formation of pseudomorphine was greater in those samples that had been autoclaved more than once. No differences in pseudomorphine content were found between samples with nitrogen-filled or air-filled headspace. The authors recommended a shelf-life of one year when stored at or below 15°C, provided the samples are not autoclaved more than once. This confirmed previous work by Roksvaag *et al.* (1980), who found that although morphine degraded quicker in glass ampoules than glass syringes; the degradation in glass ampoules remained limited.

Hor *et al.* (1997) investigated the stability of morphine sulfate in saline under simulated patient administration conditions. Two concentrations (1 mg/mL and 10 mg/mL) were monitored during slow delivery over 16 days in the dark at 32°C and 36–38% relative humidity. HPLC analysis found less than 10% change in concentration from the baseline value, and no evidence of degradation products. The high temperatures used in the study led to a small increase in drug concentration due to evaporation losses. A small decrease in pH was also seen over the study period.

Turner and Potter (1980) found that an injection of morphine sulfate 2 mg in 10 mL sodium chloride 0.9% can be packed under nitrogen and autoclaved at 115°C for 20 minutes. The authors advise against autoclaving more than once. However, Austin and Mather (1978) found no degradation after subjecting preserved morphine injection to three sterilisation cycles at 120–121°C for 30 minutes.

Taylor and Sherwood (1983) suggest that provided oxygen is excluded from the manufacture, filling and sealing, ampoules of intrathecal morphine sulfate 2 mg in 10 mL could be allocated a shelf-life of 2 years and be re-autoclaved at 115°C for 30 minutes at least twice before significant amounts of pseudomorphine could be detected. The ampoules should not be re-autoclaved within the last six months of their shelf-life.

Bioavailability data

In six healthy subjects, the absolute bioavailability of morphine from an aqueous solution, a controlled release oral tablet (MST-Continus) and a controlled release buccal tablet was approximately 23.9%, 22.4% and 18.7%, respectively, using intravenous morphine sulfate as a comparator (Hoskin *et al.*, 1989). Maximum plasma concentrations were seen at 45 minutes (oral solution), 2.5 hours (MST) and 6 hours (buccal).

Studies of oral elixir use in cancer patients reported bioavailability of between 26% and 47% (Sawe *et al.*, 1981, 1985; Gourlay *et al.*, 1986). Previous higher results shown by McQuay *et al.* (1983) have since been attributed to problems with the specificity of the assay used (Aherne and Littlejohn, 1985).

References

Aherne GW, Littlejohn P (1985). Morphine-6-glucuronide, an important factor in interpreting morphine radioimmunoassay. *Lancet* I, 221–222.

Altman L, Hopkins RJ, Ahmed S, Bolton S (1990). Stability of morphine sulphate in Cormed III (Kalex) intravenous bags. *Am J Hosp Pharm* 47: 2040–2042.

Austin KL, Mather LE (1978). Simultaneous quantitation of morphine and paraben preservatives in morphine injectables (letter). *J Pharm Sci* 67: 1510–1511.

Beaumont I (1982). Stability study of aqueous solutions of diamorphine and morphine using HPLC. *Pharm J* 229: 39–41.

British Pharmacopoeia (2007). *British Pharmacopoeia*, Vol. 2. London: The Stationery Office, pp. 1418–1420.

Budavari S, ed. (1996). *The Merck Index*, 12th edn. Rahway, NJ: Merck and Co.

Connors KA, Amidon GL, Stella ZJ (1986). *Chemical Stability of Pharmaceuticals: A Handbook for Pharmacists*, 2nd edn. New York: Wiley Interscience.
Deeks T, Davis S, Nash S (1983). Stability of an intrathecal morphine injection formulation. *Pharm J* 230: 495–497.
Duafala ME, Kleinberg ML, Nacov C, Flora KP, Hines J, Davis K, McDaniel A, Scott D (1990). Stability of morphine sulphate in infusion devices and containers for intravenous administration. *Am J Hosp Pharm* 47: 143–146.
Fennessey MR (1968). The analgesic action of morphine-*N*-oxide. *Br J Pharmacol Chemother* 34: 337.
Fennessey MR, Fearn HJ (1969). Some observations on the toxicology of morphine-*N*-oxide. *Br J Pharmacol* 21: 668.
Gourlay GK, Cherry DA, Cousins MJ (1986). A comparative study of the efficacy and pharmacokinetics or oral methadone and morphine in the treatment of severe pain in patients with cancer. *Pain* 25: 297–312.
Gove LF, Gordon NH, Miller J (1985). Prefilled syringes for self-administration of epidural opiates. *Pharm J* 234: 378–379.
Grassby PF (1991). The stability of morphine sulphate in 0.9% sodium chloride stored in plastic syringes. *Pharm J* 248: HS24–HS25.
Grassby PF, Hutchings L (1993). Factors affecting the physical and chemical stability of morphine sulphate solutions stored in syringes. *Int J Pharm Pract* 2: 39–43.
Helliwell K, Game P (1981). Stability of morphine in Kaolin and Morphine Mixture BP. *Pharm J* 227: 128–129.
Helliwell K, Jennings P (1982). Stability of morphine in ammonium chloride and morphine mixture. *Pharm J* 229: 600–601.
Hor MMS, Chan SY, Yow KL, Lim LLY, Chan E, Ho PC (1997). Stability of morphine sulphate in saline under simulated patient administration conditions. *J Clin Pharm Ther* 22: 405–410.
Hoskin PJ, Hanks GW, Aherne GW, Chapman D, Littlejohn P, Filshie J (1989). The bioavailability and pharmacokinetics of morphine after intravenous, oral and buccal administration in healthy volunteers. *Br J Clin Pharmacol* 27: 499–505.
Hung CT, Young M, Gupta PK (1988). Stability of morphine solutions in plastic syringes determined by reversed-phase ion-pair liquid chromatography. *J Pharm Sci* 77: 719–723.
Ionescu-Matiu A, Papescu A, Monicum L (1948). Contribution a l'étude de l'alteration et de lat stabilisation de la morphine et de l'apomorphine. *Ann Pharm Fr* 6: 137.
Lowey AR, Jackson MN (2008). A survey of extemporaneous dispensing activity in NHS trusts in Yorkshire, the North-East and London. *Hosp Pharmacist* 15: 217–219.
Lund W, ed. (1994). *The Pharmaceutical Codex: Principles and Practice of Pharmaceutics*, 12th edn. London: Pharmaceutical Press.
McEvoy J, ed. (1999). *AHFS Drug Information*. Bethesda, MD: American Society of Health System Pharmacists.
McQuay HJ, Moore RA, Bullingham RES, Carroll D, Baldwin D, Allen MC, Glynn Lloyd JW (1983). High systemic relative bioavailability of oral morphine in both solution and sustained release formulation. *R Soc Med Int Congr Symp Series* 64: 149–154.
Misra AL, Mule SJ (1972). Disposition and metabolism of [3*H*] pseudomorphine in the rat. *Biochem Pharmacol* 21: 103.
Orr I, Dundee J, McBride A (1982). Preservative free epidural morphine denatures in plastic syringe. *Anaesthesia* 37: 352.
Oustric-Mendes AC, Huart B, Le Hoang M-D, Perrin-Rosset M, Pailler M, Darbord JC *et al.* (1997). Study protocol: stability of morphine injected without preservative, delivered with a disposable infusion device. *J Clin Pharm Ther* 22: 2283–2290.
Preechagoon D, Sumyai V, Tonsitsirin K, Aumpon S, Pongjanyakul T (2005). Formulation development and stability testing of oral morphine solution utilizing preformulation approach. *J Pharm Pharmaceut Sci* 8: 362–369.
Rieger MM (2003). Ethylparabens. In: Rowe RC, Sheskey PJ, Weller PJ, eds. *Handbook of Pharmaceutical Excipients*, 4th edn. London: Pharmaceutical Press and Washington DC: American Pharmaceutical Association.

Roksvaag PO, Fredrikson JB, Waaler T (1980). High-performance liquid chromatographic assay of morphine and the main degradation product pseudomorphine. A study of pH, discoloration and degradation in 1 to 43 years old morphine injections. *Pharma Acta Helv* 55: 198–202.

Sawe J, Dahlstrom B, Paalzow L, Rane A (1981). Morphine kinetics in cancer patients. *Clin Pharmacol Ther* 30: 629–635.

Sawe J, Kager L, Svensson J-O, Rane A (1985). Oral morphine in cancer patients: in vivo kinetics and in vitro hepatic glucuronidation. *Br J Clin Pharmacol* 19: 495–501.

Schmidt CF, Livingstone AE (1933). A note concerning the actions of pseudomorphine. *J Pharmacol Exp Ther* 47: 473.

Strong ML, Schaaf LJ, Pankaskie MC, Robinson DH (1994). Shelf-lives and factors affecting the stability of morphine sulphate and mepiridine (pethidine) hydrochloride in plastic syringes for use in patient-controlled analgesic devices. *J Clin Pharm Ther* 19: 361–369.

Summary of Product Characteristics (2006). Oramorph. www.medicines.org.uk (accessed 28 September 2006).

Sweetman SC (2009). *Martindale: The Complete Drug Reference*, 36th edn. London: Pharmaceutical Press. ISBN 978 0 85369 840 1.

Taylor J, Sherwood MJ (1983). Intrathecal morphine injections [letter]. *Pharm J* 230: 543.

Travell J (1932). A contribution to the pharmacology of pseudomorphine. *J Pharmacol Exp Ther* 44: 123.

Trissel LA, ed. (2000). *Stability of Compounded Formulations*, 2nd edn. Washington DC: American Pharmaceutical Association.

Turner S, Potter SR (1980). Epidural morphine injection [letter]. *Pharm J* 225: 108.

United States Pharmacopeia (1999). *United States Pharmacopeia*, XXIV. Rockville, MD: The United States Pharmacopeial Convention.

Vecchio M, Walker SE, Iazetta J, Hardy BG (1988). The stability of morphine intravenous infusion solutions. *Can J Hosp Pharm* 41: 5–9, 43.

Vermeire A, Remon JP (1999). Stability and compatibility of morphine. *Int J Pharm* 187: 17–51.

Walker SE, DeAngelis C, Iazetta J (1988). Hydromorphone and morphine stability in portable infusion pump cassettes and minibags. *Can J Hosp Pharm* 41: 177–182.

Yeh SY, Lach JL (1960). Stability of morphine in aqueous solutions. *Am J Hosp Pharm* 17: 101–103.

Yeh SY, Lach J (1961). Stability of morphine in aqueous solution III: Kinetics of morphine degradation in aqueous solution. *J Pharm Sci* 50: 35–42.

Yeh SY, Lach JL (1971). Stability of morphine in aqueous solution IV: Isolation of morphine and sodium bisulfite interaction product. *J Pharm Sci* 60: 793–794.

Omeprazole oral liquid

Risk assessment of parent compound

A risk assessment survey has suggested that omeprazole has a wide therapeutic index with the potential to cause mild to moderate morbidity if inaccurate doses are administered as an extemporaneous preparation (Lowey and Jackson, 2008). Current formulations used at NHS trusts were generally of a low technical risk, based on the number and/or complexity of manipulations and calculations. Omeprazole is therefore regarded as a low-risk extemporaneous preparation. However, there may be situations where therapeutic failure could have life-threatening consequences, given the use of omeprazole in the treatment of haemorrhagic gastrointestinal conditions (Box 14.33).

Summary

The evidence surrounding the use of omeprazole suspensions is controversial and often conflicting. This in unsurprising given that the bioavailability of licensed omeprazole preparations is variable and changes over time. Omeprazole is unstable at low pH. Therefore, licensed preparations use pH-sensitive coatings in order to protect the drug from degradation in the acidic environment of the stomach. Therefore, any manipulation of these coatings may lead to an adverse effect on bioavailability. Removal of the protective coatings requires the drug to be formulated at an elevated pH, typically in sodium bicarbonate, to prevent degradation. However, once the protective coatings are removed, the drug is vulnerable to degradation by stomach acid. This may be minimised to some degree by the sodium bicarbonate vehicle, but drug loss is unpredictable. The use of licensed preparations should be recommended wherever possible in order to minimise these concerns.

Treatment with extemporaneous liquid formulations should only be carried out as a last resort under supervision of healthcare professionals who understand and acknowledge the potential problems regarding administration and bioavailability. It has been suggested that the administration of standard doses of omeprazole (in sodium bicarbonate), via the mouth or via a feeding tube, may be inadequate for routine clinical use (Balaban *et al.*, 1997; Sharma *et al.*, 2000). The residual risk associated with this practice is considerable and should be avoided wherever possible.

***Box 14.33** Preferred formula*

Note: There is a body of evidence to support the stability of this preparation. However, its bioavailability is less clear, and may be significantly and unpredictably reduced when compared to the licensed preparations. Given these concerns, the use of Losec MUPS, halved before dispersion if necessary, is considered to be preferable to extemporaneous preparation.

Using a typical strength of 2 mg/mL as an example:

- omeprazole 20 mg capsules × 5
- sodium bicarbonate 8.4% solution to 50 mL.

Method guidance

Empty the contents of the capsules and add a small volume of the sodium bicarbonate. Allow to stand for a few minutes before adding the remainder of the sodium bicarbonate. Transfer to amber glass or plastic oral syringes as required. Shake well to dissolve.

Alternative method

Pull out the plunger from a 60 mL Luer-lok syringe and attach an 18G needle or similar. Count out five omeprazole capsules and empty the contents into the syringe. Replace the plunger and draw up 50 mL of 8.4% sodium bicarbonate injection. Replace the needle with a fluid-dispensing connector and transfer the contents back and forth to a second 60 mL syringe until the granules have dissolved. Transfer to final storage container.

Shelf-life

7 days refrigerated in amber glass or plastic oral syringes (unpreserved). **Shake very well before use.**

Administration of omeprazole via feeding tubes is also problematic, as feeding tubes may become blocked by the Losec MUPS (multiple unit pellet system) micro-beads. Wherever possible, a wide-bore feeding tube should be used to allow administration of the licensed preparation, following dispersion. Where this is not possible, alternative treatment options should be considered, such as the use of lansoprazole FasTabs. A Zegerid omeprazole suspension is also available for import from the USA, where it carries a product licence.

The use of the intravenous route should be considered in very high-risk patients, for whom a reproducible plasma level is critical.

Administration by mouth

Most of the experience in the NHS with regard to extemporaneously dispensed oral liquids is associated with suspensions containing sodium bicarbonate. However, these preparations may be clinically unsuitable for some patients. The use of dispersible formulations such as Losec MUPS is preferable to extemporaneous preparation of a sodium bicarbonate-based suspension. Doses should be rounded to the nearest 5 mg where possible, to allow the use of halved MUPS tablets.

The Summary of Product Characteristics for Losec MUPS contains instructions for patients who cannot swallow tablets, including children. The dispersed tablet can be mixed with a range of beverages and semi-solid foods.

Administration via feeding tubes

The administration of omeprazole via feeding tubes is particularly problematic with respect to tube blockages (Bane, 2004). Alternative treatment options, such as the use of lansoprazole FasTabs, should be considered. Zegerid, an omeprazole suspension, is known to be licensed in the USA (J. Bane, personal communication, 2007). Currently, it appears this product is not widely used in the UK. However, it is supported by safety, efficacy and bioavailability data in adults and geriatric populations, and should be considered as a treatment option. Pharmacists are referred to the document 'Guidance for the purchase and supply of unlicensed medicinal products – notes for prescribers and pharmacists' for further information on the quality assurance of imported medicines (NHS Pharmaceutical Quality Assurance Committee, 2004).

If the use of omeprazole is essential and the Zegerid suspension is not available, the key choice is between using Losec MUPS tablets or opened capsules. The size of the feeding tube is critical in this decision. Losec MUPS provide a dispersible tablet formulation of omeprazole 'beads', and tend to be preferred where the size of the feeding tube allows. However, the dispersed beads have been reported to be large enough to block feeding teats and small-bore enteral feeding tubes (Woods and McClintock, 1993; Bane, 2004; D. Hartigan, Advanced Clinical Pharmacist, Neonatology, Leeds General Infirmary, personal communication, 2006). Losec MUPS should not therefore be administered via small-bore feeding tubes (e.g. small-bore nasogastric tubes, nasojejunal tubes). Instead, omeprazole capsules may be opened and the contents dissolved in sodium bicarbonate.

Fewer problems may be associated with administration via larger bore tubes such as a gastrostomy tube (e.g. 20 French) (Anon, 1998). Adherence to rubber plungers in syringes has also been reported (Israel and Hassall, 1998). It may be argued that the administration of sodium bicarbonate-based omeprazole liquid formulations may be more viable via the nasojejunal route, which bypasses the acidic environment of the stomach.

Specialist resources should be consulted before local guidelines are produced (e.g. Smyth, 2006; White and Bradnam, 2006). Prescribers and/or pharmacists must accept the risks of using unlicensed omeprazole liquid formulations, particularly with respect to bioavailability.

Clinical pharmaceutics

Points to consider

- Omeprazole displays optical chemistry and exists as a racemate (Agranat *et al.*, 2002). With regard to the two optical isomers of the drug, the consequences of formulating omeprazole as an oral liquid are not known. The enantiomers of omeprazole are known to show differences in terms of metabolism and activity. This complex area of research and development has been highlighted by the recent development of esomeprazole, the *S*-enantiomer of omeprazole. Further research is required with regard to oral liquid forms of omeprazole.
- Methods of administration that require large volumes of sodium bicarbonate as a vehicle and/or flush may be clinically inappropriate for some patients (e.g. congestive cardiac failure, severely fluid-restricted, chronic renal impairment, metabolic imbalances).
- As well as exposing the drug to stomach acid, chewing of omeprazole granules may yield a bitter alkaloid or quinine-like taste, resulting in poor compliance (Israel and Hassall, 1998).
- There are conflicting data with regard to the bioavailability of omeprazole suspensions. Studies suggest that the area under the curve (AUC) may be between 50% and 100% of commercially available preparations. Dose titration may be necessary, especially as children typically show a higher clearance of proton pump inhibitors than adults (Litalien *et al.*, 2005).
- When formulated in sodium bicarbonate, omeprazole suspensions may show a slight yellow colour after storage for several days (DiGiacinto *et al.*, 2000).
- New commercial preparations of omeprazole and lansoprazole in combination with sodium bicarbonate (as suspensions) have been developed over the past few years (Hepburn *et al.*, 2004; Anon, 2004; Laine *et al.*, 2004; Litalien *et al.*, 2005; J. Bane, personal communication, 2007).

Technical information

Structure

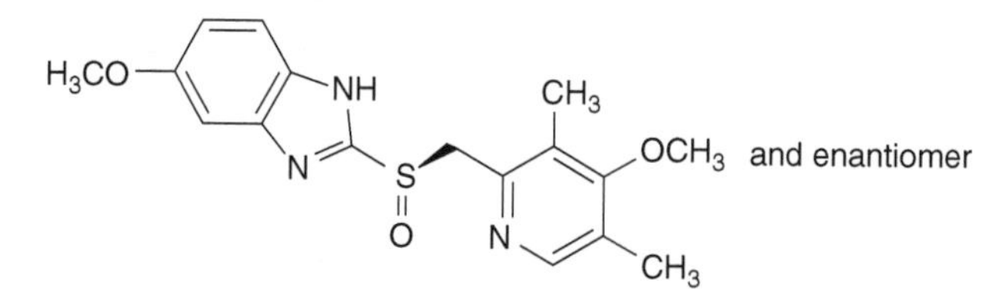

Empirical formula

$C_{17}H_{19}N_3O_3S$ (British Pharmacopoeia, 2007).

Molecular weight

345.4 (British Pharmacopoeia, 2007).

Solubility

Very slightly soluble in water (0.1–1 mg/mL) (British Pharmacopoeia, 2007).

pK_a

Pyridinium ~4 (Pilbrant and Cederberg, 1985; Qaisi *et al.*, 2006). Benzimidazole 8.8 (Pilbrant and Cederberg, 1985).

Optimum pH stability

Stability increases with increasing pH until pH 11 (Pilbrant and Cederberg, 1985), but note that enteric coatings are unstable in alkaline conditions.

Stability profile

Physical and chemical stability

Omeprazole is a white or almost white powder, very slightly soluble in water (British Pharmacopoeia, 2007). A weak base, it dissolves in dilute solutions of alkali hydroxides (Qaisi *et al.*, 2006; British Pharmacopoeia, 2007). Its degradation products include sulfonamide and benzimidazole sulfide (El-Kousy and Bebawy, 1999). Although it is often preferable to use water solutions of a drug to minimise formulation issues, omeprazole is only soluble in alkaline water solutions with physiologically unacceptable, high-pH values (Pilbrant and Cederberg, 1985).

The stability of omeprazole is a function of pH (Anon, 1997; Pesko, 1997). Omeprazole is known to be chemically unstable in acidic conditions while displaying better stability in alkaline media (Pesko, 1997). Pilbrant and Cederberg (1985) have produced a pH–degradation profile, demonstrating increasing stability with increasing pH. The maximum pH studied was pH 11; the effects of even higher pH values are not known.

Qaisi *et al.* (2006) supported these findings, showing rapid degradation at pH values of 2, 3 and 4. Electrochemical investigations suggested that degradation became slower as the solution pH was increased from pH 2 to 8, suggesting that omeprazole decomposition is acid-catalysed.

The pellets (enclosed in a capsule shell) are therefore enteric coated (Woods, 2001). The coating is designed to dissolve at a pH of over 6, allowing release of drug in the alkaline duodenum (Litalien *et al.*, 2005). Opening the capsule and mixing with a vehicle may disturb the enteric coating and expose the drug to acid, rendering it inactive (Woods and McClintock, 1993; Dunn *et al.*, 1999).

The degradation half-life of omeprazole is reported to be less than 10 minutes at pH 4, 18 hours at pH 6.5, and around 300 days at pH 11 (Pilbrant and Cederberg, 1985). Therefore, the crushing of granules and administration via a nasogastric tube in water may lead to rapid drug degradation due to the low pH in the stomach. pH values greater than 4.5 are reported to destroy the enteric coating of the granules over time (Anon, 1997). Reports suggest that the granules should not be mixed with milk, which is reported to have an approximate pH of 6.6 (Anon, 1997).

Moisture and solvents have been shown to have a deleterious effect on the stability of omeprazole and should be avoided in pharmaceutical formulations (Pilbrant and Cederberg, 1985). Stability studies conducted by Wallmark and Lindberg (1987) suggest that omeprazole is sensitive to heat and light, as well as being acid labile.

Degradation products

May include sulfonamide and benzimidazole sulfide (El-Kousy and Bebawy, 1999).

Stability in practice

There is a reasonable body of evidence to suggest that omeprazole can be formulated as a relatively stable suspension. However, the bioavailability of such suspensions is less clear. Three studies have demonstrated similar results for the stability of a 'Simplified Omeprazole Suspension' (SOS) 2 mg/mL in sodium bicarbonate 8.4%.

Phillips *et al.* (1998a,b) found no evidence of degradation over 24 weeks for samples stored in the fridge or freezer. Room temperature results showed a drop from 1.97 mg/mL after one week to 1.81 mg/mL after two weeks. After one month, only 1.57 mg/mL of omeprazole remained. No standard deviations or sample numbers were given. The HPLC analysis is allegedly based on data published elsewhere, but its specificity is not demonstrated in the brief experimental write-up, and storage temperatures are not stated. These factors limit the quality of the data presented.

Filtered versus non-filtered analysis revealed that SOS contained approximately 15–20% of omeprazole in solution, with the remaining 80–85% in suspension. The authors conclude that SOS is stable for six months if refrigerated or frozen. Stability at room temperature is restricted to 7 days. Although these results were found when the final container was clear glass, omeprazole is acknowledged to be light-sensitive, and the authors suggest a colour change may occur. Therefore, storage in amber bottles is recommended by the authors.

Quercia *et al.* (1997) have also investigated the stability of omeprazole (2 mg/mL) in an extemporaneously prepared oral liquid. Three 50 mL vials were prepared and stored at 5, 24 and –20°C. Samples were assayed and pH tested several times over 30 days. The HPLC method was modified from published sources (Amantea and Narang, 1988; Kobayashi *et al.*, 1992) and was shown to be stability-indicating.

While the omeprazole samples stored at –20°C and 5°C did not change colour during the 30-day test period, the liquid stored at 24°C gradually changed from white to brown. Following an initial pH of 8.1 for each of the samples, the pH values had increased over the 30 days by 1.0, 0.6 and 0.5 pH units at 24, 5 and –20°C, respectively. Less than 6% degradation occurred over the 30 days for the samples stored at 5 and –20°C. However, significant degradation was detected in the 24°C sample, accompanied by much wider standard deviation results. Although less than 10% degradation had occurred after two weeks, by day 18 the remaining drug content had decreased to 86.3% (±2.3%). By day 30, this figure had dropped to 84.2% (±7.6%).

The authors conclude by suggesting that the study demonstrates that omeprazole 2 mg/mL in sodium bicarbonate solution is stable for 14 days at 24°C and for 30 days at 5 and –20°C. Full details are given of the procedure used for compounding.

DiGiacinto *et al.* (2000) have studied the stability of omeprazole and lansoprazole suspensions stored in amber-coloured plastic oral syringes. The contents of 10 capsules of omeprazole 20 mg were added to 100 mL sodium bicarbonate 8.4% injection solution USP to achieve a final concentration of 2 mg/mL. Triplicate samples were prepared for room temperature

(22°C) and fridge (4°C) storage. The syringes stored at room temperature were exposed to fluorescent lighting.

All syringes were inspected and assayed over the next 60 days. Prior to assay by stability-indicated HPLC analysis, the syringe was vortexed for 2 minutes then placed on a continuous rocker for 3–5 minutes in an attempt to maintain a uniform dispersion of suspended drug particles. Omeprazole was found to be stable (>90% of initial concentration) for 14 days at 22°C and 45 days at 4°C when stored in this manner. Omeprazole samples appeared slightly yellow after one week of storage. pH values did not change significantly from baseline (starting pH approximately 8.3–8.4).

Microbiological stability was also assessed at baseline and again on days 1, 4, 7, 14, 21, 30 and 45. Samples were prepared aseptically in a vertical flow laminar airhood. Aliquots from syringes were plated in duplicate on 5% sheep blood agar plates and stored aerobically at 37°C for 24 hours. Plates were inspected for growth after 24 hours incubation. No plates were found to display growth.

Bioavailability and administration

It should be noted that the absorption of omeprazole from its licensed capsule formulation displays wide intra- and inter-individual variation (Howden, 1991; Sharma *et al.*, 2000). The systemic bioavailability of a licensed omeprazole product is approximately 35%. After repeated daily administration, this rises to around 60% (Summary of Product Characteristics, 2006). This increase in bioavailability over time makes interpretation of bioavailability data more problematic.

In common with the situation seen in adults, the absolute bioavailability of omeprazole increases with repeated dosing in children (Litalien *et al.*, 2005). The apparent clearance of omeprazole appears to be faster for children than for adults. Therefore, children may need variable and potentially considerably higher doses of omeprazole, on a milligram per kilogram basis, than for adults, to achieve the same plasma concentrations (Litalien *et al.*, 2005).

Astra Pharmaceuticals have reportedly previously recommended that crushed omeprazole granules can be administered with four doses of sodium bicarbonate solution 8 mmol/50 mL (32 mmol total) (Anon, 1998). The bicarbonate is used to increase the stomach pH and thereby decrease drug degradation. The first dose of sodium bicarbonate is given with the crushed drug granules, followed by further doses of bicarbonate 10, 20 and 30 minutes later (Anon, 1998).

This is based on a bioavailability study carried out on six healthy volunteers during the early stages of omeprazole development (Anon, 1998). The study allegedly suggested that this method produced improved results for absorption when compared to an 'unbuffered' omeprazole suspension. The dosing schedule for bicarbonate was slightly different, with an 8 mmol dose given 5 minutes before the omeprazole dose, followed by a 16 mmol dose with the omeprazole dose, then three more doses of 8 mmol 10, 20 and 30 minutes later (Anon, 1998).

Astra Pharmaceuticals have reportedly suggested that the bicarbonate solution may only be required for the first 3–5 days, after which acid suppression may be sufficient to prevent degradation of the omeprazole (Anon, 1998). There are no documented data available, however, to support these approaches. Indeed, there is some conflicting data with regard to the intragastric pH that may be seen after several days of oral omeprazole administration (Prichard *et al.*, 1985; Cederberg *et al.*, 1993). However, omeprazole is acknowledged to increase its own bioavailability to some extent; figures vary between an increase in bioavailability of 50–86% over the first one to two weeks (Howden *et al.*, 1984, 1985; Andersson *et al.*, 1990; Howden, 1991)

Pilbrant and Cederberg (1985) report the results of toxicological and phase I clinical study results. It is not known whether or not this is the study referred to by Astra Pharmaceuticals. Micronised omeprazole with a particle surface area over 2.5 m^2/g was suspended in 0.25% water solution of methylcellulose. Sodium bicarbonate was also added as a buffer. The authors report that such suspensions can be stored in the fridge for a week, or stored deep frozen for more than a year, and still retain 99% of their original potency. No data are presented.

The unbuffered suspension displayed a mean AUC of only 44% of the buffered suspension. This indicates that more than half of the dose was lost due to degradation in the stomach. The authors report that 60 mg of omeprazole in 50 mL of water with 8 mmol of sodium bicarbonate has a pH of around 8.

Other small studies have provided further evidence to support the use of omeprazole preparations formulated with sodium bicarbonate (Lind *et al.*, 1983; Londong *et al.*, 1983; Andersson *et al.*, 1990). However, such studies do not address nasogastric use and were not carried out in the clinical setting (Anon, 1998). The number and volume of sodium bicarbonate doses administered also differs between the studies.

Pilbrant and Cederberg (1985) also investigated the effect of administering antacid with enteric-coated granules of omeprazole in six healthy volunteers. No significant differences between the data were seen; co-administration of antacids had no influence on the bioavailability of omeprazole given as enteric-coated granules.

Sharma *et al.* (2000) have investigated the oral pharmacokinetics of omeprazole and lansoprazole after single and repeated doses as intact capsules or as suspensions in sodium bicarbonate. It had previously been suggested that simplified omeprazole suspension, when administered via a gastrostomy, had an effect on intragastric pH that was quantitatively smaller than that observed with either intact omeprazole capsules, or intact, non-encapsulated omeprazole granules in orange juice (Sharma *et al.*, 1999). Interestingly, a simplified lansoprazole suspension, administered via a gastrostomy, was noted to raise intragastric pH to levels that were comparable with those seen with the intact lansoprazole capsule.

Twelve healthy women were recruited and received each of four proton pump inhibitor formulations: intact omeprazole capsules 20 mg, simplified omeprazole suspension 20 mg, intact lansoprazole capsules 30 mg and simplified lansoprazole suspension 30 mg. The suspensions were prepared by opening the appropriate capsule and pouring the contents into 10 mL of 8.4% sodium bicarbonate (pH 7.8) and gently mixing until the contents were suspended. The resulting milky white liquid was then administered by mouth. pH values were 7.97 for the omeprazole suspension and 7.92 for the lansoprazole.

The number of data points collected is limited, but the results suggested that the absorption of omeprazole did improve from day 1 to day 5 for the intact capsules. The data for the omeprazole suspension did not show such a rise and the statistical test used could not detect a difference. The lansoprazole preparation showed no such improvement in bioavailability. Absorption from the simplified omeprazole suspension was lower than from the capsule, possibly in part due to the lack of improvement of the bioavailability of the omeprazole suspension over the 5 test days. Mean AUC on day 5 was 454 ng h/mL for the intact capsules compared with 265 ng h/mL for the suspension. The results showed considerable variation, and a larger sample size may have helped to provided more accurate data. Sharma *et al.* (2000) conclude that use of 20 mg simplified omeprazole suspension may be inadequate for routine clinical use.

McAndrews and Eastham (1999) have suggested that combinations of omeprazole and sodium bicarbonate require agitation during the preparation process. They also suggest that smaller volumes of sodium bicarbonate can be used to prepare the omeprazole dose (as little as 2.5 mL of sodium bicarbonate is suggested). However, such small volumes may not be sufficient to buffer intragastric pH, and enhanced drug degradation may occur if flushes of sodium bicarbonate are not co-administered.

Kaufman *et al.* (2002) suggest that the bioavailability of omeprazole and sodium bicarbonate suspension administered into the stomach and directly into the small bowel are similar, indicating adequate protection from inactivation by gastric acid.

Song *et al.* (2001) have carried out a pharmacokinetic comparison of omeprazole capsules and simplified omeprazole suspension. Seven healthy volunteers received randomly either one 20 mg delayed release capsule or simplified omeprazole suspension 20 mg in 10 mL for 7 days before crossing over (after a two-week wash-out period) to the alternative treatment for a further 7 days. The suspension was prepared 2 days before the suspension treatment phase, following the directions and formula suggested by Quercia *et al.* (1997). The doses were stored in capped 10 mL oral syringes, protected from light and at room temperature. Subjects were instructed to shake the syringe before each dose.

After one week of therapy, omeprazole absorption was faster and t_{max} was 70% shorter for the suspension. However, the overall AUC (zero to infinity) was 49% lower for the simplified omeprazole suspension. This suggests that overall bioavailability for the sodium bicarbonate-based suspension may only be half that of the commercially available capsule (Song *et al.*, 2001).

Israel and Hassall (1998) highlight the problem of drug being lost by adherence to rubber plungers in syringes, and the potential danger of sodium overload if large volumes of sodium bicarbonate are used to administer doses.

Positive and negative data and reports of the clinical use of omeprazole liquid preparations have been reported by various authors in adults and children (including simplified omeprazole suspension). However, in some cases, the suspension is prepared at the bedside by the nursing staff, by opening the capsules into a syringe with the plunger removed, before the addition of the sodium bicarbonate 8.4%. Therefore, some of the results may not directly demonstrate the successful clinical use of simplified omeprazole suspension after it has been stored for its shelf-life. The full evidence is too exhaustive for inclusion here, but should be considered before a choice of formulation is made.

The Losec MUPS formulation is now available. A 20 mg tablet is reported to contain 1000 acid-protected micropellets, measuring approximately 0.5 mm in diameter (Israel and Hassall, 1998). The coating is designed to dissolve in the alkaline pH of the duodenum. The MUPS tablet may be dispersed in water and given as a liquid. The tablets can also be cut or broken to administer smaller doses. However, part-tablet individual doses cannot be recommended for small children, as settling and aggregation of the insoluble drug is likely to lead to inaccurate dosing (Woods, 1997; Bane, 2004).

Administration of omeprazole via feeding tubes

Omeprazole is commonly administered via feeding tubes of various diameters. The diameter of the feeding tubes is measured in French units (each French unit = 0.33 mm). The smaller the diameter of the tube, the more likely a blockage is to occur.

For patients with duodenal or jejunal tubes, the drug can be delivered straight into an alkaline environment. However, jejunal tubes are often small bore in nature. It has been previously suggested that granules should be fully dissolved rather than suspended for such administration (Israel and Hassall,

1998). Blockages may be less likely with gastrostomy tubes, as such tubes tend to be of a relatively wide bore (e.g. 20 French) (Anon, 1998). However, this is not supported by any validated data. Israel and Hassall (1998) also suggest that feeding tubes with straight-angle connecting tubes are less likely to suffer blockage than tubes which have right-angle connecting tubes.

Bane (2004) investigated the administration of omeprazole in paediatric patients in an inpatient setting. Of 23 patients receiving the drug, nearly half ($n = 11$) had the drug administered via feeding tube. Of these 11 patients, 5 (45%) experienced a blockage of the tube that was attributed to omeprazole administration. The sizes of the tubes that were blocked were in the range 6–8 French. Bane also comments on the poor appreciation of potential problems by nursing staff, and the need for the production of explicit guidelines.

Peckman (1999) has suggested that a nasogastric tube size 14 French or larger is required through which to administer pellets (from an opened capsule). However, Peckman's approach has been criticised for providing a variable delivery of omeprazole, and for being labour-intensive for nursing staff (Dunn *et al.*, 1999; Woods, 2001).

Dunn *et al.* (1999) investigated the *in vitro* administration of omeprazole and lansoprazole via a nasogastric tube. Eighteen omeprazole 20 mg capsules were opened and the granules inside were counted. The granules from a single capsule were then placed in a small cup and mixed with either 30 mL of tap water, 15 mL of tap water or 15 mL of apple juice. For each assessment, a 14 French nasogastric tube was used. A flush of water or apple juice was used before the drug was administered. The drug was poured into a syringe in stages, maintaining some air between the plunger and liquid, to decrease any problems of the granules adhering to the rubber plunger. Final volume was made up to 60 mL with either water or apple juice as appropriate.

The effluent was collected and passed through a filter, with remaining granules counted by two observers. The percentage of granules successfully delivered through the tube was then calculated. The results show a massive standard error, and can be summarised as follows:

- omeprazole, 15 mL of water 53.2 (± 25.1)%
- omeprazole, 15 mL of apple juice 46.4 (± 13.3)%
- omeprazole, 30 mL of water 54.0 (± 29.3)%.

The results were described by the authors as highly variable but generally poor. Only six trials were carried out. Given the small sample size and large experimental error, no statistically significant differences could be shown between the three initial vehicles. Anecdotally, the granules appeared to stick to the cup slightly more when apple juice was used as the vehicle (Dunn *et al.*, 1999).

Zimmerman *et al.* (1997) have investigated the integrity of omeprazole granules in acidic fruit juices for up to 30 minutes. Woods (2001) has suggested that an acidic liquid such as apple juice could preserve the enteric coating of the granules and thereby improve delivery of the drug into the duodenum, where it can be absorbed at a pH greater than 6. It had also been postulated that acidic juice may also decrease the problem of partial dissolution of the enteric coating, leading to granules sticking to the syringe and tubing (Zimmerman, 1997; Dunn *et al.*, 1999). However, the data produced by Dunn *et al.* (see above) have not shown an advantage from using apple juice in preference to water.

Larson *et al.* (1996) compared the bioavailability of two 20 mg omeprazole capsules swallowed intact with that of 40 mg omeprazole granules administered with water through a nasogastric tube. Ten healthy volunteers were included in the cross-over study. There was no difference in the mean area under the plasma concentration–time curve between the two treatments. The method used for administering the granules was, however, modified during the study, after the distal portion of the nasogastric tube belonging to the first volunteer was found to contain a large amount of clumped drug granules. Following this discovery, the remaining volunteers received the drug granules in small portions (6–10 at a time), using 140 mL of water in total. As each capsule contains over 150 granules, this practice has been criticised as being labour-intensive and inappropriate for severely fluid-restricted patients (Dunn *et al.*, 1999).

Balaban *et al.* (1997) investigated the effects of nasogastric omeprazole on gastric pH in critically ill patients. In an uncontrolled study of 10 patients, the authors found that 40 mg daily of omeprazole granules via the nasogastric tube maintained intragastric pH at more than 4.0 ($n=6$), while 20 mg of omeprazole granules was not effective ($n=4$). The method of delivery involved mixing the granules with 30 mL of water before adding half to one third of the mixture to a syringe, with the plunger having been previously removed. The contents were then flushed through the nasogastric tube with 15 mL of water. This process was then repeated to complete the dose. The study by Balaban *et al.* (1997) suggests that higher doses may be required if the drug is to be administered by the nasogastric tube.

References

Agranat I, Caner H, Caldwell J (2002). Putting chirality to work: the strategy of chiral switches. *Nat Rev Drug Discov* 1: 753–768.

Amantea MA, Narang PK (1988). Improved procedure for quantitation of omeprazole and metabolites using reversed-phase high performance liquid chromatography. *J Chromatogr* 426: 216–222.

Andersson T, Andren K, Cederberg C, Lagerstrom PO, Lunborg P, Skanberg I (1990). Pharmacokinetics and bioavailability of omeprazole after single and repeated oral administration in healthy subjects. *Br J Clin Pharmacol* 29: 557–563.

Anon (1997). Compounding hotline. *Pharmacy Times* 63(9): 82.

Anon (1998). Nasogastric administration of omeprazole. *Aust J Hosp Pharm* 28: 174–176.

Anon (2004). Omeprazole/antacid-powder suspension – Santarus. *Drugs R&D* 5: 349–350.

Balaban DH, Duckworth CW, Peura DA (1997). Nasogastric omeprazole: effects on gastric pH in critically ill patients. *Am J Gastroenterol* 92: 79–93.

Bane J (2004). Drug use evaluation of omeprazole and its associated problems with administering doses less than 10 mg and/or administration via a feeding tube. Oral Presentation at Neonatal and Paediatric Pharmacists Group Annual Conference, Newcastle, 2004. www.nppg.scot.nhs.uk (accessed 13 June 2007).

British Pharmacopoeia (2007). *British Pharmacopoeia*, Vol. 2. London: The Stationery Office, pp. 1521–1524.

Cederberg C, Rohss K, Lundborg P, Olbe L (1993). Effect of once daily intravenous and oral omeprazole on 24-hour intragastric acidity in healthy subjects. *Scand J Gastroenterol* 28: 179–184.

DiGiacinto JL, Olsen KM, Bergman KL, Hole EB (2000). Stability of suspension formulations of lansoprazole and omeprazole stored in amber-colored plastic oral syringes. *Ann Pharmacother* 34: 600–605.

Dunn A, White CM, Reddy P, Quercia RA, Chow MSS (1999). Delivery of omeprazole and lansoprazole granules through nasogastric tube in vitro. *Am J Health Syst Pharm* 56: 2327–2330.

El-Kousy NM, Bebawy LI (1999). Stability-indicating methods for determining omeprazole and octylinium bromide in the presence of their degradation products. *J AOAC Int* 82: 599–606.

Hepburn B, Hardman Y, Frank WO, Goldlust B (2004). A loading-dose regimen of omeprazole-immediate release oral suspension increases bioavailability (healthy subjects) and achieves rapid control of gastric acidity in critically ill. *Gastroenterology* 126(Suppl2): 192.

Howden CW (1991). Clinical pharmacology of omeprazole. *Clin Pharmacokinet* 20: 38–49.

Howden CW, Meredith PA, Forrest JAH, Reid JL (1984). Oral pharmacokinetics of omeprazole. *Eur J Clin Pharmacol* 26: 641–643.

Howden CW, Forrest JAH, Meredith PA, Reid JL (1985). Antisecretory effect and oral pharmacokinetics following low dose omeprazole in man. *Br J Clin Pharmacol* 20: 137–139.

Israel DM, Hassall E (1998). Omeprazole and other proton pump inhibitors: pharmacology, efficacy, and safety, with special reference to use in children. *J Pediatr Gastroenterol Nutr* 27: 568–579.

Kaufman SS, Lyden ER, Brown CR, Davis CK, Andersen DA, Olsen KM *et al.* (2002). Omeprazole therapy in pediatric patients after liver and intestinal transplantation. *J Pediatr Gastroenterol Nutr* 34: 194–198.

Kobayashi K, Chiba K, Sohn DR (1992). Simultaneous determination of omeprazole and its metabolites in plasma and urine by reversed-phase high-performance liquid chromatography with an alkaline-resistant polymer C18 column. *J Chromatogr* 579: 299–305.

Laine L, Margolis B, Bagin RG, Rock J, Hepburn B, Frank W (2004). Double-blind trial of immediate-release omeprazole suspension vs intravenous cimetidine for prevention of upper gastrointestinal bleeding in critically ill patients. *Gastroenterology* 126 (Suppl2): 77–78.

Larson C, Cavuto NJ, Flockhardt DA (1996). Bioavailability and efficacy of omeprazole given orally and by nasogastric tube. *Dig Dis Sci* 41: 475–479.

Lind T, Cederberg C, Ekenved G, Haglund U, Olbe L (1983). Effects of omeprazole – a gastric proton pump inhibitor – on pentagastrin stimulated acids secretion in man. *Gut* 24: 270–276.

Litalien C, Theoret Y, Faure C (2005). Pharmacokinetics of proton pump inhibitors in children. *Clin Pharmacokinet* 44: 441–466.

Londong W, Londong V, Cedeberg C, Steffen H (1983). Dose-response study of omeprazole on meal-stimulated gastric acid secretion and gastrin release. *Gastroenterology* 85: 1373–1378.

Lowey AR, Jackson MN (2008). A survey of extemporaneous dispensing activity in NHS trusts in Yorkshire, the North-East and London. *Hosp Pharmacist* 15: 217–219.

McAndrews KL, Eastham JH (1999). Omeprazole and lansoprazole suspensions for nasogastrc administration (letter). *Am J Health Syst Pharm* 56: 81.

NHS Pharmaceutical Quality Assurance Committee (2004). Guidance for the purchase and supply of unlicensed medicinal products. Unpublished document available from regional quality assurance specialists in the UK or after registration from the NHS Pharmaceutical Quality Assurance Committee website: www.portal.nelm.nhs.uk/QA/default.aspx.

Peckman HJ (1999). Alternative method for administering proton pump inhibitors through nasogastric tubes. *Am J Health Syst Pharm* 56: 1020.

Pesko LJ (1997). Oral liquid for omeprazole. *Am Druggist* 214: 48.

Phillips JO, Metzler MH, Johnson K (1998a). The stability of simplified omeprazole suspension (SOS). *Crit Care Med* 26(Suppl A101): 221.

Phillips JO, Metzler MH, Huckfeldt RE, Olsen K (1998b). A multicenter, prospective, randomized clinical trial of continuous infusion IV ranitidine vs omeprazole suspension in the prophylaxis of stress ulcers. *Crit Care Med* 26(Suppl A101): 222.

Pilbrant A, Cederberg C (1985). Development of an oral formulation of omeprazole. *Scand J Gastroenterol* 20(Suppl 108): 113–120.

Prichard PJ, Yeomans ND, Mihaly GW, Jones DB, Buckle PJ, Smallwood RA *et al.* (1985). Omeprazole: a study of its inhibition of gastric pH and oral pharmacokinetics after morning or evening dosage. *Gastroenterology* 88: 64–69.

Qaisi AM, Tutunji MF, Tutunji LF (2006). Acid decomposition of omprazole in the absence of thiol: a differential pulse polarographic study at the static mercury drop electrode (SMDE). *J Pharm Sci* 95: 384–391.

Quercia RA, Chengde F, Xinchun Liu Chow MSS (1997). Stability of omeprazole in an extemporaneously prepared oral liquid. *Am J Health Syst Pharm* 54: 1833–1836.

Sharma VK, Vasudeva R, Howden CW (1999). Simplified lansoprazole suspension – a liquid formulation of lansoprazole – effectively suppresses intragastric acidicity when administered through a gastrostomy. *Am J Gastroenterol* 94: 1813–1817.

Sharma VK, Peyton B, Spears T, Raufman J-P, Howden CW (2000). Oral pharmacokinetics of omeprazole and lansoprazole after single and repeated doses as intact capsules or as suspensions in sodium bicarbonate. *Aliment Pharmacol Ther* 14: 887–892.

Smyth JA, ed. (2006). The NEWT Guidelines for the administration of medication to patients with enteral feeding tubes or swallowing difficulties. North East Wales NHS Trust.

Song JC, Quercia RA, Fan C, Tsikouris J, White CM (2001). Pharmacokinetic comparison of omeprazole capsules and a simplified omeprazole suspension. *Am J Health Syst Pharm* 58: 689–694.

Summary of Product Characteristics (2006). Losec MUPS. www.medicines.org.uk (accessed 14 June 2007).

Wallmark B, Lindberg P (1987). Mechanism of action of omeprazole. *Pharmacology* 1: 158–161.

White R, Bradnam V (2006). *Handbook of Drug Administration via Enteral Feeding Tubes.* London: Pharmaceutical Press.

Woods DJ (1997). Extemporaneous formulations – problems and solutions. *Paediatr Perinat Drug Ther* 1: 25–29.

Woods DJ (2001). *Formulation in Pharmacy Practice. eMixt, Pharminfotech*, 2nd edn. Dunedin, New Zealand. www.pharminfotech.co.nz/emixt (accessed 18 April 2006).

Woods DJ, McClintock AD (1993). Omeprazole administration [letter]. *Ann Pharmacother* 27: 651.

Zimmerman A, Walters JK, Katona B *et al.* (1997). Alternative methods of proton pump inhibitor administration. *Consult Pharm* 12: 990–998.

Phenobarbital and phenobarbital sodium oral liquid

Synonyms

Phenobarbitone and phenobarbitone sodium oral liquid.

Note: 438 mg of sodium phenobarbital is equivalent to 400 mg phenobarbital (referred to as the 'free acid').

Risk assessment of parent compound

A risk assessment survey completed by clinical pharmacists suggested that phenobarbital and phenobarbital sodium have a narrow therapeutic index, with the potential to cause significant morbidity if inaccurate doses were to be administered (Lowey and Jackson, 2008). Current formulations used at NHS trusts show a high technical risk, based on the number and complexity of manipulations and calculations. Phenobarbital is therefore regarded as a very high-risk extemporaneous product (Box 14.34).

***Box 14.34** Preferred formula*

Note: The extemporaneous preparation of phenobarbital oral liquid is not recommended, as this product is available from several 'Specials' manufacturers. However, the following formulae may be considering as interim measures before a 'Special' can be purchased.

Using a typical strength of 3 mg/mL as an example:

- Phenobarbital Sodium powder BP 328.5 mg (equivalent to 300 mg of phenobarbital)
- glycerol 40 mL
- sterile distilled water to 100 mL.

Shelf-life

14 days at room temperature.

Alternatives

An alternative formula if powder or glycerol are not available would be:

- phenobarbital tablets 30 mg × 10 (or equivalent weight of powder)
- Ora-Plus to 50 mL
- Ora-Sweet SF to 100 mL.

* The use of tablets is not encouraged (see below), but may be necessary for the initial supply. A suitable preparation should be purchased from a 'Specials' supplier when practicable.

Shelf-life

28 days at room temperature.

Another alternative formula would be:

- phenobarbital tablets 30 mg* 10 (or equivalent weight of powder)
- xanthan gum 0.5% with preservative ('Keltrol') to 100 mL.

* The use of tablets is not encouraged (see below), but may be necessary for the initial supply. A suitable preparation should be purchased from a 'Specials' supplier when practicable. Note also that 'Keltrol' xanthan gum formulations vary slightly between manufacturers, and may contain chloroform and ethanol. The individual specification should be checked by the appropriate pharmacist.

Shelf-life

28 days at room temperature.

Summary

Several 'Specials' suppliers are known to manufacture phenobarbital and phenobarbital sodium oral liquids. These formulations should be assessed as being appropriate for the intended age range before they are purchased. Historical formulae that contain high concentrations of alcohol are not recommended, particularly in neonates and young children.

All of the extemporaneously compounded formulae for phenobarbital or phenobarbital sodium recorded in a recent survey used pharmaceutical-grade powder rather than crushed tablets as the drug source (Lowey and Jackson, 2008). This is because tablet excipients may affect the pH of the liquid, causing precipitation of the phenobarbital. Such precipitation would be hard to see if insoluble tablet excipients are present in the liquid. As most dispensaries in the UK are unlikely to have access to pharmaceutical grade powder, or the equipment required to handle the powder appropriately, the purchase of an appropriate 'Special' preparation may be deemed necessary.

Clinical pharmaceutics

Points to consider

- Whether the drug is in suspension or solution depends on the concentration used, and whether the salt or free acid are chosen.
- Phenobarbital exhibits polymorphism (Das Gupta, 1984). In theory, changes in polymorphic form could alter a drug's particle size (Han *et al.*, 2006). The formulation of phenobarbital as a solution is therefore preferable.
- Traditional approaches have focused on the use of the soluble sodium salt as the free acid has a very bitter taste. However, this requires formulation at higher pHs, where the choice of preservative is limited. Chloroform is the usual choice for a preservative at higher pHs, and this has been classified as a class 3 carcinogen (i.e. substances that cause concern owing to possible carcinogenic effects but for which available information is not adequate to make satisfactory assessments) (CHIP3 Regulations, 2002).
- Preparations that contain stabilising agents such as alcohol and propylene glycol may not be suitable for neonates and young children.
- Formulations that contain phenobarbital in suspension are less likely to require the presence of chloroform or stabilising agents. However, such preparations may be less palatable and show poorer dosage uniformity if the bottle is not shaken thoroughly before use.
- There is no information regarding bioavailability for extemporaneously prepared phenobarbital preparations.
- Phenobarbital elixir is commercially available in some countries and official pharmacopoeial formulations also exist. Some of these (e.g. Phenobarbitone elixir BP and Phenobarbitone elixir BPC) contain over 30% ethanol which is unacceptably high, particularly for paediatric patients (Woods, 2001). Alcohol-free preparations are now recommended, particularly for neonates and infants (Woods, 2001). Some 'Specials' manufacturers still use low concentrations of alcohol, in order to improve the taste of the liquid.
- 'Keltrol' is made to slightly different specifications by several manufacturers, but typically contains ethanol and chloroform. The exact specification should be assessed before use by a suitably competent pharmacist. Ethanol is a CNS depressant. Chloroform has been classified as a class 3 carcinogen.

Technical information (adapted from British Pharmacopoeia, 2007 unless stated)

Table 14.7

Table 14.7 Technical information for phenobarbital and phenobarbital sodium oral liquid

	Phenobarbital	Phenobarbital sodium
Structure	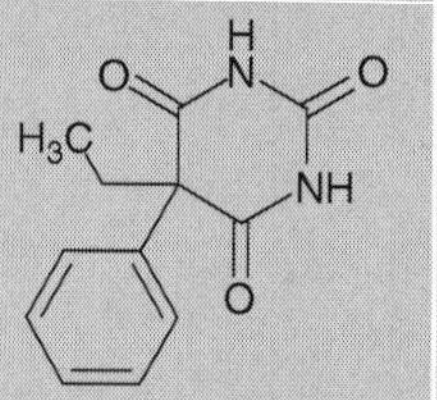	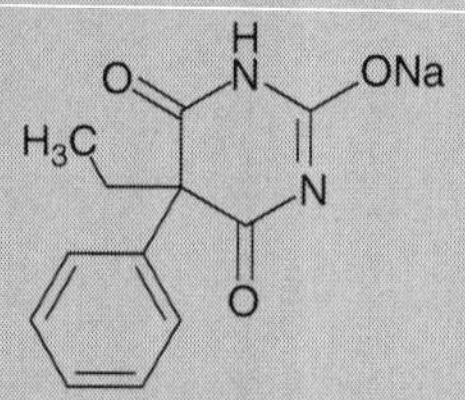
Empirical formula	$C_{12}H_{12}N_2O_3$	$C_{12}H_{11}N_2NaO_3$
Molecular weight	232.2	254.2
Solubility	Very slightly soluble in water Approximately 1 mg/mL (Trissel, 2000)	Freely soluble in carbon dioxide-free water (a small fraction may be insoluble) Solubility in water cited as 1 g/mL (Budavari, 1996)
pK_a	7.6 at 25°C (Connors *et al.*, 1986) Reported elsewhere as 7.41 (McEvoy, 1999) Also reported to have two dissociation constants of 7.3 and 11.8 (Krahl, 1940; Butler *et al.*, 1955)	
Optimum pH stability	Not known	7.5–9.0

Stability profile

Physical and chemical stability

Phenobarbital is a white, crystalline powder or colourless crystals, very slightly soluble in water. It forms water soluble compounds with alkali hydroxides and carbonates and with ammonia (British Pharmacopoeia, 2007).

Phenobarbital sodium is a white, crystalline powder, freely soluble in carbon dioxide-free water (a small fraction may be insoluble). It is hygroscopic in nature (British Pharmacopoeia, 2007).

Phenobarbital is stable in air but is reported to be relatively unstable with regard to hydrolysis (Connors *et al.*, 1986). Little decomposition is seen in the absence of water. In common with most barbiturates, phenobarbital in solution is susceptible to hydroxyl ion attack at more than one site, with initial ring cleavage occurring between the 1,2 or the 1,6 positions, leading to the formation of the diamide or ureide, respectively. The resulting products lose carbon dioxide and undergo further decomposition (Connors *et al.*, 1986).

Gardner and Goyan (1973) have suggested that the ionisation state can affect the route of degradation, with unionised drug being open to cleavage at either site, while ionised phenobarbital could only be cleaved at the 1,6 position.

Degradation of the unionised and mono-anionic species is greater at high pH values, suggesting a specific base-catalysed hydrolysis (Connors *et al.*, 1986). Above pH 9, the hydrolysis is very much accelerated. However, hydrogen ion-catalysed solvolysis has not been found (Connors *et al.*, 1986), and phenobarbital is reported to be quite stable in aqueous solutions of low pH (Chao *et al.*, 1978). At pH 7 and 80°C, the apparent first-order rate constant is $2.6 \times 10^{-6}\ s^{-1}$, corresponding to a half-life of around 74 hours. Bush and Sanders-Smith (1972) have measured the half-life of phenobarbital in 0.1 N potassium hydroxide as 46 hours (at 25°C).

Wyatt and Pitman (1979) state that as phenobarbital is an amide and imide, it is susceptible to hydrolysis and alcoholysis in solutions that contain water and alcohols or sugars.

Various approaches have been tried to improve the aqueous stability of phenobarbital, such as using a mixed solvent of water and an organic solvent (e.g. alcohol, propylene glycol, glycerin or polyethylene glycol) (Connors *et al.*, 1986). Ethanol is reported to have the most stabilising effect, followed by propylene glycol and glycerol (Connors *et al.*, 1986). The stabilising effect of ethanol may be due to a decreased dielectric constant, which slows down the reaction between ions of like charges (i.e. the ionised form of phenobarbital and the hydroxyl ions) (Das Gupta, 1984).

Phenobarbital exhibits polymorphism (Das Gupta, 1984). It is not known whether the polymorph may be altered during the process of making an oral liquid. In theory, changes in polymorphic form could alter particle size (Han *et al.*, 2006).

Phenobarbital is stable to strong oxidizing reagents (e.g. permanganate, dichromate) (Chao *et al.*, 1978). Free phenobarbital is reported to precipitate from phenobarbital sodium at lower pH, depending on the drug concentration. For example, at 3 mg/mL, free phenobarbital precipitates at pH 7.5 or below; at 20 mg/mL, it precipitates at pH 8.6 or below (Trissel, 2000, 2005).

Degradation products

In alkaline solution, hydrolysis followed by loss of carbon dioxide may yield *N*-(aminocarbonyl)-α-ethylbenzeneacetamide, which may further decompose to α-ethylbenzeneacetic acid and urea (Chao *et al.*, 1978).

Stability in practice

There is a wide body of evidence describing the stability of phenobarbital and phenobarbital sodium oral liquids. Various formulation approaches have been applied, each with their own advantages and disadvantages. The formulation chosen must be appropriate for the target population, and the choice of non-active ingredients is crucial.

Dietz *et al.* (1988) have studied the stability of 4 mg/mL phenobarbital as an elixir, emulsion, aqueous solution, and aqueous solution with propylene glycol. The emulsion was made by adding cholesterol 100 mg, phenobarbital 400 mg, span 85 (1 mL) and corn oil (30 mL) to a flask before heating and stirring. Sorenson buffer (65 mL, 2/15 M, pH 5) containing 40% v/v propylene glycol (final pH 5.4) was added to 2 mL of Tween 85. The aqueous and oily components were heated to 70°C and mixed before cooling and adjusting to volume (100 mL) with Sorenson buffer. The emulsion was then shaken, heated to 70°C and homogenised three times using a hand homogeniser. A duplicate suspension without propylene glycol was also prepared, in addition to a blank solution of propylene glycol.

For the preparation of the phenobarbital solution, 438 mg of sodium phenobarbital (equivalent to 400 mg free acid) was diluted to 100 mL with Sorenson buffer (2/15 M, pH 5), with and without the 40% v/v propylene glycol (final pH 5.4).

The elixir for comparison was a commercially available preparation (4 mg/mL, Lilly, Lot 9NN12A). All preparations were stored in amber glass medical bottles and stored for 56 weeks (temperature not stated). A stability-indicating HPLC method was used. The emulsions without propylene glycol showed a fairly rapid decline in drug content, with 95% of the starting concentration remaining after two weeks but only 66% remaining after four weeks. A precipitate also formed, which was not readily dispersed. The drug remaining after four weeks appeared to stabilise, with no more significant loss by week 12.

In contrast, the emulsion with propylene glycol, the propylene glycol solution, and the commercially available elixir retained stability (>90% of starting concentration) throughout the 56 weeks. The aqueous solution retained 90% of the initial content by week 12, but this decreased to 77% by week 55.

Although propylene glycol is generally acknowledged to be safe in small quantities (Ruddick, 1972), larger amounts may cause side-effects such as hyperosmolarity, seizures, lactic acidosis and cardiac toxicity (in both children and adults) (Martin and Finberg, 1970; Arulananatham and Genel, 1978; Cate and Hendrick, 1980; Glasgow *et al.*, 1983). Propylene glycol may also act as an osmotic laxative. Preparations that contain large quantities may not therefore be suitable for neonates and young children.

Unpublished research conducted by Walker (2006) has also confirmed the relative stability of phenobarbital formulations when co-solvents are used. Four formulations of phenobarbital sodium 10 mg/mL were tested at 5, 25 and 37°C:

Formulation A

- Phenobarbital sodium 1 g
- Glycerol BP 45 mL
- water for irrigation to 100 mL.

Formulation B

- Phenobarbital sodium 1 g
- Glycerol BP 45 mL
- Modijul blackcurrant 0.3 g
- water for irrigation to 100 mL.

Formulation C

- Phenobarbital sodium 1 g
- Glycerol BP 35 mL
- water for irrigation to 100 mL.

Formulation D

- Phenobarbital sodium 1 g
- Glycerol BP 35 mL
- Modijul blackcurrant 0.1 g
- water for irrigation to 100 mL.

Formulation B showed precipitation, probably due to the quantity of blackcurrant flavouring added; this formulation was therefore disregarded.

The other formulations showed no visible signs of deterioration over 91 days for formulation A and 54 days for formulations C and D. The starting pH of formulation A was 9.0; the starting pHs of formulations C and D were not recorded. Final pH values were 8.61, 8,84 and 8.41 for formulations A, C and D, respectively.

Accelerated stability testing allowed the calculation of shelf-lives, using a pseudo-first order reaction. Walker uses 97% of the original concentration as a cautious shelf-life indicator, given inherent experimental error. The approximate time taken to reach 97% for each formulation is listed in Table 14.8.

Table 14.8 Time taken to reach a theoretical 97% of starting concentration of phenobarbital (days)

Storage temperature	Formulation A	Formulation C	Formulation D
5°C	189	221	221
25°C	33	22	44
37°C	12	9	12

Microbiological testing was also carried out based on the test outlined in Appendix XVI of the 2004 British Pharmacopoeia ('Efficacy of antimicrobial preservation'). All three formulations tested passed the tests to justify a 14-day in-use shelf-life after opening. Results suggest that formulation A exhibited the most effective preservative system. Walker (2006) suggests that the relatively high concentrations of glycerol confer adequate antimicrobial protection.

A working shelf-life of six months has been assigned for formulation A. Walker (2006) suggests that the liquid should be used within 14 days of opening, as supported by the microbiology testing. The author observes that none of the formulations taste very pleasant, and further work is underway to improve this aspect of the formulations.

Further work on glycerol-based formulae has been carried out by Simon *et al.* (2004), who developed two oral liquid formulations for use in paediatrics. The formulations were as in Table 14.9:

These solutions, containing 5 mg/mL of phenobarbital, were stored in amber glass bottles at 25 and 32°C and assayed over 220 days using a validated HPLC method. After 150 days storage at 25°C, the loss in potency was not more than 5% of the initial concentrations (3.13% for formulation A and 3.65% for formulation B); after 220 days, however, both formulations had lost more than 10% of the initial concentration. Moreover, both formulations had lost more than 10% after 120 days storage at 32°C.

Although the starting pH values are not available, the authors state that these values did not change significantly over the study period. Although details of the HPLC system are given, the validation of the method and the sampling methodology are not available. The number of samples tested and

Table 14.9 Two oral liquid Formulations for use in paediatrics

	Formulation A	Formulation B
Phenobarbital sodium	543 mg	543 mg
Glycerol	20 mL	40 mL
Sorbitol 70%	30 mL	–
10% saccharin sodium aqueous sol.	4 mL	4 mL
Methyl hydroxybenzoate	80 mg	80 mg
Propyl hydroxybenzoate	20 mg	20 mg
Raspberry flavour (Givaudian art. No: 76525-33)	260 mg	260 mg
Lemon flavour (Givaudian art. No: 87017)	260 mg	260 mg
Distilled water (CO_2 free)	to 100 mL	to 100 mL

the accuracy of the assay are also not known. No microbiological stability data are presented; it is not possible to predict whether the preservative system would be effective without knowing the pH.

Twenty-five adults were enrolled in a taste-testing exercise, using six flavours in four combinations. A standardised questionnaire was used to assess preference, and the raspberry and lemon combination used in the formulations was chosen by the majority of the volunteers.

Simon *et al.* (2004) conclude by suggesting that both formulations appear stable for 150 days at 25°C, but that further work is required to check for paediatric approval.

Woods (2001) has described a 3 mg/mL formula taken from the Australian Pharmaceutical Formulary (14th edn):

- phenobarbital sodium 300 mg
- glycerol 20 mL
- sorbitol 30 mL
- parabens (hydroxybenzoates) 0.1%
- water to 100 mL.

However, there are no data to support the 30-day assigned shelf-life (at room temperature when protected from light). Carers should be advised to check for and report any signs of precipitation. Woods also suggests that tablets should not be used to make phenobarbital sodium mixture, as excipients may influence the pH, potentially resulting in precipitation of phenobarbital that would be difficult to detect due to the presence of insoluble tablet excipients. The author suggests that the formulation should be amended to contain ethanol 7.5% if tablets are used as the drug source, as the ethanol may help to maintain the drug in solution. The risks associated with this amount of ethanol will depend upon the individual patient scenario.

Barr and Tice (1957) found that relatively large amounts of alcohol and propylene glycol were required to solubilise phenobarbital at a concentration of 4 mg/mL. This is supported by the work done by Schmitz and Hill (1950), who found that at least 25% alcohol was needed to keep 0.4% phenobarbital in solution. In the presence of 25% glycerin, only 15% alcohol was required for the same purpose.

Barr and Tice (1957) found that the addition of sorbitol solution helped to increase drug solubility. This led to the formula below:

- phenobarbital 4 g
- alcohol 200 mL
- propylene glycol 100 mL
- sorbitol solution 350 mL
- purified water to 1,000 mL.

Orange oil was later added as a flavouring agent, as this was used in the USP formula. This necessitated an increase in the sorbitol concentration. The formulation was amended to:

- phenobarbital 4 g
- orange oil 0.25 mL
- amaranth solution 10 mL
- alcohol 200 mL
- propylene glycol 100 mL
- sorbitol solution 600 mL
- purified water to 1,000 mL.

The pH of the bright red finished product is reported to be 6.2–6.4. Higher concentrations of orange oil led to turbidity. Barr and Tice (1957) give further data on the solubility of phenobarbital in alcohol–propylene glycol–sorbitol solution and purified water mixes.

A similar formulation approach has been published by Accordino *et al.* (1985), who carried out a preliminary study of the stability of the following elixir formula:

- phenobarbital 400 mg
- imitation butterscotch essence 1 mL
- alcohol 95% 5 mL
- glycerol 50 mL
- sorbitol solution 70% 35 mL
- distilled water, freshly collected and cooled to 100 mL.

This formula had been suggested after concerns were raised with regard to the palatability of the formula previously published by Barr and Tice (1957). Five retained samples of the above formulation were stored in amber glass bottles in an air-conditioned pharmacy at 22–24°C, and analysed using a specific HPLC method over a period of 3–32 months.

Details of the dilutions and analysis are supplied. Each of the samples tested retained between 97% and 104% of the starting drug concentration, suggesting minimal (if any) degradation. The authors suggest a shelf-life of at least 2 years if stored at 22–24°C, but do acknowledge the need for further testing, including accelerated stability testing. Microbiological analysis was not carried out, and the overall number of samples taken was relatively low. Details of the sampling procedures are not stated.

Wyatt and Pitman (1979) studied the stability of three formulations:

Formulation 1

Initial pH 7.75:

- phenobarbital 300 mg
- alcohol (90%) 40 mL

- glycerol 30 mL
- aromatic syrup 20 mL
- purified water to 100 mL.

Formulation 2

Initial pH 4.6:

- phenobarbital 300 mg
- alcohol (90%) 10 mL
- glycerol 30 mL
- aromatic syrup 20 mL
- syrup 30 mL
- purified water to 100 mL.

Formulation 3

Initial pH 4.6:

- sodium phenobarbital 300 mg
- green colour 2.4 mL
- aromatic syrup 20 mL
- syrup to 100 mL.

The formulations were stored in 100 mL clear glass bottles with metal screw tops (with plastic liners), and stored at 30 or 40°C. Aliquots were subsequently removed for testing and diluted with sodium hydroxide to pH 11.5 for spectrophotometric analysis. The assay was estimated to be accurate to ±2%. No attempt was made to quantify rate constants as the pH was not controlled, and the time period studied equated to less than one half-life.

The shelf-lives (t_{90}) at 30°C for the three formulations were estimated to be 54 days (formulation 1), 102 days (formulation 2) and 104 days (formulation 3). Degradation was found to be increased for all formulations when stored at 40°C. However, even at 40°C, all three formulations maintained shelf-lives in excess of 30 days.

Formulation 3 degraded the quickest, probably due to base-catalysis. This is in line with normal behaviour of amides, imides and other condensation products of carboxylic acids (Wyatt and Pitman, 1979).

Wyatt and Pitman (1979) also report carrying out similar experiments at 4°C. However, a crystalline precipitate was found in each formulation, being most pronounced in formulation 2. Aliquots of supernatant contained less drug than expected, suggesting that the precipitate contained phenobarbital. Shaking of the mixtures led to the drug redissolving.

The authors conclude that each formulation could be stored at 30 or 40°C, protected from light. They also suggest that 75% of the alcohol in formulation 1 with syrup might produce a more palatable formulation for children without affecting the chemical stability of phenobarbital. The storage of phenobarbital formulations in the refrigerator may lead to the precipitation of phenobarbital.

Several 'Specials' suppliers manufacture phenobarbital or phenobarbital sodium preparations, including the following formula supplied by P.S. Bendell of the Torbay Pharmacy Manufacturing Unit, Pharmacy Department Torbay Hospital (P.S. Bendall, personal communication, 2006):

- phenobarbital sodium powder 150 g
- sorbitol 20 kg
- sodium saccharin 75 g
- orange oil (terpeneless) 25 mL
- lemon oil (terpeneless) 5 mL
- methyl hydroxybenzoate 31.5 g
- absolute alcohol 1,000 mL
- freshly distilled water to 50 L.

NHS Quality Control Laboratory analysis has been carried out on this and similar formulae using a stability-indicating HPLC analysis. Data on file suggest that a shelf-life of 12 months at room temperature is justifiable (F. Haines-Nutt, Torbay Quality Control, Torbay Hospital, personal communication, 2006). The appropriateness of the excipients should be assessed by an appropriately competent person before such a product is purchased. The preparation is relatively low in alcohol (2% v/v), sucrose-free and chloroform-free, but does contain hydroxybenzoate; this is a widely used antimicrobial preservative in cosmetics and oral and topical pharmaceutical formulations. Concern has been expressed over its use in infant parenteral products as bilirubin binding may be affected (Rieger, 2003). This is potentially hazardous in hyperbilirubinaemic neonates (Loria *et al.*, 1976). However, there is no proven link to similar effects when used in oral products. The World Health Organization has suggested an estimated total acceptable daily intake for methyl-, ethyl- and propyl-hydroxybenzoates at up to 10 mg/kg body weight (WHO/FAO, 1974).

Alcohol-free and chloroform-free 'Specials' are also available, such as the one made by St Mary's Pharmaceutical Unit (V. Fenton-May, St Mary's Pharmaceutical Unit, personal communication, 2006). A 10 mg/mL phenobarbital sodium mixture is formulated as follows:

- phenobarbital sodium 50 g
- orange tincture 78 mL
- Syrup BP 870 mL
- glycerol 1,218 mL
- preservative solution to 5 L.

Full details of the formulation are available from the unit. Stability data are available for a previous formulation that contained chloroform, but the quality controller is satisfied that the mixture is stable, provided that the pH is alkaline (typically pH 8.5–9) (V. Fenton-May, personal communication, 2006).

Commercial 'Specials' manufacturers are also known to produce alcohol-free batch products, such as a 3 mg/mL phenobarbital sodium preparation that uses powder as the starting material (A. Twitchell, Nova Laboratories Ltd, personal communication, 2006). The exact formula may be discussed with the manufacturer, but is known to contain a raspberry flavour, glycerol and hydroxybenzoate preservative. No xanthan gum is used in the preparation. A shelf-life of six months has been allocated, with a stability-indicating HPLC assay reported to demonstrate minimal degradation of phenobarbital sodium after 12 months.

Related products

A 10% w/v phenobarbital sodium aqueous solution showed 7% decomposition in four weeks at 20°C (Larsen and Jensen, 1970). Nahata and Hipple (1986) also studied the stability of phenobarbital sodium 10 mg/mL in sodium chloride 0.9%. At 4°C, the solution was physically compatible with no loss of phenobarbital over 28 days. The pH of the diluted product was 8.5. However, Woods (2001) has suggested that different brands and concentrations may result in precipitation and that therefore the physical stability of such dilutions should be closely monitored.

It should be noted that some brands of injection will be formulated with ethanol and/or propylene glycol (Woods, 2001).

Bioavailability data

There are no data available specific to the extemporaneous preparations.

References

Accordino A, Chambers RA, Don JL (1985). Preliminary study of the stability of phenobarbitone elixir PMH. *Aust J Hosp Pharm* 15: 229–230.

Arulananatham K, Genel M (1978). Central nervous system toxicity associated with the ingestion of propylene glycol. *J Pediatr* 93: 515–516.

Barr M, Tice LF (1957). An improved formula for phenobarbital elixir and pentobarbital elixir. *Am J Pharm* 129: 332–342.

British Pharmacopoeia (2007). *British Pharmacopoeia*, Vol. 2. London: The Stationery Office, pp. 1618–1620.

Budavari S, ed. (1996). *The Merck Index*, 12th edn. Rahway, NJ: Merck and Co.

Bush MT, Sanders-Smith E (1972). In: Woodbury DM, Penry JK, Schmidt RP, eds. *Antiepileptic Drugs*. New York: Raven Press, pp. 292–302.

Butler TC, Ruth JM, Tucker GF (1955). The second ionisation of 5,5-disubstituted derivatives of barbituric acid. *J Am Chem Soc* 77: 1486–1488.

Cate J, Hedrick R (1980). Propylene glycol intoxication and lactic acidosis. *N Engl J Med* 303: 1237.

Chao MKC, Albert KS, Fusari SA (1978). Phenobarbital. In: Florey K, ed. *Analytical Profiles of Drug Substances*, Vol. 7. London: Academic Press, pp. 359–399.

CHIP3 Regulations (2002). Chemicals (Hazard Information and Packaging for Supply) Regulations. SI 2002/1689.

Connors KA, Amidon GL, Stella ZJ (1986). *Chemical Stability of Pharmaceuticals: A Handbook for Pharmacists*, 2nd edn. New York: Wiley Interscience.

Das Gupta V (1984). Effect of ethanol, glycerol, and propylene glycol on the stability of phenobarbital sodium. *J Pharm Sci* 73: 1661–1662.

Dietz NJ, Cascella PJ, Houglum JE, Chappell GS, Sieve RM (1988). Phenobarbital stability in different dosage forms: alternatives for elixirs. *Pharm Res* 5: 803–805.

Gardner LA, Goyan JE (1973). Mechanism of phenobarbital degradation. *J Pharm Sci* 62: 1026.

Glasgow A, Boeckx R, Miller M, MacDonald M, August G, Goodman S (1983). Hyperosmolarity in small infants due to propylene glycol. *Pediatrics* 72: 353–355.

Han J, Beeton A, Long PF, Wong I, Tuleu C (2006). Physical and microbiological stability of an extemporaneous tacrolimus suspension for paediatric use. *J Clin Pharm Ther* 31: 167–172.

Krahl ME (1940). The effect of variation in ionic strength and temperature of the apparent dissociation constants of thity substituted barbituric acids. *J Phys Chem* 44: 449.

Larsen SS, Jensen VG (1970). Studies on stability of drugs in frozen systems II. The stabilities of hexobarbital sodium and phenobarbital sodium in frozen aqueous solutions. *Dan Tidsskr Farm* 44: 21–31.

Loria CJ, Excehverria P, Smith AL (1976). Effect of antibiotic formulations in serum protein: bilirubin interaction of newborn infants. *J Pediatr* 89: 479–482.

Lowey AR, Jackson MN (2008). A survey of extemporaneous dispensing activity in NHS trusts in Yorkshire, the North-East and London. *Hosp Pharmacist* 15: 217–219.

Martin G, Finberg L. Propylene glycol: a potentially toxic vehicle in liquid dosage form. *J Pediatr* 77: 877–878.

McEvoy J, ed (1999). *AHFS Drug Information*. Bethesda, MD: American Society of Health System Pharmacists.

Nahata MC, Hipple TF (1986). Stability of phenobarbitone sodium diluted in 0.9% sodium chloride injection. *Am J Hosp Pharm* 43: 384–385.

Rieger MM (2003). Ethylparabens. In: Rowe RC, Sheskey PJ, Weller PJ, eds. *Handbook of Pharmaceutical Excipients*, 4th edn. London: Pharmaceutical Press and Washington DC: American Pharmaceutical Association.

Ruddick J (1972). Toxicology, metabolism, and biochemistry of 1,2-propanediol. *Toxicol Appl Pharmacol* 21: 102–111.

Schmitz RE, Hill JS (1950). The stability of phenobarbital in hydro-alcoholic and glycerol-hydro-alcoholic solution. *J Am Pharm Assoc* (Practical Pharmacy Edition) 2: 500–501.

Simon C, Griffiths W, Ing H, Sadeghipour F, Bonnabry P. (2004). Development, stability and flavour acceptability of two oral liquid formulations of phenobarbital for use in paediatrics. Presented at the 9th Congress of the European Association of Hospital Pharmacists, Palacio de Congresos, Sevilla, Spain 17–19th March, 2004. www.hcuge.ch/Pharmacie/rd/posters.htm

Trissel LA, ed. (2000). *Stability of Compounded Formulations*, 2nd edn. Washington DC: American Pharmaceutical Association.

Trissel LA (2005). *Handbook on Injectable Drugs*, 13th edn. Bethesda, MD: American Society of Health System Pharmacists.

Walker RDS (2006). The formulation and stability of phenobarbital liquid. Submitted as research project, Module 8. MSc in Pharmaceutical Technology and Quality Assurance, September 2006. University of Leeds.

WHO/FAO (1974). Toxicological evaluation of certain food additives with a review of general principles and of specifications. Seventeenth report of the Joint FAO/WHO Expert Committee on Food Additives. World Health Organ. Tech. Rep. Ser. No. 539.

Woods DJ (2001). *Formulation in Pharmacy Practice. eMixt, Pharminfotech*, 2nd edn. Dunedin, New Zealand. www.pharminfotech.co.nz/emixt (accessed 18 April 2006).

Wyatt KA, Pitman IH (1979). Stability of elixirs and mixtures of phenobarbitone. *Aust J Hosp Pharm* 9: 37–38.

Phenoxybenzamine hydrochloride oral liquid

Risk assessment of parent compound

A survey of clinical pharmacists suggested that phenoxybenzamine has a moderate to narrow therapeutic index, with the potential to cause serious morbidity if inaccurate doses are administered (Lowey and Jackson, 2008). Current formulations used at NHS trusts were generally of a moderate technical risk, based on the number and/or complexity of manipulations and calculations. Phenoxybenzamine hydrochloride oral liquid is therefore regarded as a high-risk extemporaneous item (Box 14.35).

Summary

Phenoxybenzamine hydrochloride oral suspension is currently prepared in a limited number of NHS centres (Lowey and Jackson, 2008). There is a limited

Box 14.35 *Preferred formula*

Using a typical strength of 1 mg/mL as an example:

- phenoxybenzamine 10 mg capsules × 3
- Ora-Plus 15 mL
- Ora-Sweet SF to 30 mL.

Method guidance

Open the capsules and grind the contents to a fine, uniform powder in a pestle and mortar. A small amount of Ora-Plus may be added to form a paste, before adding further portions of Ora-Plus up to 50% of the final volume. Transfer to a measuring cylinder. The Ora-Sweet SF can be used to wash out the pestle and mortar before making the suspension up to 100% volume. Transfer to an amber medicine bottle.

Shelf-life

3 days refrigerated in amber glass.

body of data with regard to liquid formulations. The drug is known to be relatively unstable with optimum stability shown to be at pH<3 in solution (Adams and Kostenbauder, 1985). A suspending vehicle with a low pH has been used in an attempt to minimise drug loss, and a short shelf-life has been assigned. Alternative treatments or methods of administering the drug should also be considered.

Future research may focus on the possibility of formulating the drug at higher pHs. Although this will make the drug come out of solution, it may help to improve overall stability. The use of a suspending agent with a pH above the pK_a of the drug may prove useful. Further work is required. Until this work is carried out, the shelf-life for the above formulation is restricted to 3 days.

Clinical pharmaceutics

Points to consider

- At typical paediatric concentrations (e.g. 1 mg/mL), the drug is likely to be completely in solution and therefore open to aqueous degradation mechanisms.
- Capsules can reportedly be opened and contents administered (Guy's and St Thomas' and Lewisham Hospitals, 2005). However, phenoxybenzamine hydrochloride in powder form should not be allowed to come into contact with eyes or skin, as it may cause irritation (Moffat, 1986). Phenoxybenzamine has also been characterised as a potential carcinogen (Budavari, 1996), and has been found to be carcinogenic in rodents (Martindale, 2005). Therefore, a local risk assessment must be carried out before any capsules are opened. Appropriate safety equipment should be available.
- Possible alternative practices include opening the capsule(s) and mixing with water for each individual dose. However, this practice carries health and safety concerns. Suitable protective equipment should be worn, as contact sensitisation may occur (BNF, 2007).

Technical information

Structure

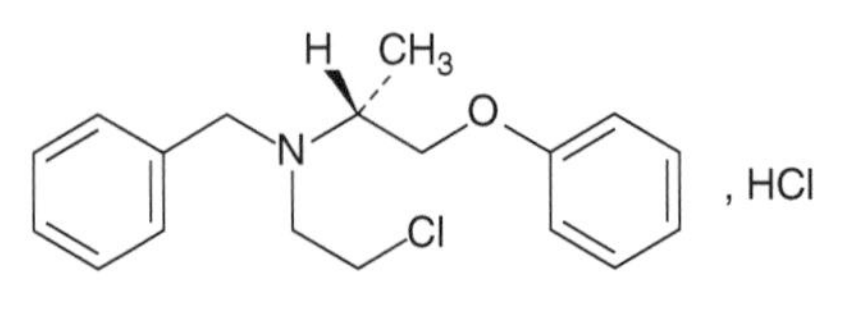

and enantiomer

Empirical formula

$C_{18}H_{22}ClNO,HCl$ (British Pharmacopoeia, 2007).

Molecular weight

340.3 (British Pharmacopoeia, 2007).

Solubility

40 mg/mL (Trissel, 2000).

pK_a

Around 5.0–5.1 (Adams and Kostenbauder, 1985).

Optimum pH stability

Less than 3 (Adams and Kostenbauder, 1985).

Stability profile

Physical and chemical stability

Phenoxybenzamine hydrochloride is sparingly soluble in water (Trissel, 2000). However, given the strength of typical paediatric formulations (1 mg/mL), the drug is likely to be almost completely in solution, and therefore susceptible to degradation. Phenoxybenzamine is also soluble in ethanol (167 mg/mL) and propylene glycol (Trissel, 2000).

Phenoxybenzamine hydrochloride has been reported to be stable in acidified non-aqueous solutions but unstable in neutral or alkaline solutions (Martindale, 2005). However, this may be due to insolubility. The formulation of phenoxybenzamine at higher pHs as a suspension warrants further investigation.

Adams and Kostenbauder (1985) also suggest that the stability of phenoxybenzamine hydrochloride in aqueous solution is pH dependent, with the drug showing better stability when pH values are less than 3.

Investigating the degradation kinetics of phenoxybenzamine hydrochloride in 1:1 absolute ethanol–water mixtures in the pH range 1.5–8.0, Adams and Kostenbauder (1985) have proposed that the drug degrades through a pH-dependent cyclisation reaction. At a pH less than or equal to 4.5, the reaction may be suppressed because a large proportion of the drug in solution has an additional proton (p$K_a = 5.0$ in the solvent system) (Lim *et al.*, 1997).

Degradation products

No information available.

Stability in practice

Lim *et al.* (1997) investigated the stability of 19 aqueous and non-aqueous formulations of phenoxybenzamine solution (2 mg/mL). The formulations consisted of:

1. alcohol
2. propylene glycol
3. glycerin
4. 0.15% citric acid in water
5. 0.5% propylene glycol and 0.15% citric acid in water
6. 1.0% propylene glycol and 0.15% citric acid in water
7. 5.0% propylene glycol and 0.15% citric acid in water
8. 10% propylene glycol and 0.15% citric acid in water
9. 20% propylene glycol and 0.15% citric acid in water
10. 1.0% alcohol and 0.15% citric acid in water
11. 3.0% alcohol and 0.15% citric acid in water
12. 5.0% alcohol and 0.15% citric acid in water
13. 1.0% alcohol, 1% propylene glycol and 0.15% citric acid in water
14. 5.0% alcohol, 1% propylene glycol and 0.15% citric acid in water
15. 25% syrup, 1% propylene glycol and 0.15% citric acid in water
16. 50% syrup, 1% propylene glycol and 0.15% citric acid in water
17. 75% syrup, 1% propylene glycol and 0.15% citric acid in water
18. 1% propylene glycol and 0.15% citric acid
19. propylene glycol (to be diluted 1:4 v/v with syrup).

(The syrup preparations contained 66.7% food-grade sucrose in distilled water with 0.08% w/v methylparaben and 0.02% w/v propylparaben as preservatives.)

Phenoxybenzamine hydrochloride was obtained as drug powder. A traditional pestle and mortar approach was used. Drug concentration for formulations 1–18 was 2 mg/mL with a volume of 10 mL. For the stock solution (formulation 19), a 10 mg/mL concentration was chosen, with a volume of 20 mL. More details of the preparation process are stated.

Two variations of preparation 18 were prepared, using phenoxybenzamine hydrochloride from dibenyline capsules, and using pharmaceutical syrup rather than syrup prepared with food-grade sucrose in the vehicle.

All formulations were made in triplicate and stored in amber medicine bottles at 4°C, except for formulation 1, which was stored in foil-covered cylinders with stoppers due to volatility concerns.

Samples of 1 mL were removed after shaking, on days 0, 1, 2, 3 and 4. Formulations 5–8 and 10–18 were also sampled on days 7 and 10. Formulation 19 was sampled on days 0, 2, 4, 7, 10, 14 and 30. A stability-indicating HPLC method was used to assay drug concentration. There were no apparent changes in turbidity, colour or odour for any of the formulations during their specified storage periods.

In general, degradation was found to be rapid: of the 19 formulations, only six retained 90% or more of the starting concentration at 7 days. However, formulations numbers 1 to 4 were not tested after day 4, at which time both formulations 1 and 2 retained more than 90% of the starting concentration. Of the remaining formulations, numbers 6, 7, 10, 11, 19A and 19B retained more than 90% of the original drug concentration. (Formulation 19A = undiluted stock solution; 19B = stock solution 1 hour after dilution 1:4 v/v with syrup.)

Solutions prepared in syrup were extremely unstable, with significant degradation occurring very rapidly after compounding, with the baseline concentration check showing only 1.76 mg/mL of the expected 2 mg dose (88%). The authors suggest that the degradation may be catalysed by alkoxide ions derived from the hydroxyl group in the sucrose molecule.

The initial pH of formulations 1–19 varied between 1.7 and 3.0. Final pH readings did not exceed 3.5 and did not change by more than 0.8 units. As previous reports have postulated that rapid decomposition occurs at pH values over 4.5 (Nedergard, 1969; Adams and Kostenbauder, 1985), Lim *et al.* (1997) suggest that pH may not be the sole factor influencing the stability of phenoxybenzamine hydrochloride. Despite using citric acid to acidify the formulations, phenoxybenzamine hydrochloride was still found to be very unstable in aqueous media.

Other factors that may have affected stability were the polarity of the solvent and the hygroscopic nature of the drug during storage.

Lim *et al.* (1997) also formulated a 10 mg/mL stock solution of phenoxybenzamine hydrochloride in propylene glycol that was shown to be stable for 30 days when stored at 4°C. The authors suggest the dilution of this stock solution in syrup immediately before administration, in order to reduce the concentration of propylene glycol delivered. Even after dilution with syrup, the final concentration was 20% (v/v). Although propylene glycol is generally acknowledged to be safe in small quantities (Ruddick, 1972), larger amounts may cause side-effects such as hyperosmolarity, seizures, lactic acidosis and cardiac toxicity (in both children and adults) (Martin and Finberg, 1970; Arulananatham and Genel, 1978; Cate and Hedrick, 1980; Glasgow *et al.*, 1983). Propylene glycol and polyethylene glycol may also act as osmotic laxatives.

Although the addition of alcohol and propylene glycol may help by reducing the polarity of the aqueous system, both of these agents should be avoided where possible in paediatric formulations, due to their potentially toxic nature (Martin and Finberg, 1970; American Academy of Pediatrics Committee on Drugs, 1984).

No significant differences were seen when comparing drug obtained from powder or from opened capsules.

Other related preparations

Nedergaad (1969) studied the degradation of a parenteral solution containing: phenoxybenzamine chloride 5 g, 99% ethanol (197.5 g), propylene glycol (260.0 g) and 1 N hydrochloric acid (0.5 g). Final pH of the solution was approximately 2.5. The ethanol–propylene glycol medium is used due to the poor solubility of phenoxybenzamine in water.

Autoclaving the solution in clear glass vials at 120°C for 20 minutes had no effect on the drug content. Analysis of the vials used UV-spectrometric data every second day for the first two weeks, then weekly. Thin-layer chromatography was also carried out to add further evidence.

Negergaard (1969) suggests that the solution is 'stable' for around two and a half months, and that the solution did not need to be protected from light or stored at a lower temperature. However, no data are given, and the author acknowledges that the UV technique failed to detect the degradation, as the major degradant gave a curve similar to that of phenoxybenzamine itself.

The stability was reportedly pH-dependent, with the drug being most stable at pH 2–3 and decomposition much more rapid at pH above 5. The limitations of the type of technology used in this work, including the failure and the UV methodology to detect the major degradant should be noted.

Bioavailability data

No data available.

References

Adams WP, Kostenbauder HB (1985). Phenoxybenzamine stability in aqueous ethanolic solution I: application of potentiometric pH stat analysis to determine kinetics. *Int J Pharm* 25: 293–312.

American Academy of Pediatrics Committee on Drugs (1984). Ethanol in liquid preparations intended for children. *Pediatrics* 73: 405–407.

Arulananatham K, Genel M (1978). Central nervous system toxicity associated with the ingestion of propylene glycol. *J Pediatr* 93: 515–516.

British Pharmacopoeia (2007). *British Pharmacopoeia*, Vol. 2. London: The Stationery Office, pp. 1623–1624.

Budavari S, ed. (1996). *The Merck Index*, 12th edn. Rahway, NJ: Merck and Co.

Cate J, Hedrick R (1980). Propylene glycol intoxication and lactic acidosis. *N Engl J Med* 303: 1237.

Glasgow A, Boeckx R, Miller M, MacDonald M, August G, Goodman S (1983). Hyperosmolarity in small infants due to propylene glycol. *Pediatrics* 72: 353–355.

Guy's and St Thomas' and Lewisham Hospitals (2005). *Paediatric Formulary*. London: Guy's and St Thomas' and Lewisham NHS Trust.

Joint Formulary Committee (2010). *British National Formulary*, No. 59. London: British Medical Association and Royal Pharmaceutical Society of Great Britain.

Lim L-Y, Tan L-L, Chan EWY, Yow KL, Chan SY, Ho PCL (1997). Stability of phenoxybenzamine hydrochloride in various vehicles. *Am J Health Syst Pharm* 54: 2073–2078.

Lowey AR, Jackson MN (2008). A survey of extemporaneous dispensing activity in NHS trusts in Yorkshire, the North-East and London. *Hosp Pharmacist* 15: 217–219.

Martin G, Finberg L (1970). Propylene glycol: a potentially toxic vehicle in liquid dosage forms. *J Pediatr* 77: 877–878.

Moffat AC, Osselton MD, Widdop B (2003). *Analysis of Drugs and Poisons*, 3rd edn. London: Pharmaceutical Press. ISBN 978 0 85369 4731.

Nedergaad M (1969). Phenoxybenzamine in solutions. *J Hosp Pharm* 27: 174–176.

Ruddick J (1972). Toxicology, metabolism, and biochemistry of 1,2-propanediol. *Toxicol Appl Pharmacol* 21: 102–111.

Sweetman SC (2009). *Martindale: The Complete Drug Reference*, 36th edn. London: Pharmaceutical Press.

Trissel LA, ed. (2000). *Stability of Compounded Formulations*, 2nd edn. Washington DC: American Pharmaceutical Association.

Potassium acid phosphate oral liquid

Synonyms

Potassium dihydrogen phosphate oral liquid and monobasic potassium phosphate oral liquid.

Risk assessment of parent compound

A risk assessment survey completed by clinical pharmacists suggested that potassium phosphate has a medium therapeutic index. However, inaccurate doses were thought to be associated with significant morbidity (Lowey and Jackson, 2008). Current formulations used at NHS trusts were generally of a moderate technical risk, based on the number and/or complexity of manipulations and calculations. Potassium acid phosphate is therefore regarded as a high-risk extemporaneous preparation (Box 14.36).

Summary

Current best practice appears to involve manipulation of the licensed injectable preparation, which is drawn up from ampoules and transferred to an appropriately sized amber glass bottle (Lowey and Jackson, 2008).

Box 14.36 *Preferred formula*

Using a typical strength of potassium acid phosphate 13.6% (1 mmol of potassium and 1 mmol of phosphate per 1 mL), as an example:

- potassium acid phosphate 13.6% injection 5 × 10 mL ampoules.

Method guidance

Open the glass ampoules with care, following local procedures. Use a filter needle or filter straw with a 50 mL syringe to draw up the contents of the five ampoules. Slowly transfer the contents of the syringe into a 50 mL amber medicine bottle.

Shelf-life

7 days refrigerated in amber glass (unpreserved).

The unpreserved nature of the presentation restricts shelf-life to 7 days in the fridge.

Oral potassium phosphate solutions are available from 'Specials' manufacturers. The formulations and their supporting evidence should be assessed by a suitably competent pharmacist before use. The purchase of a 'Special' product should be considered in preference to an extemporaneous preparation under normal circumstances.

Clinical pharmaceutics

Points to consider

- The solution is unpreserved and a restricted shelf-life of 7 days in the fridge is therefore imposed.
- There are no data available with regard to oral bioavailability.

Other notes

- Potassium acid phosphate (potassium dihydrogen phosphate) 13.6% contains 1 mmol/mL potassium and 1 mmol/mL phosphate.

Technical information

Empirical formula

KH_2PO_4 (British Pharmacopoeia, 2007).

Molecular weight

136.1 (British Pharmacopoeia, 2007).

Solubility

Freely soluble in water (British Pharmacopoeia, 2007). Soluble in approximately 4.5 parts water (Budavari, 1999).

Optimum pH stability

Not relevant.

Stability profile

Physical and chemical stability

Potassium acid phosphate is a white, crystalline powder or colourless crystals, freely soluble in water and practically insoluble in alcohol

(British Pharmacopoeia, 2007). Each gram represents approximately 7.3 mmol of potassium and of phosphate (Martindale, 2005).

Potassium acid phosphate is chemically stable in solution. A 1% solution in water has a pH of about 4.5 (Martindale, 2005). Potassium acid phosphate powder should be stored in airtight containers (Martindale, 2005).

Degradation products

No information.

Stability in practice

No data.

Bioavailability data

No information.

References

British Pharmacopoeia (2007). *British Pharmacopoeia*, Vol. 2. London: The Stationery Office, pp. 1696–1697.

Budavari S, ed. (1996). *The Merck Index*, 12th edn. Rahway, NJ: Merck and Co.

Lowey AR, Jackson MN (2008). A survey of extemporaneous dispensing activity in NHS trusts in Yorkshire, the North-East and London. *Hosp Pharmacist* 15: 217–219.

Sweetman SC (2009). *Martindale: The Complete Drug Reference*, 36th edn. London: Pharmaceutical Press.

Primidone oral liquid

Risk assessment of parent compound

A risk assessment survey has suggested that primidone has a medium therapeutic index with the potential to cause moderate to severe morbidity if inaccurate doses are administered as an extemporaneous preparation (Lowey and Jackson, 2008). The technical risk associated with the preparation of primidone suspensions was regarded as moderate, based on the number and complexity of calculations and/or physical manipulations. Primidone oral liquid is therefore regarded as a high-risk extemporaneous preparation (Box 14.37).

Summary

A primidone oral suspension was previously available commercially. It is possible that this suspension (or a similar one) may be marketed again by a specialist company in the near future.

Box 14.37 *Preferred formula*

Using a typical strength of 50 mg/mL as an example:

- primidone tablets 250 mg × 20
- xanthan gum 0.4% ('Keltrol') or similar to 100 mL.

Method guidance

Tablets can be ground to a fine, uniform powder in a pestle and mortar. A small amount of 'Keltrol' may be added to form a paste, before adding further portions of up to approximately 75% of the final volume. Transfer to a measuring cylinder. The remaining 'Keltrol' can be used to wash out the pestle and mortar before making the suspension up to 100% volume. Transfer to an amber medicine bottle.

Shelf-life

28 days in amber glass at room temperature. **Shake the bottle.**

Note: 'Keltrol' xanthan gum formulations vary slightly between manufacturers, and may contain chloroform and ethanol. The individual specification should be checked by the appropriate pharmacist.

A recent survey of NHS trusts showed the use of 'Keltrol'-based formulae (Lowey and Jackson, 2008). Other suspending agents may be equally appropriate, given the poor solubility of primidone. Although no data are available to support the shelf-life above, theory would suggest minimal drug degradation is likely to occur over 28 days. Furthermore, Mysoline oral suspension was previously marketed with a long shelf-life. Therefore, there would appear to be evidence that primidone can be successfully formulated as an oral liquid.

Primidone may be a candidate with a longer shelf-life to buy from a 'Specials' manufacturer. The purchaser should take steps to scrutinise and approve the formula and stability data provided by the manufacturer.

Clinical pharmaceutics

Points to consider

- At a typical strength of 50 mg/mL, almost all of the drug will be in suspension. **Shake the bottle to ensure uniformity of drug throughout the suspension.**
- Two polymorphic forms of primidone have been reported. However, no differences have been seen in dissolution or solubility properties (Al-Badr and El-Obeid, 1988).
- 'Keltrol' is made to slightly different specifications by several manufacturers, but typically contains ethanol and chloroform. The exact specification should be assessed before use by a suitably competent pharmacist. Ethanol is a CNS depressant. Chloroform has been classified as a class 3 carcinogen (i.e. substances that cause concern owing to possible carcinogenic effects but for which available information is not adequate to make satisfactory assessments) (CHIP3 Regulations, 2002).
- Scored tablets may be crushed/dispersed in water for full-tablet or half-tablet doses (C. Norton, Lead Pharmacist for Clinical Trials and Adult Psychiatry and WMLRN Medicines for Children Pharmacist Adviser, Birmingham Children's Hospital, personal communication, 2007).
- Primidone is odourless or almost odourless, with a slightly bitter taste (Lund, 1994).
- There are no bioavailability data available for primidone extemporaneous formulations.

Technical information

Structure

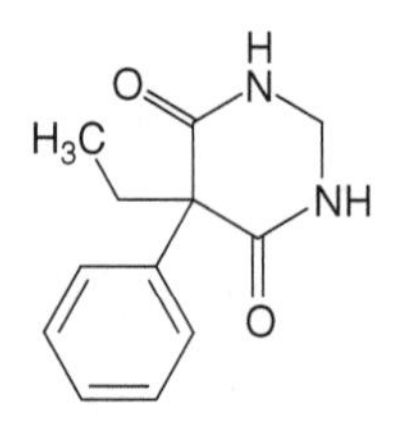

Empirical formula

$C_{12}H_{14}N_2O_2$ (British Pharmacopoeia, 2007).

Molecular weight

218.3 (British Pharmacopoeia, 2007).

Solubility

Solubility in water is reported to be 500 micrograms/mL at 'room temperature' (Daley, 1973) and 600 micrograms/mL at 37°C (Lund, 1994).

pK_a

Not known.

Optimum pH stability

Not known.

Stability profile

Physical and chemical stability

Primidone is a white or almost white, crystalline powder, very slightly soluble in water and slightly soluble in alcohol (British Pharmacopoeia, 2007).

Daley (1973) described primidone as 'quite stable', as predicted from the absence of reactive functional groups, low solubility and high melting point and suggested that 'no reports of degradation under ordinary conditions could be found, although the drug will react under drastic chemical conditions'.

Primidone powder should be preserved in well-closed containers (United States Pharmacopeia, 2003). A monograph for primidone oral suspension in the United States Pharmacopeia (50 mg/mL) states that it should be preserved in tight, light-resistant containers (Lund, 1994; United States Pharmacopeia, 2003). The final pH of the suspension should be between 5.5 and 8.5 (United States Pharmacopeia, 2003).

An equilibrium solubility of 564 micrograms/mL was determined for primidone in distilled water at pH 5.5 and 37°C. This is reported to decrease to 172 micrograms/mL in 500 mg/mL citric acid (Summers and Enever, 1976; Lund, 1994). This poor solubility should limit the amount of drug that is susceptible to aqueous degradation mechanisms.

Two polymorphs of primidone have been reported. However, no differences were recorded between either the solubility or dissolution rate of the two forms in distilled water, pH 5.5 at 37°C (Summers and Enever, 1976; Lund, 1994).

Schott and Royce (1985) have investigated the effect of non-ionic surfactants on aqueous primidone suspensions.

Degradation products

Include phenobarbital (phenobarbitone) (El Bayoumi *et al.*, 1999).

Stability in practice

The Mysoline suspension was formally marketed by Wyeth-Aventis, then by Astra Zeneca. It was reported to carry an expiration date of 5 years (Al-Badr and El-Obeid, 1988). Mysoline tablets are now marketed by Acorus Therapeutics.

A formula for primidone oral suspension (50 mg/mL) is stated in the Pharmaceutical Codex (Lund, 1994), with a diluent containing carmellose sodium (50 cP) 1% w/v, sucrose 20% w/v, methyl-hydroxybenzoate 0.15% w/v, propyl-hydroxybenzoate 0.015% w/v, in freshly boiled and cooled water to 100% (Lund, 1994).

Other excipients that have been reported to have been used in primidone suspensions include aluminium magnesium silicate, ammonia solution (diluted), cetosteryl alcohol/ethylene oxide condensate, citric acid, methylcellulose, quinoline yellow (E104), saccharin sodium, sodium alginate, sodium benzoate, sodium citrate, sodium hypochlorite solution, sorbic acid, sorbitan monolaurate, Sunset Yellow FCF (E110), syrup, vanilla flavouring (Lund, 1994).

Bioavailability data

The bioequivalence of generic and branded Mysoline tablets has been questioned. Wylie *et al.* (1987) reported an increase in seizure frequency when changing patients to the generic tablets (Bolar, USA). Variations have been documented in plasma concentrations of phenobarbitone (an active metabolite of primidone) in 12 patients taking two different batches of Mysoline tablets (Bielman *et al.*, 1974). The two batches displayed different *in vitro* dissolution and disintegration times. However, plasma concentrations of primidone itself were not significantly different (Lund, 1994).

The tablet formulation was changed following publication of the US Pharmacopeia (2003), which contained new guidelines for tablet dissolution. Meyer *et al.* (1998) compared the *in vitro* dissolution and *in vivo* bioavailability of the old tablet formulation, the new tablet formulation, and the Mysoline suspension. Although the new tablet formulation was found to dissolve much more quickly than the old formulation *in vitro*, there was no significant difference observed between the three formulations in terms of overall bioequivalence. The authors concluded that the *in vitro* dissolution test for immediate release dosage forms was not predictive of *in vivo* bioavailability.

References

Al-Badr AA, El-Obeid A (1988). Primidone. In: Florey K, ed. *Analytical Profiles of Drug Substances*, Vol. 17. London: Academic Press, pp. 749–795.

Bielmann P, Levac TH, Langlois Y, Tetreault L (1974). Bioavailability of primidone in epilepsy patients. *Int J Clin Pharmacol Ther Toxicol* 9: 132–137.

British Pharmacopoeia (2007). *British Pharmacopoeia*, Vol. 2. London: The Stationery Office, pp. 1728–1729.

CHIP3 Regulations (2002). Chemicals (Hazard Information and Packaging for Supply) Regulations. SI 2002/1689.

Daley RD (1973). Primidone. In: Florey K, ed. *Analytical Profiles of Drug Substances*, Vol. 2. London: Academic Press, pp. 409–437.

El Bayoumi AA, Amer SM, Moustafa NM, Tawakkol MS (1999). Spectrodensiometric determination of clorazepate dipotassium, primidone and chlorzoxazone each in presence of its degradation product. *J Pharm Biomed Anal* 20: 727–735.

Lowey AR, Jackson MN (2008). A survey of extemporaneous dispensing activity in NHS trusts in Yorkshire, the North-East and London. *Hosp Pharmacist* 15: 217–219.

Lund W, ed. (1994). *The Pharmaceutical Codex: Principles and Practice of Pharmaceutics*, 12th edn. London: Pharmaceutical Press.

Meyer MC, Straughn AB, Mhatre RM, Shah VP, Williams RL, Lesko LJ (1998). Lack of in vivo/in vitro correlations for 50 mg and 250 mg primidone tablets. *Pharm Res* 15: 1085–1089.

Schott H, Royce AE (1985). Effect of non-ionic surfactants on aqueous primidone suspensions. *J Pharm Sci* 74: 957–962.

Summers MP, Enever RP (1976). Preparation and properties of solid dispersion system containing citric acid and primidone. *J Pharm Sci* 65: 1613–1617.

United States Pharmacopeia (2003). *United States Pharmacopeia*, XXVI. Rockville, MD: The United States Pharmacopeial Convention.

Wylie E, Pippenger CE, Rothner AD (1987). Increased seizure frequency with generic primidone. *J Am Med Assoc* 258: 1216–1217.

Pyrazinamide oral liquid

Risk assessment of parent compound

A risk assessment survey completed by clinical pharmacists suggested that pyrazinamide has a moderate therapeutic index, with the potential to cause significant morbidity if inaccurate doses were to be administered (Lowey and Jackson, 2008). Current formulations used at NHS trusts show a high technical risk, based on the number and complexity of manipulations and calculations. Pyrazinamide is therefore regarded as a high-risk extemporaneous product (Box 14.38).

Summary

The body of available evidence appears to indicate that pyrazinamide is stable in a number of different formulations. A 1:1 mixture of Ora-Sweet SF and Ora-Plus is easy to prepare, and avoids the use of syrup in paediatric populations.

Xanthan gum-based preparations are currently widely used in the NHS (Lowey and Jackson, 2008).

Clinical pharmaceutics

Points to consider

- At typical paediatric strengths (50–100 mg/mL), the majority of the drug is likely to be in suspension. **Shake the bottle to ensure uniformity of drug throughout the suspension.**
- Ora-Plus and Ora-Sweet SF are free from chloroform and ethanol.
- Ora-Plus and Ora-Sweet SF contain hydroxybenzoate preservative systems.
- Ora-Sweet SF has a cherry flavour.
- 'Keltrol' is made to slightly different specifications by several manufacturers, but typically contains ethanol and chloroform. The exact specification should be assessed before use by a suitably competent pharmacist. Ethanol is a CNS depressant. Chloroform has been classified as a class 3 carcinogen (i.e. substances that cause concern owing to possible carcinogenic effects but for which available information is not adequate to make satisfactory assessments) (CHIP3 Regulations, 2002).

Box 14.38 *Preferred formula*

Using a typical strength of 100 mg/mL as an example:

- pyrazinamide 500 mg tablets × 20
- Ora-Plus to 50 mL
- Ora-Sweet SF to 100 mL.

Method guidance

Tablets can be ground to a fine, uniform powder in a pestle and mortar. A small amount of Ora-Plus may be added to form a paste, before adding further portions of Ora-Plus up to 50% of the final volume. Transfer to a measuring cylinder. The Ora-Sweet can be used to wash out the pestle and mortar before making the suspension up to 100% volume. Transfer to an amber medicine bottle.

Shelf-life

28 days at room temperature in amber glass. **Shake the bottle.**

Alternative

An alternative formula would be:

- pyrazinamide 500 mg tablets × 20
- 'Keltrol' 0.4% (or similar) to 100 mL.

Shelf-life

28 days at room temperature in amber glass.

Note: 'Keltrol' xanthan gum formulations vary slightly between manufacturers, and may contain chloroform and ethanol. The individual specification should be checked by the appropriate pharmacist.

- There are no known licensed pyrazinamide tablets available in the UK. Use of an imported tablet preparation is widespread (Lowey and Jackson, 2008). For information on the quality assessment of imported products, pharmacists are referred to the document 'Guidance for the purchase and supply of unlicensed medicinal products – notes for prescribers and pharmacists' for further information (NHS Pharmaceutical Quality Assurance Committee, 2004).
- No bioavailability data are available for the extemporaneously prepared suspensions.

- Pyrazinamide has a bitter taste; some trusts have been known to add sweetening agents to improve the palatability (e.g. saccharin, sorbitol, peppermint water).
- Scored tablets can be used to administer whole-tablet or half-tablet doses after crushing or dispersing under water (A. Fox, Principal Pharmacist, Southampton University Hospitals NHS Trust, personal communication, 2007).

Technical information

Structure

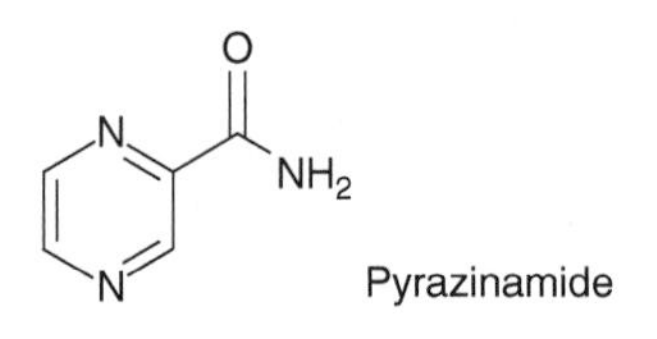

Empirical formula

$C_5H_5N_3O$ (British Pharmacopoeia, 2007) .

Molecular weight

330.3 (British Pharmacopoeia, 2007).

Solubility

15 mg/mL (Budavari, 1996).

pK_a

0.5 (Martindale, 2005)

Optimum pH stability

Unknown.

Stability profile

Physical and chemical stability

Pyrazinamide is a white, crystalline powder that is sparingly soluble in water (British Pharmacopoeia, 2007). This limited solubility should help to minimise the extent to which pyrazinamide will be vulnerable to degradation in an extemporaneous formulation. There is no published information describing the degradation pathway.

When in the solid state, pyrazinamide is reported to exhibit good stability. Felder and Pitre (1983) report no apparent degradation of bulk sample, either in wet or dry atmosphere. Unpublished data are reported to suggest that pyrazinamide is also stable when exposed to natural light (Felder and Pitre, 1983).

Mariappan *et al.* (2000) have investigated the breakdown of various anti-TB agents under acid conditions, alone and in combination. Stability-indicating HPLC analysis was used to quantify the breakdown of drug(s) after storage in either simulated gastric fluid (pH 1.2) or in 0.1 M hydrochloric acid at 37°C for 50 minutes. Pyrazinamide showed negligible amounts of degradation whether stored alone or with the other anti-tuberculosis agents in either solution.

Degradation products

May include pyrazinoic acid (P. Cowin, Deputy QA Manager, Pharmacy, Charing Cross, 2006).

Stability in practice

There is a significant body of evidence to support the stability of extemporaneous preparations of pyrazinamide as an oral liquid.

Allen and Erickson (1998) investigated the stability of pyrazinamide formulated as a 10 mg/mL suspension. The stability of the following three different formulations was investigated when the suspensions were stored in amber clear polyethylene terephthalate prescription bottles in the dark at 5 and 25°C:

- 1:1 mixture of Ora-Sweet and Ora-Plus
- 1:1 mixture of Ora-Sweet Sugar Free (SF) and Ora-Plus
- cherry syrup (cherry syrup concentrate diluted 1:4 with simple syrup).

The suspensions were prepared by manually crushing 500 mg tablets and mixing thoroughly with the appropriate vehicle in aliquots. Three bottles of each formulation were stored at 5°C and three bottles at 25°C.

No significant changes were noted in physical appearance or odour over the test period (60 days). The initial pH values were 4.4 to 4.5 and did not change by more than 0.5 pH units. A stability-indicating HPLC methodology was used to assess changes in drug concentration. Taking into account standard deviations, the mean drug concentration remained above 94% for all of the suspensions, regardless of the vehicle chosen.

The sampling methodology is not detailed and microbiological testing was not carried out. Actual mean initial drug concentrations ranged from 10.4 to 11 mg/mL. The authors suggest assigning a 60-day expiry to these formulations.

Nahata *et al.* (1995) provide further data for the stability of pyrazinamide in two syrup-based formulations. Crushed 500 mg tablets were added to either simple syrup, or a vehicle containing a 1:1 mixture of 1% methylcellulose and simple syrup. The theoretical final concentration of the suspensions was 100 mg/mL; 60 mL of each of the two formulations were stored in 10 plastic and 10 amber glass bottles, and then equally divided between storage at 4 and 25°C.

Samples were collected in duplicate on days 0, 7, 15, 30, 45 and 60 and frozen at −70°C until analysed. Some details of the sampling methodology are given, using 5 mL aliquots removed from the middle of each bottle after shaking for 20 minutes using an automated wrist-action shaker. A stability-indicating HPLC assay was used to quantify drug concentration, adapting the method described by Gaitonde and Pathak (1990) to quantify pyrazinamide and isoniazid in plasma and urine.

The mean drug concentration remained above 90% for both formulations throughout the two-month study period for all the test conditions and storage devices. pH remained within the range 5.80–5.84 for the simple syrup formulation and from 4.78 to 4.83 for the methylcellulose and simple syrup formulation.

Woods (2001) suggests the use of the methylcellulose and syrup formulation, with the minor alteration of the addition of 0.1% parabens. This is justified by the stability shown by the formulation above described by Allen and Erickson (1998), which also contains parabens.

Seifart *et al.* (1991) investigated the stability of suspensions of pyrazinamide alone, and in the presence of other anti-tuberculosis agents. A 100 mg/mL suspension was formulated by crushing tablets in a mortar and adding to a suspending agent as follows:

- pyrazinamide tablets 500 mg × 400
- compound of tragacanth powder 40 g
- aqua chloroph. conc (concentrated chloroform water) 50 mL
- water to 2 L.

Ascorbic acid was added to a sample of each of the suspensions to give a concentration of 20 micrograms/mL.

Before analysis via HPLC, samples were agitated to ensure uniformity throughout the sample. An aliquot of 1 mL was dissolved in 10 mL of methanol and then further diluted to 100 mL with a 50:50 solution of methanol and water. Analysis occurred over 28 days for suspensions stored at 4, 24 and 40°C.

Results were compared against a randomly chosen standard. Each batch of samples was analysed in quadruplicate to exclude dilution or sampling errors. Inter-day variation was less than 1.5%, with less than 1.0% variation within the samples based on the area under the curve.

The suspensions stored in the absence of ascorbic acid were reported to retain more than 98% of the starting concentration of pyrazinamide at day 28, regardless of storage temperature (standard deviation not quoted). However, in the presence of ascorbic acid, pyrazinamide was found to be much more unstable at day 28, with values ranging from 89% at 4°C to 56% at 40°C.

An unpublished study, documented on the St Mary's website, was carried out by P. Cowin in 1989 at Charing Cross Hospital Pharmacy Department (St Mary's Stability Database, 2006). The formula for the 100 mg/mL suspension used was:

- pyrazinamide tablets 500 mg × 20
- sorbitol solution BPC 25 mL
- saccharin solution BPC 2 mL
- concentrated peppermint water BP 2.5 mL
- 'Keltrol' 1%* 50 mL
- distilled water to 100 mL.

*Contains 0.2% w/v Nipasept (hydroxybenzoates) as preservative.

HPLC investigations were designed to detect the degradation product pyrazinoic acid. Suspensions were stored at room temperature and in the fridge and examined over six months. No significant degradation was seen at either temperature. Since the assay, the formula has been switched to use pyrazinamide powder.

The author suggests that the sorbitol and saccharin (added as a sugar-free sweetening system) could be removed if needed. Similarly, the peppermint water, designed to mask the drug's bitter taste, could be omitted if needed (P. Cowin, personal communication, 2006). Flavoured versions of xanthan gum vehicle are available if needed. Cowin also suggests that a 28-day expiry could be allocated for storage either in the fridge or at room temperature, if the formula was prepared extemporaneously.

Bioavailability data

No data available.

References

Allen LV, Erickson MA (1998). Stability of bethanechol chloride, pyrazinamide, quinindine, rifampicin, and tetracycline hydrochloride in extemporaneously compounded oral liquids. *Am J Health Syst Pharm* 55: 1804–1809.

British Pharmacopoeia (2007). *British Pharmacopoeia*, Vol. 2. London: The Stationery Office, pp. 1770–1771.

Budavari S, ed. (1996). *The Merck Index*, 12th edn. Rahway, NJ: Merck and Co.

CHIP3 Regulations (2002). Chemicals (Hazard Information and Packaging for Supply) Regulations. SI 2002/1689.

Felder E, Pitre D (1983). Pyrazinamide. In: Florey K, ed. *Analytical Profiles of Drug Substances*, Vol 12. London: Academic Press, pp. 433–462.

Gaitonde CD, Pathak PV (1990). Rapid liquid chromatographic method for the estimation of isoniazid and pyrazinamide in plasma and urine. *J Chromatogr* 532: 418–423.

Lowey AR, Jackson MN (2008). A survey of extemporaneous dispensing activity in NHS trusts in Yorkshire, the North-East and London. *Hosp Pharmacist* 15: 217–219.

Mariappan TT, Singh B, Singh S (2000). A validated reversed-phase (C18) HPLC method for simultaneous determination of rifampicin, isoniazid and pyrazinamide in USP dissolution medium and simulated gastric fluid. *Pharm Pharmacol Commun* 6: 345–349.

Nahata MC, Morosco RS, Peritore SP (1995). Stability of pyrazinamide in two suspensions. *Am J Health Syst Pharm* 52: 1558–1560.

NHS Pharmaceutical Quality Assurance Committee (2004). Guidance for the purchase and supply of unlicensed medicinal products. Unpublished document available from regional quality assurance specialists in the UK or after registration from the NHS Pharmaceutical Quality Assurance Committee website: www.portal.nelm.nhs.uk/QA/default.aspx

Seifart HI, Parkin DP, Donald PR (1991). Stability of isoniazid, rifampin and pyrazinamide in suspensions used for the treatment of tuberculosis in children. *Paediatr Infect Dis J* 10: 827–831.

St Mary's Online Stability Database (2006). http://www.stmarys.demon.co.uk/ (no longer available) (accessed 7 April 2006).

Sweetman SC (2009). *Martindale: The Complete Drug Reference*, 36th edn. London: Pharmaceutical Press.

Trissel LA, ed. (2000). *Stability of Compounded Formulations*, 2nd edn. Washington DC: American Pharmaceutical Association.

Woods DJ (2001). *Formulation in Pharmacy Practice. eMixt, Pharminfotech*, 2nd edn. Dunedin, New Zealand. www.pharminfotech.co.nz/emixt (accessed 18 April 2006).

Pyridoxine hydrochloride (vitamin B6) oral liquid

Risk assessment of parent compound

A risk assessment survey completed by clinical pharmacists suggested that pyridoxine hydrochloride has a wide therapeutic index. Inaccurate doses were thought likely to lead to minor morbidity (Lowey and Jackson, 2008). Current formulations used at NHS trusts were generally of a moderate technical risk, based on the number and/or complexity of manipulations and calculations. Pyridoxine hydrochloride is therefore regarded as a low- to medium-risk extemporaneous preparation (Box 14.39).

Summary

Most of the evidence and experience in the NHS to support the formulation of pyridoxine as an oral liquid is based on the use of xanthan gum suspending agents. The exact formulation of xanthan gum suspending agents varies between manufacturers, and should be checked by a suitably competent pharmacist. These 'Keltrol' preparations may or may not contain ethanol, chloroform and flavourings. pH values also vary.

Clinical pharmaceutics

Points to consider

- At typical strengths (1–30 mg/mL), almost all of the drug should be in solution. Dosage uniformity should not be an issue.
- 'Keltrol' is made to slightly different specifications by several manufacturers, but typically contains ethanol and chloroform. The exact specification should be assessed before use by a suitably competent pharmacist. Ethanol is a CNS depressant. Chloroform has been classified as a class 3 carcinogen (i.e. substances that cause concern owing to possible carcinogenic effects but for which available information is not adequate to make satisfactory assessments) (CHIP3 Regulations, 2002).
- There are no available data with regard to the bioavailability of these preparations.

Box 14.39 *Preferred formula*

Using a typical strength of 10 mg/mL as an example:

- pyridoxine hydrochloride 50 mg tablets × 20
- xanthan gum 0.5% ('Keltrol' or similar) to 100 mL.

Method guidance

Tablets can be ground to a fine powder in a pestle and mortar. A small amount of suspending agent may be added to form a paste, before adding further up to approximately 75% of the final volume. Transfer to a measuring cylinder. The remaining suspending agent can be used to wash out the pestle and mortar before making the suspension up to 100% volume. Transfer to an amber medicine bottle.

Shelf-life

28 days at room temperature in amber glass. **Protect from light.**

Alternative

Using an example of 10 mg/mL as an example:

- pyridoxine hydrochloride 100 mg/mL injection × 10 mL
- preserved syrup to 100 mL.

Shelf-life

28 days at room temperature in amber glass. **Protect from light.**

Other notes

Some tablets may be quartered or halved to allow smaller doses to be given (Guy's and St Thomas' and Lewisham Hospitals, 2005).

Technical information

Structure

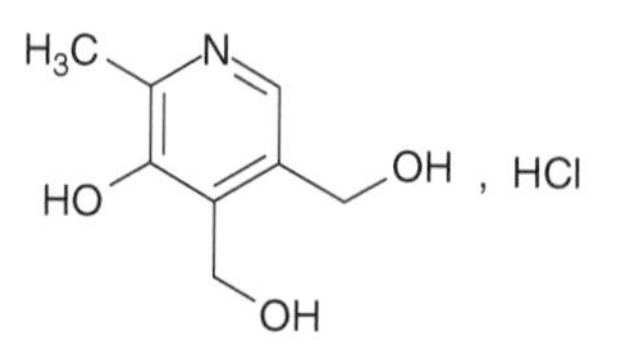

Empirical formula

$C_8H_{11}NO_3$,HCl (British Pharmacopoeia, 2007).

Molecular weight

205.6 (British Pharmacopoeia, 2007).

Aqueous solubility

200–222 mg/mL (Trissel, 2000).

pK_a

5.0 (−N=) and 9.0 (−OH) (Moffat, 1986).

Optimum pH stability

Not known.

Stability profile

Physical and chemical stability

Pyridoxine hydrochloride is a white or almost white, crystalline powder, freely soluble in water, slightly soluble in alcohol (British Pharmacopoeia, 2007). It has a slightly bitter taste (Trissel, 2000).

Pyridoxine is one of three similar compounds that have been referred to as vitamin B6; the other two compounds are pyridoxamine and pyridoxal. Pyridoxamine is not used in pharmaceutical preparations (Higuchi and Hansen, 1961; Aboul-Enein and Loutfy, 1984) and the use of pyridoxal is rare (P. North-Lewis, Paediatric Liver Pharmacist, St James' Hospital, Leeds Teaching Hospitals NHS Trust, personal communication, 2007). Pyridoxine is the most stable of the three forms (Aboul-Enein and Loutfy, 1984). Acidic aqueous solutions of vitamin B6 may be heated at 120°C for 30 minutes without decomposition (Aboul-Enein and Loutfy, 1984).

The drug is reported to undergo decomposition slowly when exposed to light, particularly at higher pHs (Shepherd and Labadarios, 1986; Trissel, 2000). Higuchi and Hansen (1961) have suggested that pyridoxine is destroyed by UV radiation in neutral or alkaline solutions, but not in acidic solution.

Pyridoxine hydrochloride is incompatible with oxidising agents, iron salts and alkaline solutions (McEvoy, 1999; Trissel, 2000). Pyridoxine is unstable towards various food, drug and cosmetic colours, due to the formation of complex addition products between the colours and pyridoxine (Epley and Hall, 1947; Aboul-Enein and Loutfy, 1984). Studies are available describing the addition of pyridoxine to bread, flour, cereals, and during canning and freezing of sweetcorn.

Mizuno *et al.* (1981) have investigated the effect of dyes on the photodecomposition of pyridoxine and pyridoxamine. Erythrosine, eosine Y, rose bengal, mercurochrome, methylene blue, azure A and azure B accelerated the decomposition of pyridoxine and pyridoxamine at pH 5–9, but fluorescein, rhodamine B, brilliant blue and acid red did not. The degradation in the presence of erythrosine was greatest at pH 9. The photodecomposition of pyridoxine and pyridoxamine in the presence of dyes was depressed by aminopyrine, sulpyrine and tryptophan.

In 1944, Hochberg *et al.* presented data on the effect of light, heat, acids, alkali and oxidizing agents on pyridoxine. The authors demonstrated rapid destruction of pyridoxine by light in neutral and alkaline solutions by various methods of measurement. Little loss was seen, however, in 0.1 N hydrochloric acid. Similar results were obtained when irradiations were performed in an atmosphere of nitrogen, indicating that the destruction was not due to photolytic oxidation.

Heating the vitamin for 1 hour at 100°C with 5 N hydrochloric acid, sulfuric acid or sodium hydroxide did not affect pyridoxine. Treatment with nitric acid, however, led to oxidative degradation. Hochberg *et al.* (1944) state that pyridoxine may be autoclaved for 15 minutes with 1 N hydrochloric acid or sodium hydroxide, or for 30 minutes in 4 N sulfuric acid, without deleterious effect. Manganese dioxide has no effect on pyridoxine, but potassium permanganate destroys the vitamin (Hochberg *et al.*, 1944).

The authors conclude that the vitamin is unstable when irradiated in aqueous solutions at pH 6.8 or above. At pH 1, pyridoxine is almost unaffected, and the destruction is not due to photolytic oxidation.

Cunningham *et al.* (1945) found the same instability in neutral and alkaline solution.

A 5% aqueous solution has a pH between 2.4 and 3.0, and the injection has a pH between 2.0 and 3.8 (McEvoy, 1999; Trissel, 2000; Martindale, 2005). Pyridoxine hydrochloride should be stored protected from light (British Pharmacopoeia, 2007). The injection should not be frozen (McEvoy, 1999).

Pyridoxine is contained in the Pabrinex IM and IV high-potency liquid injection preparations (BNF, 2007).

Degradation products

No information available.

Stability in practice

The evidence to support the stability of pyridoxine oral liquids is limited. However, unpublished data suggest that reasonable shelf-lives should be achievable.

Nahata and Hipple (2004) have described the formulation of a 1 mg/mL pyridoxine oral solution. One millilitre of pyridoxine hydrochloride 100 mg/mL injection was mixed with 99 mL of syrup and stored in a glass bottle. The authors recommend a shelf-life of 30 days based on experience. No data are available to support the shelf-life.

P. Nixon (personal communication, 1999) has carried out unpublished stability work on a pyridoxine hydrochloride 30 mg/mL mixture in diluent A ('Keltrol'). The mixture was stored at room temperature (not stated) for 28 days, and sampled on days 0, 14 and 28. Details of the stability-indicating HPLC method are available, but no details are supplied with regard to the sampling methodology or number of duplicate samples.

The initial concentration was only 92.7% of expected, but little or no degradation occurred over the 28-day test period (91.5% and 91.3% remaining on day 14 and day 28, respectively). Nixon suggests that the product appears to be chemically stable for 28 days. No changes in pH or physical appearance were observed (pH range 3.37–3.40).

Related preparations

Allwood (1984) reviewed the compatibility and stability of total parenteral nutrition (TPN) mixtures in flexible containers. Pyridoxine was found to be light sensitive, although the amount of degradation was far less than losses associated with vitamin A or riboflavin. No losses were reported unless the solution was exposed to direct sunlight when destruction of more than 80% of the added pyridoxine occurred in 8 hours. Under normal circumstances, therefore, pyridoxine losses have been anticipated to be small.

Chen *et al.* (1983) have also investigated the stability of B vitamins in mixed parenteral nutrition solution. Three identical TPN solutions were prepared under aseptic conditions, and vitamins were added. The bags were then protected from light, and baseline samples were taken. One infusion bag was placed in a windowless room with fluorescent lights, another by a window containing non-tinted glass, and a third bag was placed in direct sunlight. Samples were taken from the bags at 2-hourly intervals, after the contents had been mixed by inverting the bag.

Under direct sunlight, 86% of pyridoxine hydrochloride was destroyed in 8 hours, determined by a microbiological method. Fluorescent light and indirect sunlight had little effect on pyridoxine hydrochloride concentration over 8 hours.

Bioavailability data

No information available.

References

Aboul-Enein HY, Loutfy MA (1984). Pyridoxine hydrochloride. In: Florey K, ed. *Analytical Profiles of Drug Substances*, Vol. 13. San Diego: Academic Press, pp. 447–486.
Allwood MC (1984). Compatibility and stability of TPN mixtures in big bags. *J Clin Hosp Pharm* 9: 181–198.
British Pharmacopoeia (2007). *British Pharmacopoeia*, Vol. 2. London: The Stationery Office, pp. 1772–1773.
Chen MF, Boyce HW, Jr Triplett L (1983). Stability of the B vitamins in mixed parenteral nutrition solution. *J Parenter Enter Nutr* 7: 462–464.
CHIP3 Regulations (2002). Chemicals (Hazard Information and Packaging for Supply) Regulations. SI 2002/1689.
Cunningham E, Snell EE (1945). The vitamin B6 group - VI The comparative stability of pyridoxine, pyridoxamine and pyridoxal. *J Biol Chem* 158: 491–495.
Epley HC, Hall AG (1947). An experimental study of the stability of certain factors of vitamin B complex toward various food, drug and cosmetic colors. *Am J Pharm* 119: 309–314.
Guy's and St. Thomas' and Lewisham Hospitals (2005). *Paediatric Formulary*. London: Guy's and St Thomas' and Lewisham NHS Trust.
Higuchi T, Hansen EB (1961). *Pharmaceutical Analysis*. New York: Interscience Publishers, p. 668.
Hochberg M, Melnick D, Oser BL (1944). On the stability of pyridoxine. *J Biol Chem* 155: 129–136.
Joint Formulary Committee (2010). *British National Formulary*, No. 59. London: British Medical Association and Royal Pharmaceutical Society of Great Britain.
Lowey AR, Jackson MN (2008). A survey of extemporaneous dispensing activity in NHS trusts in Yorkshire, the North-East and London. *Hosp Pharmacist* 15: 217–219.
Sweetman SC (2009). *Martindale: The Complete Drug Reference*, 36th edn. London: Pharmaceutical Press.
McEvoy J, ed (1999). *AHFS Drug Information*. Bethesda, MD: American Society of Health System Pharmacists.
Mizuno N, Fujiwara A, Morita E (1981). Effect of dyes on the photodecomposition of pyridoxine and pyridoxamine. *J Pharm Pharmacol* 33: 373–376.
Moffat AC, Osselton MD, Widdop B (2003). *Analysis of Drugs and Poisons*, 3rd edn. London: Pharmaceutical Press. ISBN 978 0 85369 4731.
Nahata MC, Hipple TF (2004). *Pediatric Drug Formulations*, 5th edn. Cincinnati, OH: Harvey Whitney Books.
Shepherd GS, Labadarios D (1986). Degradation of vitamin B6 standard solutions. *Clin Chim Acta* 160: 307–312.
Trissel LA, ed. (2000). *Stability of Compounded Formulations*, 2nd edn. Washington DC: American Pharmaceutical Association.

Quinine sulfate oral liquid

Risk assessment of parent compound

A survey of clinical pharmacists has suggested that quinine sulfate has a moderate therapeutic index with the potential to cause severe morbidity if inaccurate doses are given (Lowey and Jackson, 2008). Current formulations used at NHS trusts were generally of a low to moderate technical risk, based on the number and/or complexity of manipulations and calculations. Quinine sulfate is therefore regarded as a high-risk extemporaneous preparation (Box 14.40).

***Box 14.40** Preferred formula*

Using a typical strength of 40 mg/mL as an example:

- quinine sulfate 200 mg tablets × 20
- 'Keltrol' 1% 50 mL
- distilled water 10 mL
- raspberry syrup 16.5 mL
- preserved syrup to 100 mL.

Method guidance

The water is used to soften the tablets, which can be ground to a fine, uniform powder in a pestle and mortar. After the addition of the raspberry syrup, the 'Keltrol' can be added sequentially. Transfer to a measuring cylinder. The syrup can be used to wash out the pestle and mortar before making the suspension up to 100% volume. Transfer to an amber medicine bottle.

Shelf-life

7 days refrigerated in amber glass.

Note: 'Keltrol' xanthan gum formulations vary slightly between manufacturers, and may contain chloroform and ethanol. The individual specification should be checked by the appropriate pharmacist.

Summary

There is a paucity of high-quality data to support the formulation of quinine sulfate as an oral liquid. Alternative treatment options should be considered, including ward-based manipulations for whole-tablet doses. However, there are no obvious routes of rapid degradation and further research may allow the future extension of the suggested shelf-life. Moreover, quinine is known to be present in tonic water, suggesting inherent stability. Current best practice would appear to comprise the use of xanthan gum-based suspending agents. Various methods have been tried to disguise the very bitter taste associated with quinine.

Clinical pharmaceutics

Points to consider

- At typical paediatric strengths (20–40 mg/mL), most of the drug will be in suspension. **Shake the bottle to ensure uniformity of drug throughout the suspension.**
- Quinine displays optical chemistry; quinidine is an isomer of quinine. Isomerisation should be limited by the relatively low solubility of quinine sulfate, but may become an issue if very low-strength liquids are prepared (e.g. less than 5 mg/mL). The use of quinine bisulfate is not recommended due to the increased solubility (1 in 8 of water), which may lead to more extensive problems with isomerisation.
- 'Keltrol' is made to slightly different specifications by several manufacturers, but typically contains ethanol and chloroform. The exact specification should be assessed before use by a suitably competent pharmacist. Ethanol is a CNS depressant. Chloroform has been classified as a class 3 carcinogen (i.e. substances that cause concern owing to possible carcinogenic effects but for which available information is not adequate to make satisfactory assessments) (CHIP3 Regulations, 2002).
- Quinine sulfate has a very bitter taste (Lund, 1994)
- No data are available on bioavailability.

Technical information

Structure

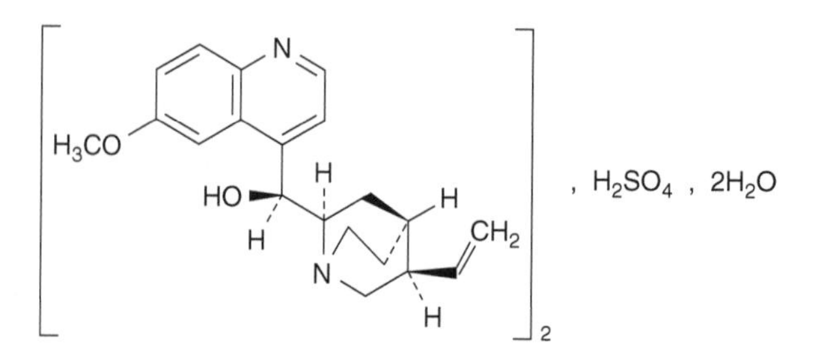

Empirical formula

$(C_{20}H_{24}N_2O_2)_2H_2SO_4,2H_2O$ (British Pharmacopoeia, 2007).

Molecular weight

783 (British Pharmacopoeia, 2007).

Solubility

Slightly soluble in water (1 in 810 or 1.24 mg/mL) (Lund, 1994; British Pharmacopoeia, 2007).

pK_a

4.1, 8.5 (at 20°C) (Clarke, 1986; Lund, 1994).

Optimum pH stability

Not known.

Stability profile

Physical and chemical stability

Quinine sulfate is a white or almost white crystalline powder or fine, colourless needles. It should be protected from light (British Pharmacopoeia, 2007). Quinine sulfate becomes brown on exposure to light (Lund, 1994).

The relatively low solubility of quinine sulfate will help to limit aqueous degradation kinetics.

Quinine sulfate solutions precipitate quinine in the presence of alkalis and their carbonates (Lund, 1994). Quinine salts are also reported to be incompatible with acetates, benzoates, citrates, iodides, salicylates and tartrates (Lund, 1994).

Quinine is an optical isomer of quinidine (Lund, 1994). Any optical isomerisation should again be limited by the low solubility of quinine sulfate.

Degradation products

No information.

Stability in practice

There are no published reports of the stability of extemporaneous preparations of quinine sulfate. Information from NHS trusts is summarised below.

Unpublished stability data dating from 1993–1994 was supplied by Norfolk and Norwich Hospital Pharmacy QC Laboratory (R. Shaw, personal communication, 2006). It describes the testing of quinine sulfate 40 mg/mL in a standard suspending agent:

- xanthan gum NF XVII ('Keltrol') 500 mg
- citric acid monohydrate BP 1.29 g
- sodium phosphate BP 2.75 g
- aspartame 200 mg
- Nipasept 100 mg
- antifoam emulsion (Dow Corning M30) 2 drops
- purified water to 100 mL.

The suspending agent was buffered to pH 4 before use, and quinine sulfate powder was used as the source of drug. HPLC analysis suggested no evidence of extra peaks due to degradation over 79 days. The assay results showed a result consistently above 100%, rising to 110.4% on day 79. However, the final formula adopted by the hospital was reported to be based on quinine dihydrochloride to the following formula:

- quinine dihydrochloride injection or powder qs
- standard suspending agent (see above) to 100 mL.

The formula reportedly carried a four-week expiry when stored in a sealed container at room temperature and protected from light, but no supporting data are available.

The following formulation is reported to be physically stable for 28 days (R. Tomlinson, Senior Pharmacist, Huddersfield Royal Infirmary, personal communication, 2006):

- quinine sulfate powder qs
- vanillin 0.05%
- syrup 75%
- neutral suspending base 100%.

Neutral suspending base is sodium carboxymethylcellulose sodium 100 g, methyl hydroxybenzoate 5 g, propyl hydroxybenzoate 0.5 g, glycerol 250 mL, preserved syrup 2 L, distilled water to 5 L, with hydrochloric acid for pH adjustment.

Other NHS centres have been reported to use the following formula (Lowey and Jackson, 2008):

- quinine sulfate 200 mg tablets × 20
- 'Keltrol' 1% 50 mL
- distilled water 10 mL
- raspberry syrup 16.5 mL
- syrup (preserved) to 100 mL.

A 7-day expiry in the fridge has been used, but no evidence to support this shelf-life could be found. The water is used to soften the tablets before the addition of the raspberry syrup and then the 'Keltrol'. It should be noted that 'Keltrol' is now available from several suppliers to varying specifications, typically containing chloroform and ethanol. The appropriateness of the formulation should be assessed before use by a suitably competent pharmacist.

The Pharmaceutical Codex (Lund, 1994) suggests that the following excipients have been used in suspensions of quinine sulfate: cocoa powder, compound tragacanth powder, methyl and propyl hydroxybenzoates, syrup, vanilla essence.

Strategies such as doubling the flavour concentration, increasing the sugar content, adding excess citric acid to fruit-flavoured syrups and increasing the viscosity of the solutions have been tried to improve the palatability of quinine oral liquids (not solely quinine sulfate preparations) (Lund, 1994).

Related studies – other salts

Lerkiatbundit *et al.* (1993) has studied the stability of quinine dihydrochloride injection (not sulfate) at a concentration of 1.2 mg/mL in three common intravenous solutions (sodium chloride 0.9%, dextrose 5% and dextrose 5% in sodium chloride 0.9%). A stability-indicating HPLC method was used to assay drug concentrations. All solutions were prepared in glass bottles to avoid sorption problems with plastic containers. All solutions were stored at 27°C under normal lighting conditions in the laboratory to simulate in-use conditions.

Over a 24-hour period, less than 10% degradation occurred in any of the three solutions. However, the authors still recommend use of such a solution as soon as possible, as concentrations did decrease over the test period. The final concentrations of drug expressed as percentages of the initial concentration were 98.96 (± 1.01), 94.9 (± 3.32) and 90.40 (± 0.29) for the dextrose 5%, dextrose 5% in normal saline, and normal saline bottles, respectively.

A quinine dihydrochloride injection is currently available from 'Specials' manufacturers in the UK with a typical shelf-life of 2 years (C. Akinola, Cardinal Healthcare, personal communication, 2006).

Bioavailability data

No known data.

References

British Pharmacopoeia (2007). *British Pharmacopoeia*, Vol. 1. London: The Stationery Office, pp. 783–785 and Vol. 2, pp. 2090–2092.

CHIP3 Regulations (2002). Chemicals (Hazard Information and Packaging for Supply) Regulations. SI 2002/1689.

Lerkiatbundit S (1993). Stability of quinine hydrochloride in commonly used intravenous solutions. *J Clin Pharm Ther* 18: 5: 343–345.

Lowey AR, Jackson MN (2008). A survey of extemporaneous dispensing activity in NHS trusts in Yorkshire, the North-East and London. *Hosp Pharmacist* 15: 217–219.

Lund W, ed. (1994). *The Pharmaceutical Codex: Principles and Practice of Pharmaceutics*, 12th edn. London: Pharmaceutical Press.

Moffat AC, Osselton MD, Widdop B (2003). *Analysis of Drugs and Poisons*, 3rd edn. London: Pharmaceutical Press. ISBN 978 0 85369 4731.

Sildenafil citrate oral liquid

Risk assessment of parent compound

A recent survey of clinical pharmacists suggested that sildenafil has a moderate to narrow therapeutic index (Lowey and Jackson, 2008). Given its use in paediatric cardiology centres, it carries the potential to cause serious morbidity or death if inaccurate dosages are administered. Current formulations used at NHS trusts were generally of a moderate technical risk, based on the number and/or complexity of manipulations and calculations. A risk assessment carried out by clinical pharmacists suggested that sildenafil is regarded as medium-risk extemporaneous preparation. This risk may be considered to be higher in certain clinical scenarios (Box 14.41).

Summary

Evidence exists to suggest that sildenafil citrate is stable in a number of suspending agents. Given the relative stability of this drug in aqueous formulations, sildenafil citrate may be a candidate for purchase from a 'Specials' manufacturer. If made in the appropriate conditions with batch release and quality control testing, the shelf-life may be able to be extended over time.

Clinical pharmaceutics

Points to consider

- At lower strengths such as 1 mg/mL all the drug is likely to be in solution. At higher strengths such as 5 mg/mL the majority of the drug will be held in solution, but some drug may settle. Therefore, the formula should include a suspending agent and be shaken before use.
- Ora-Plus and Ora-Sweet SF are free from chloroform and ethanol.
- Ora-Plus and Ora-Sweet contain hydroxybenzoate preservative systems.
- There are no bioavailability data available.

Technical information

Structure

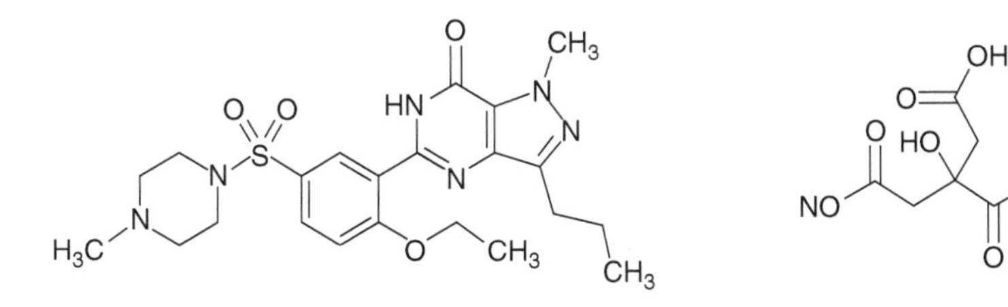

Box 14.41 *Preferred formula*

Using a typical strength of 2 mg/mL as an example:

- sildenafil citrate tablets 100 mg × 1
- Ora-Plus 25 mL
- Ora-Sweet or Ora-Sweet SF to 50 mL.

Method guidance

The tablet can be ground to a fine, uniform powder in a pestle and mortar. A small amount of Ora-Plus may be added to form a paste, before adding further portions of Ora-Plus up to 50% of the final volume. Transfer to a measuring cylinder. The Ora-Sweet can be used to wash out the pestle and mortar of any remaining drug before making the suspension up to 100% volume. Transfer to an amber medicine bottle.

Shelf-life

28 days in amber glass at room temperature. **Shake the bottle.**

Alternative

An alternative 2 mg/mL formula would be:

- sildenafil citrate tablets 100 mg × 1
- methylcellulose 1% 25 mL
- simple syrup to 50 mL.

Shelf-life

7 days in amber glass at room temperature. **Shake the bottle.**

Note: The shelf-life is restricted due to concerns regarding microbiological stability.

Empirical formula

$C_{22}H_{30}N_6O_4S.C_6H_8O_7$ (Moffat *et al.*, 2004).

Molecular weight

666.7 (Moffat *et al.*, 2004).

Solubility

3.5 mg/mL (Pfizer Information Sheet, 2006).

pK_a

8.7 (Cooper *et al.*, 1997).

Optimum pH stability

Not known.

Stability profile

Physical and chemical stability

Sildenafil citrate is a white to off-white crystalline powder with a solubility of 3.5 mg/mL in water (Pfizer Information Sheet, 2006). Sildenafil has basic functional groups with a pK_a value of 8.7, although a weak acidic moiety is also present (Cooper *et al.*, 1997).

Development of a stability-indicating HPLC method by Daraghmeh *et al.* (2001) has revealed some basic information regarding the degradation of sildenafil citrate samples. Various samples of the drug in solid state and in the mobile phase (a 1:1 mixture of ammonium acetate (pH 7.0 0.2 M) and acetonitrile) were exposed to direct sunlight and the dark for 6 days. The solid state showed almost a full recovery without any significant degradation. However, the solutions showed a degradation of around 0.7% with an almost complete recovery. Samples kept in the dark showed no signs of degradation.

Further sildenafil citrate sample solutions kept in the dark at 40°C for 6 days did not show any significant levels of degradation, suggesting that the degradation that took place in sunlight may be due solely to the light effect.

Daraghmeh *et al.* (2001) also showed specificity of their assay by inducing degradation of sildenafil acetate by treating samples with 1.5% hydrogen peroxide at room temperature for 60 minutes, or with 0.1 M hydrochloric acid at 65°C for 12 days, or 0.1 M sodium hydroxide at 65°C for 12 days. Although treatment with hydrogen peroxide produced significant degradation (35%), the treatment at 65°C with acid or alkali yielded less than 1% degradation.

Segall *et al.* (2000) described a reversed-phase HPLC method for the determination of sildenafil acetate in the presence of its oxidative-induced degradation products.

Degradation products

No information.

Stability in practice

Two main studies provide evidence for the stability of sildenafil when formulated as an oral liquid.

Nahata *et al.* (2006) have investigated the formulation of sildenafil citrate oral suspensions for the treatment of pulmonary hypertension in children. Twenty-five milligram tablets were ground to a fine powder in a pestle and mortar before adding to a vehicle consisting of either:

- 1:1 mixture of Ora-Plus and Ora-Sweet or
- 1:1 mixture of methylcellulose 1% and simple syrup.

The theoretical concentration of the preparation was 2.5 mg/mL. Enough of each suspension was made to produce fill 10 plastic prescription bottles, each containing 30 mL. Five bottles were stored at 4°C, and five at 25°C. Samples were removed after the suspensions had been subject to a wrist-action shaker for 10 minutes and then allowed to stand for 2 minutes. Samples were withdrawn from the centre of the suspension at mid-level. The schedule consisted of 500 microlitre samples collected from each bottle on days 0, 7, 14, 28, 42, 56 70 and 91. A previously published stability-indicating HPLC methodology was used (Daraghmeh *et al.*, 2001). pH measurements were also carried out daily, and the liquid was observed for any changes in colour and turbidity against black and white backgrounds.

The authors found that the mean concentration of drug exceeded 98% of the initial concentration in each formulation at both temperatures throughout the test period of 91 days. Typical values for the standard deviation were less than 3%. No significant changes were detected in pH, colour, odour or turbidity. Nahata *et al.* (2006) suggested that both formulations appeared uniformly suspended.

Tuleu *et al.* (2003) have also produced data to support the stability of sildenafil in a 1:1 mixture of Ora-Plus and Ora-Sweet SF. A 2 mg/mL suspension was prepared using crushed Viagra tablets suspended in a 1:1 mixture of Ora-Plus and Ora-Sweet. The authors mimicked use by a patient by opening the suspensions twice a day and removing 1 mL. HPLC methodology was used to assay drug content on a weekly basis for four weeks. Particle size was monitored using a Malvern MasterSizer based on laser diffraction, and rheology was assessed using a Cari Med Rheometer. The authors also reported that the suspension was examined under a light microscope coupled to polarisers at ×40, ×100 and ×400 magnifications.

No significant changes in particle size or drug content were detected over the four-week test period. Rheology was reported to be unaffected. However, some non-crystalline 'objects' were observed under the microscope that may justify further investigation. Tuleu *et al.* (2003) also reported that the suspension remained physically and microbiologically 'stable' for six months when stored at 4°C.

Sildenafil citrate has been reported to be physically compatible with diluent A suspending agent ('Keltrol' or xanthan gum) (Nova Laboratories Ltd, personal communication, 2005). Shelf-lives of up to 28 days at room temperature have been reported for these preparations, but no data are currently available in the public domain (Lowey and Jackson, 2008).

Given the apparent stability of sildenafil citrate in oral liquid formulations, it may be prudent to investigate the commissioning of a suitable formulation from a 'Specials' manufacturer.

The use of sildenafil powders has been documented in a case report in the literature (Hon *et al.*, 2005). However, no validation data are available.

Bioavailability data

No information, although Nahata *et al.* (2006) suggest that the bioavailability of a well-suspended liquid is expected to be comparable to that of a tablet. The researchers also suggest that there has been successful anecdotal use of the Ora-based suspension. However, it is important to note that no bioavailability, safety or efficacy data are available.

Sildenafil has also been used enterally as part of a treatment strategy in a case report of severe neonatal pulmonary hypertension (Filan *et al.*, 2006).

References

Cooper JDH, Muirhead JE, Taylor JE, Baker PR (1997). Development of an assay for the simultaneous determination of sildenafil (viagra) and its metabolite (UK-103,320) using automated sequential trace enrichment of dialysates and high performance liquid chromatography. *J Chromatogr B* 701: 87–95.

Daraghmeh N, Al-Omari M, Badwan AA, Jaber AM (2001). Determination of sildenafil citrate and related substances in commercial products and tablet dosage form using HPLC. *J Pharm Biomed Anal* 25: 483–492.

Filan PM, McDougall PN, Shekerdemian LS (2006). Combination pharmacotherapy for severe neonatal pulmonary hypertension. *J Paediatr Child Health* 42: 219–220.

Hon KL, Cheung KL, Siu KL, Leung TF, Yam MC, Fok TF, Ng PC (2005). Oral sildenafil for treatment of severe pulmonary hypertension in an infant. *Biol Neonate* 88: 109–112.

Lowey AR, Jackson MN (2008). A survey of extemporaneous dispensing activity in NHS trusts in Yorkshire, the North-East and London. *Hosp Pharmacist* 15: 217–219.

Moffat AC, Osselton MD, Widdop B (2004). *Clarke's Analysis of Drugs and Poisons*, 3rd edn. London: Pharmaceutical Press.

Nahata MC, Morosco RS, Brady MT (2006). Extemporaneous sildenafil oral suspensions for the treatment of pulmonary hypertension in children. *Am J Health Syst Pharm* 63: 254–257.

Pfizer Information Sheet (2006). Viagra tablets. www.pfizer.com/pfizer/download/uspi_viagra.pdf (accessed 4 August 2006).

Segall AI, Vitale MF, Perez VL, Palacios ML, Pizzorno MT (2000). Reversed-phase HPLC determination of sildenafil citrate in the presence of its oxidative-induced degradation products. *J Liq Chromatogr Relat Technol* 23: 1377–1386.

Tuleu C, Long P, Wong I, Cope J, Haworth SG (2003). Safe use of the extemporaneous preparation of sildenafil (Viagra) for children with pulmonary hypertension. Research based at: Centre for Paediatric Pharmacy Research, The School of Pharmacy, London. Winner of the 2003 ManMed Research Award. Presented at the 10th Annual NPPG Conference, Newcastle. www.nppg.org.uk (accessed 4 August 2006).

Sodium bicarbonate oral liquid

Synonyms

Sodium hydrogen carbonate.

Risk assessment of parent compound

A risk assessment survey completed by clinical pharmacists suggested that sodium bicarbonate has a wide therapeutic index. In general, inaccurate doses were thought likely to lead to relatively minor morbidity only (Lowey and Jackson, 2008). Current formulations used at NHS trusts were generally of a moderate technical risk, based on the number and/or complexity of manipulations and calculations. Due to the potential technical complexity of the calculations involved (i.e. conversion of units – mmol/mL, percentages, etc.), sodium bicarbonate is regarded as a high-risk extemporaneous preparation (Box 14.42).

Box 14.42 *Preferred formulae*

Using a typical strength of 8.4% (1 mmol/mL) as an example:

- sodium bicarbonate 8.4% Polyfusor – decant into an appropriately sized amber glass bottle.

Shelf-life

7 days refrigerated in amber glass.

Alternative

An alternative formula would be:

- sodium bicarbonate BP 8.4 g
- purified water to 100 mL.

Expiry (if unpreserved)

7 days refrigerated in amber glass

Note: The addition of a suitable preservative may allow extension of the shelf-life.

Summary

The majority of practical experience in the NHS with regard to extemporaneous preparation of sodium bicarbonate comprises the use of licensed intravenous products in as an oral solution (e.g. Polyfusor).

However, several NHS and non-NHS 'Specials' manufacturers are known to produce sodium bicarbonate solutions. The formula should be checked and agreed by a suitably competent pharmacist. Some 'Specials' manufacturers produce terminally sterilised products which are preservative-free. If a preservative system is included, it should be chosen with the target patient population in mind, as selected preservatives may show toxicity in some populations.

Clinical pharmaceutics

Points to consider

- There should not be any dosage uniformity issues as sodium bicarbonate is soluble in water at typical concentrations.
- Unpreserved sodium bicarbonate solutions should be stored in the fridge with a restricted shelf-life. Most centres restrict expiry to 7 days for an unpreserved solution (Lowey and Jackson, 2008).
- Sodium bicarbonate has a slightly salty alkaline taste (Trissel, 2000; Martindale, 2005)
- There are no available data with regard to bioavailability.

Other notes

- Sodium bicarbonate solutions are known to be available from several 'Specials' manufacturers. The exact formula should be checked and approved by a suitably competent pharmacist before purchase.

Technical information

Empirical formula

$NaHCO_3$ (British Pharmacopoeia, 2007).

Molecular weight

84.0 (British Pharmacopoeia, 2007).

Solubility

Approximately 83 mg/mL (Trissel, 2000).

pK_a

Not known.

Optimum pH stability

Not known.

Stability profile

Physical and chemical stability

Sodium bicarbonate is a white crystalline powder, soluble in water and practically insoluble in alcohol (British Pharmacopoeia, 2007). When heated in the dry state or in solution, it gradually changes into sodium carbonate (British Pharmacopoeia, 2007). When added to acid in aqueous solution, sodium bicarbonate evolves carbon dioxide gas bubbles and effervescence (Trissel, 2000).

Trissel (2000) suggests that sodium bicarbonate slowly decomposes in moist air to sodium carbonate, carbon dioxide and water. The decomposition is accelerated by heat or agitation (Gable, 2003). It is stable in dry air (Trissel, 2000). Sodium carbonate is a much more alkaline compound; a pH check of heat-sterilised products as a quality control measure has therefore been suggested (Trissel, 2000).

Aqueous solutions of sodium bicarbonate may be sterilised by filtration or autoclaving. Decomposition via decarboxylation on autoclaving is minimised by passing carbon dioxide through the solution in its final container (Gable, 2003). The sealed container should not be opened for 2 hours after it has returned to ambient temperature, to allow reformation of the bicarbonate (Gable, 2003).

Sodium bicarbonate solutions in glass have been reported to develop deposits of small glass particles, and sediments of calcium carbonate with traces of magnesium or other metal carbonates have been found in autoclaved injections (Gable, 2003). This is thought to occur due to impurities or to the extraction of calcium and magnesium ions from the glass. Problems with sedimentation may be retarded by the inclusion of 0.01–0.02% disodium edetate (Hadgraft and Hewer, 1964; Hadgraft, 1966; Smith, 1966; Gable, 2003).

Aqueous solutions of sodium bicarbonate are slightly alkaline to litmus or phenolphthalein (Budavari, 1996). The pH of a freshly prepared 0.1 M solution at 25°C is 8.3 (Budavari, 1996; Trissel, 2000). The USP states that sodium bicarbonate injection has a pH of 7.0–8.5, and that the tablets and powder should be stored in well-closed containers at a controlled room temperature (United States Pharmacopeia, 2003).

Sodium bicarbonate injection should be stored at controlled room temperature and protected from freezing (Trissel, 2000; United States Pharmacopeia, 2003).

Degradation products

Include sodium carbonate, carbon dioxide and water (Trissel, 2000).

Stability in practice

A simple formula has been endorsed by Nahata *et al.* (2004) to prepare an 8.4% (1 mmol/mL) solution:

- sodium bicarbonate powder 84 g
- purified water to 1,000 mL.

An expiry of 30 days is suggested at room temperature. However, unpreserved solutions usually carry shorter expiry dates and are stored in the fridge.

A recent survey of NHS trusts in the North-East, Yorkshire and London found that most trusts decanted a Polyfusor preparation into an amber glass bottle with a 7-day expiry in the fridge (Lowey and Jackson, 2008). Chemical stability is not likely to be an issue given the stability of sodium bicarbonate, and the prolonged shelf-lives of the Polyfusor and other sodium bicarbonate preparations.

Bioavailability data

No information available.

References

British Pharmacopoeia (2007). *British Pharmacopoeia*, Vol. 2. London: The Stationery Office, p. 1871.

Budavari S, ed. (1996). *The Merck Index*, 12th edn. Rahway, NJ: Merck and Co.

Gable CG (2003). Sodium bicarbonate. In: Rowe RC, Sheskey PJ, Weller PJ, eds. *Handbook of Pharmaceutical Excipients*, 4th edn. London: Pharmaceutical Press and Washington DC: American Pharmaceutical Association.

Hadgraft JW (1966). Unsatisfactory infusions of sodium bicarbonate [letter]. *Pharm J* i: 603.

Hadgraft JW, Hewer BD (1964). Molar injection of sodium bicarbonate [letter]. *Pharm J* 192: 544.

Lowey AR, Jackson MN (2008). A survey of extemporaneous dispensing activity in NHS trusts in Yorkshire, the North-East and London. *Hosp Pharmacist* 15: 217–219.

McEvoy J, ed. (1999). *AHFS Drug Information*. Bethesda, MD: American Society of Health System Pharmacists.

Nahata MC, Pai VB, Hipple TF (2004). *Pediatric Drug Formulations*, 5th edn. Cincinnati: Harvey Whiteney Books.

Smith G (1966). Unsatisfactory infusions of sodium bicarbonate [letter]. *Lancet* i: 658.

Sweetman SC (2009). *Martindale: The Complete Drug Reference*, 36th edn. London: Pharmaceutical Press.

Trissel LA, ed. (2000). *Stability of Compounded Formulations*, 2nd edn. Washington DC: American Pharmaceutical Association.

United States Pharmacopeia (2003). *United States Pharmacopeia*, XXVI. Rockville, MD: The United States Pharmacopeial Convention.

Sodium chloride oral solution

Risk assessment of parent compound

A risk assessment survey completed by clinical pharmacists suggested that sodium chloride has a wide to medium therapeutic index. In general, inaccurate doses were deemed likely to lead to only relatively minor morbidity (Lowey and Jackson, 2008). Current formulations used at NHS trusts were generally of a moderate to high technical risk, based on the number and/or complexity of manipulations and calculations. Due to the potential technical complexity of the calculations involved, sodium chloride is regarded as a high-risk extemporaneous preparation (Box 14.43).

Summary

The shelf-life is limited by concerns over microbial contamination. A suitable preservative may be added to extend the expiry. 'Specials' are commonly available from NHS production centres with added preservative systems.

Clinical pharmaceutics

Points to consider

- At typical strengths (1–5 mmol/mL), sodium chloride will be in solution. Dosage uniformity is not an issue.
- Dilution of licensed injection and Polyfusor preparations will result in the formation of unpreserved systems. As such, these formulations must be stored in the fridge with a limited shelf-life.

Other notes

- Hydroxybenzoate preservatives have been used to preserve some formulations. Hydroxybenzoates (also known as parabens) are a group of widely used antimicrobial preservatives in cosmetics, and oral and topical pharmaceutical formulations. Concern has been expressed over their use in infant parenteral products as bilirubin binding may be affected (Rieger, 2003). This is potentially hazardous in hyperbilirubinaemic neonates (Loria *et al.*, 1976). However, there is no proven link to similar effects when used in oral products. The World Health Organization has

***Box 14.43** Preferred formula*

Note: 'Specials' are available from reputable manufacturers in the UK. Extemporaneous preparation should only occur where there is a demonstrable clinical need.

Using a typical strength of 1 mmol/mL as an example:

- sodium chloride BP 5.8 g
- sterile water to 100 mL.

Shelf-life

7 days refrigerated in amber glass.

Alternatives

Using a strength of 1 mmol/mL:

- sodium chloride 30% injection (5 mmol/mL) 20 mL
- sterile water to 100 mL.

Shelf-life

7 days refrigerated in amber glass.

Using a strength of 5 mmol/mL:

- sodium chloride 30% injection or Polyfusor (5 mmol/mL).

Transfer into an appropriately sized amber bottle (use filter straw or filter needle if ampoules are used).

Shelf-life

7 days refrigerated in amber glass.

suggested an estimated total acceptable daily intake for methyl-, ethyl- and propyl-hydroxybenzoates at up to 10 mg/kg body weight (WHO/FAO, 1974). The solubility of methyl hydroxybenzoate is decreased in aqueous sodium chloride solutions (McDonald and Lindstrom, 1974).

Technical information

Empirical formula

NaCl (British Pharmacopoeia, 2007).

Molecular weight

58.44 (British Pharmacopoeia, 2007).

Solubility

Approximately 357 mg/mL (Martindale, 2005).

pK_a

Not applicable.

Optimum pH stability

Not known.

Stability profile

Physical and chemical stability

Sodium chloride is a white crystalline powder or colourless crystals or white pearls, freely soluble in water, practically insoluble to ethanol (British Pharmacopoeia, 2007). A saturated aqueous solution has a pH of approximately 6.7–7.3 (Owen, 2003).

The solid material is stable and should be stored in well-closed container in a cool, dry place (Owen, 2003).

Solutions of some sodium salts, including sodium chloride, when stored may cause separation of solid particles from glass containers, and solutions containing such particles must not be used (Martindale, 2005). However, sodium chloride solutions are generally regarded as stable, and may be sterilised by autoclaving or filtration (Owen, 2003).

Sodium chloride solutions should be protected from excessive heat and freezing (Trissel, 2005). Below 0°C, salt may precipitate as a dihydrate (Owen, 2003).

Degradation products

None.

Stability in practice

Sodium chloride is inherently chemically stable in solution.

The St Mary's Stability Database (2006) references an analysis of a 10% w/v sodium chloride solution by Frary. No decomposition of the solution was detected after 7 years storage at room temperature. An expiry of 5 years at room temperature was suggested.

Bioavailability data

No information available.

References

British Pharmacopoeia (2007). *British Pharmacopoeia*, Vol. 2. London: The Stationery Office, pp. 1879–1880.

Loria CJ, Excehverria P, Smith AL (1976). Effect of antibiotic formulations in serum protein: bilirubin interaction of newborn infants. *J Pediatr* 89: 479–482.

Lowey AR, Jackson MN (2008). A survey of extemporaneous dispensing activity in NHS trusts in Yorkshire, the North-East and London. *Hosp Pharmacist* 15: 217–219.

McDonald C, Lindstrom RE (1974). The effect of urea on the solubility of methyl *p*-hydroxybenzoate in aqueous sodium chloride solution. *J Pharm Pharmacol* 26: 39–45.

Owen SC (2003). Sodium chloride. In: Rowe RC, Sheskey PJ, Weller PJ, eds. *Handbook of Pharmaceutical Excipients*, 4th edn. London: Pharmaceutical Press and Washington DC: American Pharmaceutical Association.

Rieger MM (2003). Ethylparabens. In: Rowe RC, Sheskey PJ, Weller PJ, eds. *Handbook of Pharmaceutical Excipients*, 4th edn. London: Pharmaceutical Press and Washington DC: American Pharmaceutical Association.

St Mary's Online Stability Database (2006). http://www.stmarys.demon.co.uk/ (no longer available) (accessed 7 April 2006).

Sweetman SC (2009). *Martindale: The Complete Drug Reference*, 36th edn. London: Pharmaceutical Press.

Trissel LA (2005). *Handbook on Injectable Drugs*, 13th edn. Bethesda, MD: American Society of Health System Pharmacists.

WHO/FAO (1974). Toxicological evaluation of certain food additives with a review of general principles and of specifications. Seventeenth report of the Joint FAO/WHO Expert Committee on Food Additives. World Health Organ. Tech. Rep. Ser. No. 539.

Sodium phenylbutyrate oral liquid

Synonym

Sodium 4-phenylbutyrate.

Risk assessment of parent compound

A risk assessment survey completed by clinical pharmacists suggested that sodium phenylbutyrate has a moderate therapeutic index. Inaccurate doses were thought to be potentially associated with serious morbidity (Lowey and Jackson, 2008). Formulations in use at the NHS trusts surveyed were of a low technical risk, based on the number and/or complexity of manipulations and calculations. Sodium phenylbutyrate is therefore regarded as a moderate-risk extemporaneous preparation (Box 14.44).

Summary

Sodium phenylbutyrate is available as granules and tablets (Ammonaps formulations) and is licensed for use within the EU. Unless there is valid reason for not doing so, licensed preparations should be used in preference to unlicensed preparations (MHRA, 2006).

There is a lack of data in the public domain to support the stability of sodium phenylbutyrate oral liquids. However, from first principles, there are no obvious routes of degradation. For patients who cannot use the licensed products, there are several powder-based preparations that are readily available from several 'Specials' manufacturers. The 'Specials' manufacturers should be able to provide supporting stability data for their formulation(s) and shelf-life. This data should be assessed by the requesting pharmacist prior to procurement.

Box 14.44 *Preferred formula*

Using a typical strength of 200 mg/mL as an example:

- sodium phenylbutyrate powder 20 g
- sterile water to 100 mL.

Shelf-life

7 days refrigerated in amber glass.

As licensed preparations and 'Specials' are available for this product, extemporaneous preparation should only be used as an interim measure until a supply can be sourced from a reputable manufacturer.

Clinical pharmaceutics

Points to consider

- At typical concentrations (200 mg/mL), all of the drug should be in solution. Therefore, dosage uniformity should not be an issue.
- Sodium phenylbutyrate has a strongly salty taste (Summary of Product Characteristics, 2007).
- Unpreserved solutions should be stored in the refrigerator and discarded 7 days after opening, unless local data are available to support a longer shelf-life.
- There are no data available with regard to bioavailability of unlicensed sodium phenylbutyrate oral liquids.
- The unlicensed injection has been reported to have been used orally (Guy's and St Thomas' and Lewisham Hospitals, 2005).

Other notes

- Sodium phenylbutyrate is a prodrug of sodium phenylacetate (Martindale, 2005).

Technical information

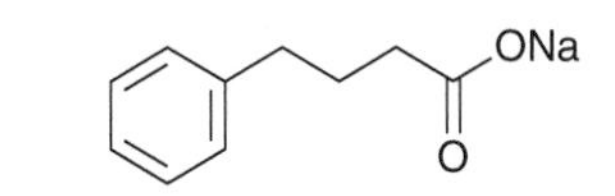

Empirical formula

$C_{10}H_{11}O_2Na$ (Summary of Product Characteristics, 2007).

Molecular weight

186 (Summary of Product Characteristics, 2007).

Solubility

Soluble in water; up to 500 mg/mL (EMEA, 2005).

pK_a

Not known.

Optimum pH stability

Not known.

Stability profile

Physical and chemical stability

Sodium phenylbutyrate is an off-white crystalline substance which is soluble in water and has a strong salty taste (Summary of Product Characteristics, 2007).

Ammonaps tablets and granules, available through Orphan Europe, should not be stored above 30°C (EMEA, 2005). A marketing authorisation valid throughout Europe was granted by the EMEA in 1999. Following submission of acceptable stability data to the EMEA, a 2-year shelf-life has been granted to both formulations if stored below 30°C.

Degradation products

No data.

Stability in practice

There are no known stability data available in the public domain for the unlicensed preparations.

Bioavailability data

No known data for the unlicensed preparations.

References

EMEA (2005). EMEA scientific discussion on ammonaps. http://www.emea.europa.int/humandocs/Humans/EPAR/ammonaps/ammonaps.htm (accessed 24 October 2006).

Guy's and St Thomas' and Lewisham Hospitals (2005). *Paediatric Formulary*. London: Guy's and St Thomas' and Lewisham NHS Trust.

Lowey AR, Jackson MN (2008). A survey of extemporaneous dispensing activity in NHS trusts in Yorkshire, the North-East and London. *Hosp Pharmacist* 15: 217–219.

MHRA (Medicines and Healthcare products Regulatory Agency) (2006). The supply of unlicensed relevant medicinal products for individual patients. Guidance note 14 (revised 2006). www.mhra.gov.uk (accessed 25 October 2006).

Summary of Product Characteristics (2007). Buphenyl. www.buphenyl.com (accessed 01 June 2007).

Sweetman SC (2009). *Martindale: The Complete Drug Reference*, 36th edn. London: Pharmaceutical Press.

St Mark's Solution

Risk assessment of parent compound

A risk assessment survey completed by clinical pharmacists suggested that St Mark's Solution has a wide therapeutic index, with the potential to cause significant side-effects if inaccurate doses were to be administered (Lowey and Jackson, 2008). Current formulations used at NHS trusts generally show a low to moderate technical risk, based on the number and/or complexity of manipulations and calculations. St Mark's Solution is therefore generally regarded as moderate-risk extemporaneous preparation (Box 14.45).

Summary

There is no information in the public domain to support the formulation or stability of the powder or subsequent solution. However, it has been in use for some time in several NHS centres, and theory would suggest that stability is not likely to be a problem. Each component is stable in solution and there are no reported incompatibilities between the ingredients.

Purchase of the powder is now possible from 'Specials' manufacturers.

Clinical pharmaceutics

Points to consider

- Current best practice is to reconstitute the 26 g powder with half to one litre of tap water before administration as a drink that is sipped

Box 14.45 *Preferred formula*

- Glucose 20 g
- sodium citrate 2.5 g
- sodium chloride 3.5 g

(Total 26 g as a powder).

Shelf-life

One year as a powder; discard 24 hours after reconstitution with 1 litre of tap water. (500 mL may be used in some situations.)

throughout the day. Any remaining drink is discarded after 24 hours (J. Eastwood, Specialist Pharmacist, St Mark's Hospital, personal communication, 2007).

- It has been reported that sodium bicarbonate has been used in place of the sodium citrate component (C. Brady, Senior Technician, County Durham and Darlington Acute Hospitals NHS Trust, personal communication, 2007).

Technical information

Table 14.10

Table 14.10 Technical information for St Mark's solution

	Glucose	Sodium citrate	Sodium chloride
Structure	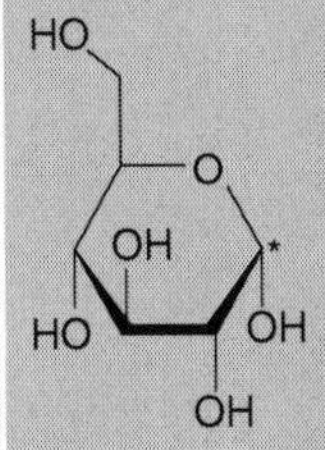 HO, OH, HO, OH, OH, O (glucose ring structure) and epimer at C*, H_2O	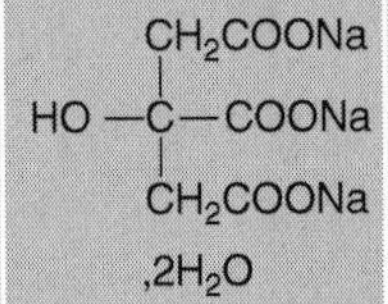 HO—C(CH$_2$COONa)(COONa)(CH$_2$COONa) ,$2H_2O$	NaCl
Empirical formula	$C_6H_{12}O_6,H_2O$	$C_6H_5Na_3O_7,2H_2O$	NaCl
Molecular weight	198.2	294.1	58.44
Solubility	Freely soluble in water (>100 mg/mL)	Freely soluble in water (>100 mg/mL)	Approximately 357 mg/mL

Adapted from British Pharmacopoeia (2007) and Martindale (2005).

Stability profile

Physical and chemical stability

Glucose is a white, crystalline powder with sweet taste, freely soluble in water, sparingly soluble in alcohol (British Pharmacopoeia, 2007).

Sodium citrate is a white, crystalline powder or white, granular crystals, slightly deliquescent in moist air (i.e. it is inclined to absorb moisture from the air until it dissolves). It is freely soluble in water, practically insoluble in alcohol. It should be stored in an airtight container (British Pharmacopoeia, 2007).

Sodium chloride is a white crystalline powder or colourless crystals or white pearls, freely soluble in water, practically insoluble to ethanol (British Pharmacopoeia, 2007). A saturated aqueous solution has a pH of approximately 6.7–7.3 (Rowe *et al.*, 2003). The solid material is stable and should be stored in a well-closed container in a cool, dry place (Rowe *et al.*, 2003).

Solutions of some sodium salts, including sodium chloride, when stored may cause separation of solid particles from glass containers, and solutions containing such particles must not be used (Martindale, 2005). However, sodium chloride solutions are generally regarded as stable, and may be sterilised by autoclaving or filtration (Rowe *et al.*, 2003).

Sodium chloride solutions should be protected from excessive heat and freezing (Trissel, 2005). Below 0°C, salt may precipitate as a dihydrate (Rowe *et al.*, 2003).

Degradation products

No data.

Stability in practice

There are no reports of the stability of the powder or the resultant solution in the public domain.

Bioavailability data

No data.

References

British Pharmacopoeia (2007). *British Pharmacopoeia*, Vols 1 and 2. London: The Stationery Office.

Lowey AR, Jackson MN (2008). A survey of extemporaneous dispensing activity in NHS trusts in Yorkshire, the North-East and London. *Hosp Pharmacist* 15: 217–219.

Rowe RC, Sheskey PJ, Weller PJ, eds (2003). *Handbook of Pharmaceutical Excipients*, 4th edn. London: Pharmaceutical Press and Washington DC: American Pharmaceutical Association.

Sweetman SC (2009). *Martindale: The Complete Drug Reference*, 36th edn. London: Pharmaceutical Press.

Trissel LA (2005). *Handbook on Injectable Drugs*, 13th edn. Bethesda, MD: American Society of Health System Pharmacists.

Tacrolimus oral mouthwash

Note: See other stability summary for tacrolimus *oral suspension*.

Risk assessment of parent compound

A risk assessment survey completed by clinical pharmacists suggested that tacrolimus has a medium therapeutic index, with the potential to cause minor side-effects if inaccurate doses were to be administered (Lowey and Jackson, 2008). Current formulations used at NHS trusts for the mouthwash show a low technical risk, based on the number and complexity of manipulations and calculations. Tacrolimus mouthwash is therefore regarded as a low- to medium-risk extemporaneous product.

Note: This risk assessment applies to the mouthwash only. The use of tacrolimus as an oral suspension may be associated with a much higher risk (Box 14.46).

Summary

Tacrolimus mouthwash is used at a limited number of NHS centres in the UK. While there is reported to be some published and unpublished data to support its clinical use (e.g. Olivier *et al.*, 2002), there are no direct stability data available. Therefore, a cautious shelf-life of 7 days is recommended.

Future developments of the formulation may include investigation of the use of suspending agents rather than water, given the poor water solubility of the drug. Note that the handling of tacrolimus is associated with health and

Box 14.46 *Preferred formula – mouthwash*

Using a typical strength of 1 microgram/mL as an example:

- tacrolimus 1 mg capsule × 1
- sterile water to 1 litre.

Shelf-life

7 days at room temperature in amber glass. **Shake the bottle before use. Do not filter. Not to be swallowed.**

safety considerations, and should only be carried out in suitable facilities with appropriate protective equipment for the operator(s).

Clinical pharmaceutics

Points to consider

- The mouthwash should not be filtered, as this may remove some of the active drug.
- Although tacrolimus is very poorly soluble, the very low strength of this preparation should mean that most or all of the drug is in solution. However, some drug may be bound to insoluble excipients. As a precaution, the mouthwash should be shaken thoroughly before use. Future formulation developments may consider whether the use of a suspending agent is desirable.

Other notes

- Tacrolimus is not compatible with PVC plastics (Taormina *et al.* 1992; Summary of Product Characteristics, 2006).

Technical information

Structure

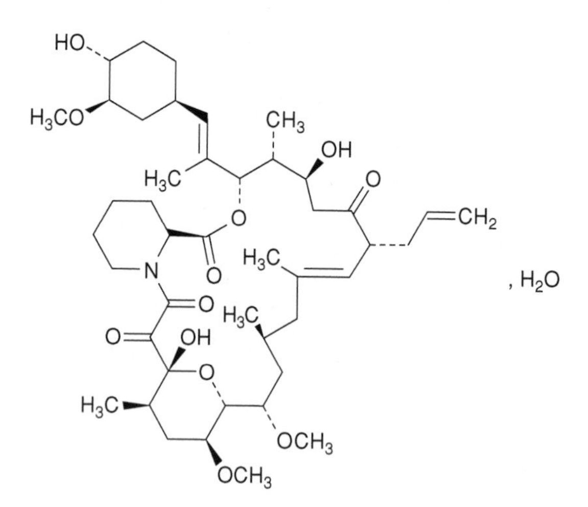

Empirical formula

$C_{44}H_{69}NO_{12},H_2O$ (Martindale, 2005).

Molecular weight

822 (Martindale, 2005).

Solubility

1–2 micrograms/mL in water (Yamashita *et al.*, 2003; Han *et al.*, 2006); 2–5 micrograms/mL in water (Namiki *et al.*, 1995).

pK_a

Not known.

Optimum pH stability

2–6 (Namiki *et al.*, 1995).

Stability profile

Physical and chemical stability

Tacrolimus is reported to be highly lipophilic and insoluble in water, but soluble in alcohol (Woods, 2001). It occurs as white crystals or crystalline powder (Trissel, 2000).

Tacrolimus exhibits maximum stability at pH 2–6; higher pH environments substantially increase degradation rates (Namiki *et al.*, 1995).

Details of the degradation process for tacrolimus are not known to have been published. However, first principles would suggest that the primary route of degradation may be hydrolysis.

Tacrolimus is known to exhibit tautomerism. Further research is needed to establish the relevance of this phenomenon in oral liquid formulations.

Tacrolimus capsules require storage at controlled room temperature (<25°C). After opening of the blister packs, the capsules are reported to be stable for 12 months. Once the aluminium wrapper is opened, the capsules in the blister strips should be kept in a dry place (Summary of Product Characteristics, 2006).

Degradation products

No information.

Stability in practice

No data available.

Bioavailability data

No data are available when formulated as a mouthwash.

References

Han J, Beeton A, Long PF, Wong I, Tuleu C (2006). Physical and microbiological stability of an extemporaneous tacrolimus suspension for paediatric use. *J Clin Pharm Ther* 31: 167–172.

Lowey AR, Jackson MN (2008). A survey of extemporaneous dispensing activity in NHS trusts in Yorkshire, the North-East and London. *Hosp Pharmacist* 15: 217–219.

Namiki Y, Fujiwara A, Kihara N, Oda S, Hane K, Yasuda T (1995). Factors affecting tautomeric phenomenon of a novel potent immunosuppressant (FK506) on the design for injectable formulation. *Drug Dev Ind Pharm* 21: 809–822.

Olivier V, Lacour J-P, Mousnier A, Garraffo R, Monteil RA, Ortonne J-P (2002). Treatment of chronic erosive oral lichen planus with low concentrations of topical tacrolimus. *Arch Dermatol* 138: 1335–1338.

Summary of Product Characteristics (2006). Prograf Capsules. www.medicines.org.uk (accessed 20 October 2006).

Sweetman SC (2009). *Martindale: The Complete Drug Reference*, 36th edn. London: Pharmaceutical Press.

Taormina D, Abdallah HY, Venkatamanan R, Logue L, Burckart GJ, Ptachcinski RJ *et al.* (1992). Stability and sorption of FK 506 in 5% dextrose injection and 0.9% sodium chloride injection in glass, polyvinyl chloride, and polyolefin containers. *Am J Hosp Pharm* 49: 119–123.

Trissel LA, ed. (2000). *Stability of Compounded Formulations*, 2nd edn. Washington DC: American Pharmaceutical Association.

Woods DJ (2001). *Formulation in Pharmacy Practice. eMixt, Pharminfotech*, 2nd edn. Dunedin, New Zealand. www.pharminfotech.co.nz/emixt (accessed 18 April 2006).

Yamashita K, Nakate T, Okimoto K, Ohike A, Tokunaga Y, Ibuki R *et al.* (2003). Establishment of new preparation method for solid dispersion formulation of tacrolimus. *Int J Pharm* 267: 79–91.

Tacrolimus oral suspension

Note: See other stability summary for tacrolimus *mouthwash*.

Risk assessment of parent compound

Tacrolimus has a narrow therapeutic index, with the potential to cause death if inaccurate doses are administered. Current formulations used at NHS trusts are generally of a moderate to high technical risk, based on the number and/or complexity of manipulations and calculations. Tacrolimus oral suspension is therefore regarded as a very high-risk extemporaneous preparation (Box 14.47).

Summary

There are several studies to support the stability of tacrolimus when formulated as an oral suspension. However, the bioavailability and efficacy of these suspensions are more controversial. It would appear a sensible step to convert the patient from oral suspension to capsules when practicable. Close monitoring of plasma drug levels should be considered, particularly given the high-risk nature of the patients treated with such preparations.

Purchase of the suspension from a 'Specials' manufacturer may be possible. Note that the method of manufacture may alter particle size and could lead to differences in bioavailability. The formulation should be reviewed by a suitably competent pharmacist before use.

Note that the handling of tacrolimus is associated with health and safety considerations, and should only be carried out in suitable facilities with appropriate protective equipment for the operator(s).

Clinical pharmaceutics

Points to consider

- Due to the very poor aqueous solubility of tacrolimus, the vast majority of drug is likely to be suspension rather than solution. **Shake the bottle before use to improve dosage uniformity.**
- Concerns have been raised with regard to poor bioavailability and erratic plasma concentrations for the oral suspension (Van Mourik *et al.*,

Box 14.47 *Preferred formula – oral suspension*

Note: 'Specials' are now available from reputable manufacturers. Extemporaneous preparation should only occur where there is a demonstrable clinical need.

Using a typical strength of **1 mg/mL** as an example:

- tacrolimus 5 mg capsules × 20
- Ora-Plus 50 mL
- Ora-Sweet to 100 mL.

Method guidance

The contents of the capsules can be ground to a fine, uniform powder in a pestle and mortar. A small amount of Ora-Plus may be added to form a paste, before adding further portions of Ora-Plus up to 50% of the final volume. Transfer to a measuring cylinder. The Ora-Sweet can be used to wash out the pestle and mortar of any remaining drug before making the suspension up to 100% volume. Transfer to an amber medicine bottle.

Shelf-life

28 days at room temperature in amber glass. **Shake the bottle before use.**

Alternative

Using a typical strength of **500 micrograms/mL:**

- tacrolimus 5 mg capsules × 10
- Ora-Plus 50 mL
- Syrup BP to 100 mL.

Shelf-life

28 days at room temperature in amber glass. **Shake the bottle before use.**

1999; Reding *et al.*, 2002; Han *et al.*, 2006). There are conflicting data with regard to the bioavailability of the suspension, with estimates ranging from around 50% to 100% of the bioavailability of the capsule (McGhee *et al.*, 1997; Reding *et al.*, 2002). Close monitoring of plasma levels is recommended.

- Ora-Plus and Ora-Sweet are free from ethanol and chloroform. Ora-Plus is free from sucrose.
- The particle size of the suspension may vary according to the method of manufacture. Suspensions made with a pestle and mortar may differ in their bioavailability from those made with automated equipment.
- The combination of Ora-Plus and Ora-Sweet is preferred, as the combination of Ora-Plus and Syrup BP (or Simple Syrup NF as in the American literature) has not been widely validated as an effective suspending vehicle. Note that Syrup BP is sucrose 667 g with purified water sufficient to produce 1,000 mL. One or more suitable antimicrobial preservatives may be added. Syrup NF (simple syrup NF) is sucrose 850 g with purified water sufficient to produce 1,000 mL.

Other notes

- Tacrolimus is not compatible with PVC plastics (Summary of Product Characteristics, 2006).
- Preparation of tacrolimus powder papers is time-consuming and carries a high potential for error (Jacobson *et al.*, 1997).
- The intravenous formulation should be used with caution to compound an oral liquid due to the presence of dehydrated alcohol and hydrogenated castor oil as stabilizing agents (Elefante *et al.*, 2006).

Technical information

Structure

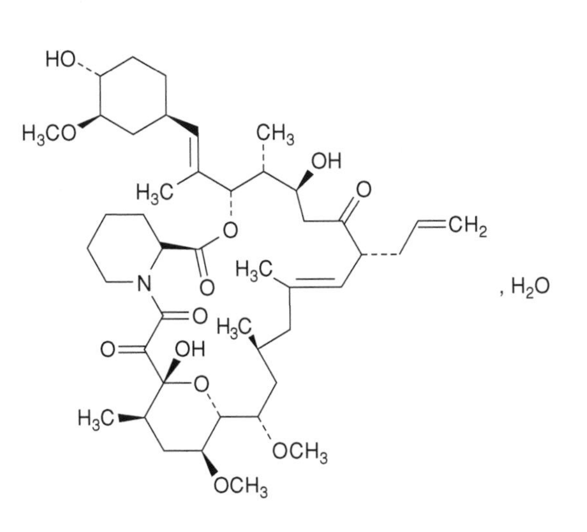

Empirical formula

$C_{44}H_{69}NO_{12},H_2O$ (Martindale, 2005).

Molecular weight

822 (Martindale, 2005).

Solubility

Reported separately as: 1–2 micrograms/mL in water (Yamashita *et al.*, 2003; Han *et al.*, 2006); 2–5 micrograms/mL in water (Namiki *et al.*, 1995).

pK_a

Not known.

Optimum pH stability

2–6 (Namiki *et al.*, 1995).

Stability profile

Physical and chemical stability

Tacrolimus is reported to be highly lipophilic and insoluble in water, but soluble in alcohol (Woods, 2001). It occurs as white crystals or crystalline powder (Trissel, 2000).

Tacrolimus exhibits maximum stability at pH 2–6; higher pH environments substantially increase degradation rates (Namiki *et al.*, 1995).

Details of the degradation process for tacrolimus are not known to have been published. However, first principles would suggest that the primary route of degradation may be hydrolysis.

Tacrolimus is known to exhibit tautomerism. Further research is needed to establish the relevance of this phenomenon in oral liquid formulations.

Tacrolimus capsules require storage at controlled room temperature (<25°C). After opening of the blister packs, the capsules are reported to be stable for 12 months. Once the aluminium wrapper is opened, the capsules in the blister strips should be kept in a dry place (Summary of Product Characteristics, 2006).

Degradation products

No information available.

Stability in practice

There are three main studies which describe the physical and chemical stability of tacrolimus in vehicles consisting of Ora-Plus and a syrup preparation, and Ora-Plus and Ora-Sweet.

Jacobson *et al.* (1997) have investigated the stability of tacrolimus 500 micrograms/mL in an extemporaneously compounded liquid. The contents of six 5 mg tacrolimus capsules were suspended in 60 mL of a 1:1 mixture of Ora-Plus and Simple Syrup NF (sucrose 850 g with purified water to 1,000 mL). Six identical suspensions were prepared and divided equally between storage in glass and plastic prescription bottles. All the samples were stored at room temperature, defined as 24–26°C.

A 1 mL sample was withdrawn from each bottle on days 0, 7, 15, 30, 45 and 56, and analysed using a stability-indicating assay. The lack of microbiological testing is justified by the authors due to the presence of effective preservatives in both Ora-Plus and Simple Syrup NF. Ora-Plus contains a hydroxybenzoate preservative system. The preservative in the Simple Syrup NF preparation used by Jacobson *et al.* (1997) is not known.

At least 98% of the initial tacrolimus concentration remained in all the suspensions throughout the 56 days. No significant changes were detected in colour, odour or pH (starting mean pH 4.6). The authors describe a sweet flavour with a slightly bitter aftertaste. The plastic in the bottles was polypropylene resin, not PVC.

Following concerns regarding physical stability and particle size (see below), Han *et al.* (2006) studied the physical and microbiological stability of a formulation very similar to that described by Jacobson *et al.* (1997), although they used a BP grade syrup preparation in place of Simple Syrup NF. Note that Syrup BP contains 667 g sucrose, purified water to 1,000 mL, and that one or more suitable antimicrobial preservatives may be added. The authors refer to the syrup preparation as 'simple syrup of BP grade'. The lead author has confirmed that the formulation used Syrup BP with preservatives (J. Han, personal communication, 2007).

Although the preparation process was essentially the same as that described by Jacobson *et al.* (1997), one extra homogenisation step was added. The contents of the tacrolimus capsules were mixed into Ora-Plus using a pestle and mortar, followed by stepwise addition of the syrup. After thorough mixing, the suspensions were further homogenised using a high shear mixer (Model L4RT; Silverson Machine Ltd) at a speed of 7,500 rpm for 2 minutes. The suspensions were made on behalf of the authors by a UK Specials manufacturer (Specials Laboratories Ltd, Low Prudhoe, UK) under good manufacturing practice (GMP) conditions. Three batches of the suspension were packed into 50 mL amber glass bottles with screw-cap seals and supplied for testing. Analysis began less than 24 hours after production.

Allowing for normal experimental error and drift, no significant changes were detected in particle size, using laser diffraction and microscopical examination. The particle size measured was the mean particle size, including drug and excipient components. None of the samples showed any microbiological contamination.

Ora-Plus contains a blend of suspending agents, which may have helped to form a structured gel-like matrix of high viscosity reducing the coalescence and sedimentation of the particles (Ofner *et al.*, 1996). Han *et al.* (2006) suggest that cellulosic derivatives in the Ora-Plus and the high viscosity of the media may have suppressed crystal growth. The low pH of the formulation and the antimicrobial nature of tacrolimus may have contributed to the lack of microbial growth.

In conclusion, the reported low bioavailability and erratic plasma concentration could not be correlated to physical or microbiological instability. The authors suggest that poor dosing (e.g. impractically small volumes) or poor shaking may warrant further investigation as potential causes of erratic plasma concentrations.

The method of manufacture of the suspension may also be a factor, as suspensions made with automated equipment may display different physical characteristics to those suspensions made by hand using only the traditional pestle and mortar approach.

Elefante *et al.* (2006) studied the long-term stability of a 1 mg/mL tacrolimus suspension, after the 500 micrograms/mL suspension described above led to administration errors and the confusion of milligrams with millilitres. The contents of six 5 mg capsules were placed in an amber plastic bottle and 5 mL of water was added. The bottle was agitated until the powder dispersed to form a slurry. Equal volumes of Ora-Plus and Ora-Sweet were added to a final volume of 30 mL. Study samples of 3 mL were prepared in triplicate and stored at room temperature (23–26°C) in capped plastic bottles.

The samples were analysed by a stability-indicating HPLC assay on days 0, 78 and 132. Although statistical analysis found no difference between the results on these three days, the range of values at day 132 extended down to 91.3% of the original concentration. If more data points had been available, the statistical package used may have detected a trend. The study is limited by the number of data points, and by the use of 3 mL samples volumes, which would not mimic the practical usage of the suspension.

The authors note the difficulty associated with degrading the tacrolimus during assay validation, in addition to the long-term stability of the suspension over four months at room temperature.

Bioavailability data

Oral absorption of tacrolimus, including licensed preparations, is reported to be erratic. Oral bioavailability varies widely, with typical mean oral bioavailability from the capsules quoted at 20–25% (Summary of Product Characteristics, 2006). The variation in bioavailability may not be highly important for the mouthwash preparation, but may become crucial if an oral suspension is formulated. Given the known variation in bioavailability from the capsules, it follows that oral suspensions derived from the capsules will also display a marked variation in plasma levels.

Although Jacobson *et al.* (1997) acknowledge that the bioavailability of the Ora-Plus/Simple Syrup NF formulation was not assessed formally, the authors suggest that its bioavailability should not differ significantly from that of the capsule. The authors also state that adults and children have shown satisfactory plasma levels using this formulation. The exact procedure for making the suspension is presented, using a pestle and mortar.

McGhee *et al.* (1997) have subsequently reported the successful use of the formulation described by Jacobson *et al.* (1997) in 20 paediatric patients. Nineteen of the patients were solid organ recipients (heart, $n = 5$; heart–lung, $n = 4$; lung, $n = 4$; liver–intestine, $n = 4$; liver, $n = 3$; intestine, $n = 2$; multivisceral, $n = 1$) and one patient had dermatomyositis. The suspension was administered orally or via an enteral tube (nasogastric, nasojejunal, gastrostomy, gastrojejunal or jejunostomy tube). McGhee *et al.* (1997) suggest that all patients were able to maintain therapeutic drug levels (normal range quoted as 5–20 ng/mL) at standard doses comparable to those used with capsules.

Nine patients continued the suspension as outpatients, with no loss of efficacy. The authors conclude that the suspension is an effective method of delivering tacrolimus to infants and children who cannot swallow, as well as patients who require administration via an enteral tube.

Reding *et al.* (2002) found the bioavailability of the suspension to be approximately 50% lower than that of the capsule, when used in paediatric liver transplant recipients. The oral suspension was administered to 15 paediatric liver transplant recipients (mean age 1.4 years), and the data compared to corresponding data with tacrolimus capsules. Graft and survival rates were 100%, with acute rejection and steroid-resistant rejection encountered in 9/15 and 3/15 patients, respectively.

The authors suggest that the oral absorption of tacrolimus was lower for the oral suspension than for the capsules. However, the comparator data for patients taking capsules is derived from a separate study based on a different cohort of patients (Wallemacq and Reding, 1993). This difference may make some of the comparisons inaccurate. Moreover, no direct correlation could be shown between pharmacokinetic parameters and rejection, and a similar rejection incidence was found with both formulations (60% versus 55% for the suspension and capsules, respectively). The authors conclude by stating that the oral suspension may be useful for dose titration in low body weight recipients in the early post-transplant phase. However, prompt conversion to capsules is recommended. Overall bioavailability for the oral suspension is estimated at 12.9%, compared with a figure of 25% described in the separate cohort of patients taking capsules rather than suspension (Reding *et al.*, 2002).

Van Mourik *et al.* (1999) have reported erratic trough levels in the first few days of administration in a group of 20 children. The authors suggest that this finding may be linked to an increased risk of acute graft rejection, which was approximately double that of a control group treated with granules or capsules. It is unclear whether this difference is due to hepatic metabolism or intestinal absorptive barriers of the children, or due to formulation factors including poor physical stability (particle size changes, especially crystal growth during storage) or poor dosing (Han *et al.*, 2006).

The work carried out by Han *et al.* (2006) did not find any significant change in particle size over eight weeks for the 500 micrograms/mL tacrolimus suspension in a 1:1 mixture of Ora-Plus and Syrup BP. The authors suggest that dosing techniques and poor shaking may warrant further investigation (Han *et al.*, 2006).

References

Elefante A, Nuindi J, West K, Dunford L, Abel S, Paplham P *et al.* (2006). Long-term stability of a patient-convenient 1 mg/ml suspension of tacrolimus for accurate maintenance of stable therapeutic levels. *Bone Marrow Transplant* 37: 781–784.

Han J, Beeton A, Long PF, Wong I, Tuleu C (2006). Physical and microbiological stability of an extemporaneous tacrolimus suspension for paediatric use. *J Clin Pharm Ther* 31: 167–172.

Jacobson PA, Johnson CE, West NJ, Foster JA (1997). Stability of tacrolimus in an extemporaneously compounded oral liquid. *Am J Health Syst Pharm* 54: 178–180.

McGhee B, McCombs JR, Boyle G, Webber S, Reyes JR (1997). Therapeutic use of an extemporaneously prepared oral suspension of tacrolimus in paediatric patients. *Transplantation* 64: 941.

Namiki Y, Fujiwara A, Kihara N, Koda S, Hane K, Yasuda T (1995). Factors affecting tautomeric phenomenon of a novel potent immunosuppressant (FK506) on the design for injectable formulation. *Drug Dev Ind Pharm* 21: 809–822.

Ofner CM, Schnaare RL, Schwartz JB (1996). Oral aqueous suspensions. In: Lierberman HA, Rieger MM, Banker GS, eds. *Pharmaceutical Dosage Forms – Disperse Systems*, Vol. 2. New York: Marcel Dekker, pp. 149–181.

Olivier V, Lacour J-P, Mousnier A, Garraffo R, Monteil RA, Ortonne J-P (2002). Treatment of chronic erosive oral lichen planus with low concentrations of topical tacrolimus. *Arch Dermatol* 138: 1335–1338.

Reding R, Sokal E, Paul K, Janssen M, Evrard V, Wilmotte L *et al.* (2002). Efficacy and pharmacokinetics of tacrolimus oral suspension in pediatric liver transplant recipients. *Pediatr Transplant* 6: 124–126.

Summary of Product Characteristics (2006). Prograf Capsules. www.medicines.org.uk (accessed 20 October 2006).

Sweetman SC (2009). *Martindale: The Complete Drug Reference*, 36th edn. London: Pharmaceutical Press.

Taormina D, Abdallah HY, Venkatamanan R, Logue L, Burckart GJ, Ptachcinski RJ *et al.* (1992). Stability and sorption of FK 506 in 5% dextrose injection and 0.9% sodium chloride injection in glass, polyvinyl chloride, and polyolefin containers. *Am J Hosp Pharm* 49: 119–123.

Trissel LA, ed. (2000). *Stability of Compounded Formulations*, 2nd edn. Washington DC: American Pharmaceutical Association.

Van Mourik IDM, Holder G, Beath S, McKiernan PUB, Kelly D (1999). Increased risk of acute graft rejection in paediatric liver transplant recipients on tacrolimus suspension. Paper presented at the 3rd International Congress on Immunosuppression, San Diego, USA.

Wallemacq P, Reding R (1993). FK 506 (Tacrolimus), a novel immunosuppressant in organ transplantation: clinical, biomedical and analytical aspects. *Clin Chem* 39: 2219–2228.

Woods DJ (2001). *Formulation in Pharmacy Practice. eMixt, Pharminfotech*, 2nd edn. Dunedin, New Zealand. www.pharminfotech.co.nz/emixt (accessed 18 April 2006).

Yamashita K, Nakate T, Okimoto K, Ohike A, Tokunaga Y, Ibuki R *et al.* (2003). Establishment of new preparation method for solid dispersion formulation of tacrolimus. *Int J Pharm* 267: 79–91.

Thiamine hydrochloride oral liquid

Risk assessment of parent compound

A risk assessment survey completed by clinical pharmacists suggested that thiamine hydrochloride has a wide therapeutic index, with the potential to cause minor side-effects if inaccurate doses were to be administered (Lowey and Jackson, 2008). Current formulations used at NHS trusts generally show a moderate technical risk, based on the number and/or complexity of manipulations and calculations. Thiamine hydrochloride is therefore regarded as a low- to moderate-risk extemporaneous preparation (Box 14.48).

Summary

Previous NHS extemporaneous formulations of thiamine hydrochloride have been based on the use of xanthan gum ('Keltrol') suspending agents. However, the exact formulation pH of such suspending agents is variable depending on the supplier, and typically contains ethanol and chloroform. The exact details should be checked by a suitably competent pharmacist

***Box 14.48** Preferred formula*

Using a typical strength of 20 mg/mL as an example:

- thiamine hydrochloride 100 mg tablets × 20
- Ora-Plus 50 mL
- Ora-Sweet to 100 mL.

Method guidance

Tablets can be ground to a fine, uniform powder in a pestle and mortar. A small amount of Ora-Plus may be added to form a paste, before adding further portions of Ora-Plus up to 50% of the final volume. Transfer to a measuring cylinder. The Ora-Sweet can be used to wash out the pestle and mortar before making the suspension up to 100% volume. Transfer to an amber medicine bottle.

Shelf-life

28 days at room temperature in amber glass. **Shake the bottle.**

before use. The pH of 'Keltrol' preparations may be close to the threshold pH of 6, above which degradation is exponentially increased. The lower pH of the 'Ora' preparations may help to limit degradation.

Although the published data below is based on powdered thiamine hydrochloride as a starting material, a 28-day shelf-life may be justifiable for suspensions made from tablets. A slight change in colour and taste may occur during the shelf-life, but this should not adversely affect stability.

The commissioning of this product as a 'Special' may be a future option, as stability would appear to be relatively good. 'Specials' manufacturers should be able to provide justification for the shelf-life attributed.

Clinical pharmaceutics

Points to consider

- Thiamine hydrochloride should be in solution at all usual strengths (20–100 mg/mL), due to its very good solubility.
- The exact formulation of 'Keltrol' varies according to the manufacturer, but typically contains ethanol and chloroform. The pH of the formulation may be crucial for some drugs, such as thiamine. The exact specification should be checked by a suitably competent pharmacist before use.
- Ora-Plus and Ora-Sweet SF are free from chloroform and ethanol.
- Ora-Plus and Ora-Sweet SF contain hydroxybenzoate preservative systems.
- Ora-Sweet SF has a cherry flavour which may help to mask the bitter taste of the drug.
- There are no known bioavailability data.
- Thiamine has a characteristic odour described as 'meat-like' (Al-Rashood *et al.*, 1989).

Technical information

Structure

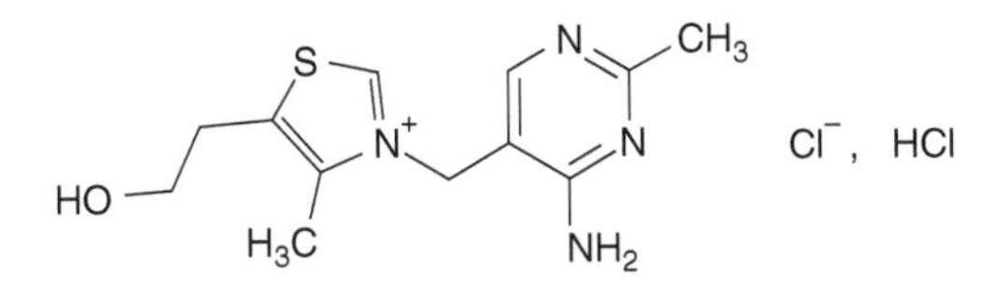

Empirical formula

$C_{12}H_{17}ClN_4OS,HCl$ (British Pharmacopoeia, 2007).

Molecular weight

337.3 (British Pharmacopoeia, 2007).

Solubility

1 g/mL (Al-Rashood *et al.*, 1989; Martindale, 2005).

pK_a

5.0 and 9.5 at 25°C (Connors *et al.*, 1986).

Optimum pH stability

Approximately pH 2; the pH of aqueous solutions should be lower than 6 for acceptable stability (Connors *et al.*, 1986; Al-Rashood *et al.*, 1989). Described elsewhere as 3.5 (Windheuser and Higuchi, 1962).

Stability profile

Physical and chemical stability

Thiamine hydrochloride is a white or almost white, crystalline powder or colourless crystals (British Pharmacopoeia, 2007). It has a faint characteristic odour (Martindale, 2005).

When powders or tablets are stored under dry conditions, stability is described as satisfactory over a period of years (Connors *et al.*, 1986). In the dry form, thiamine is reported to be stable at 100°C for 24 hours (Al-Rashood *et al.*, 1989).

Thiamine is known to be increasingly unstable in solution as the pH rises and is decomposed by oxidising or reducing agents (Kearney *et al.*, 1995). In aqueous solution, it can be sterilised at 110°C, but rapid destruction of thiamine occurs if the pH of the solution is above 5.5 (Allwood, 1984; Al-Rashood *et al.*, 1989). The pH of maximum stability is reported to be around 2 (Al-Rashood *et al.*, 1989; Martindale, 2005). In water, the stability after six months storage at pH 2 and 37°C is reported to be 100% (Al-Rashood *et al.*, 1989).

A 2.5% solution in water has a pH of 2.7 – 3.3 (Martindale, 2005). A 1% solution in water has a pH of 3.13, and a 0.1% solution in water has a pH of 3.58 (Al-Rashood *et al.*, 1989). Sterile solutions of pH 4 or less are reported to lose activity only very slowly, but neutral or alkaline solutions degrade rapidly, especially when open to the air (Martindale, 2005). The anhydrous powder quickly absorbs about 4% of water to form a hydrate when exposed to air (Connors *et al.*, 1986; Martindale, 2005). Thiamine undergoes several hydrolytic reactions, and exhibits maximal stability near pH 2 (Connors *et al.*, 1986). Metal ion catalysis is prevented with the use of sodium edetate.

The complex pH–rate profile shows an exponential increase in degradation at around pH 7 and above. The activation energies for thiamine hydrolysis are 27.4 kcal/mol, 30.6 kcal/mol and 25 kcal/mol at pH 3.0, 5.0 and 6.37, respectively (Connors *et al.*, 1986).

Thiamine is incompatible with oxidising agents, and reducing agents such as sulfites (Connors *et al.*, 1986; Martindale, 2005). The molecule is cleaved by sulfite or bisulfite, and the combination of sulfite and a nucleophile causes rapid loss of thiamine. The mechanism of degradation of thiamine in the presence of sulfite was first reported in 1939 (Williams and Spies, 1988), and the hydrolytic cleavage increases with increasing pH to a maximum at pH 6 (Kearney *et al.*, 1995).

When oxygen and sulfite are avoided, the primary routes of degradation are hydrolytic, with the nature of the reaction dependent on pH (Connors *et al.*, 1986). The hydrolysis is first order. Connors *et al.* (1986) describe various mechanisms, and state that thiamine is capable of existing in several ionic forms. Protonated thiamine loses a proton to give the unprotonated quaternary form (pK_a 5.0 at 25°C). The unprotonated form reacts with hydroxide to eventually form a thiol.

At high pHs, thiamine is converted via an acid/base equilibrium process to a yellow substance (average pK_a for reaction = 11.6), the colour of which fades as a subsequent reaction yields the thiol (Connors *et al.*, 1986).

Thiamine may also be oxidised, forming degradation products that include the strongly fluorescent thiochrome (Connors *et al.*, 1986).

Thiamine hydrochloride powder should be stored in a non-metallic container, protected from light (British Pharmacopoeia, 2007). Thiamine hydrochloride BP is formulated at pH 2.8–3.4 (British Pharmacopoeia, 2007). The degradation of thiamine tablets is minimised by the exclusion of moisture (Connors *et al.*, 1986).

Degradation products

May include the strongly fluorescent thiochrome (Connors *et al.*, 1986).

Stability in practice

There are two main publications that discuss the stability of thiamine in oral liquid formulation. They provide encouraging data to support the formulation of thiamine hydrochloride as an extemporaneous product.

Ensom and Decarie (2005) have studied the stability of thiamine 100 mg/mL in extemporaneously compounded suspensions. The suspensions were prepared from thiamine powder in a 1:1 mixture of Ora-Sweet and Ora-Plus.

Six 50 mL test liquids were prepared and divided between storage at 4 and 25°C. The 4°C samples were refrigerated and not exposed to light except during analysis, but the 25°C samples were exposed to fluorescent light in the laboratory. The suspensions were checked each week (up to 91 days) for odour, pH, taste, colour, viscosity, formation of precipitates and ease of resuspension. One-millilitre samples were withdrawn after shaking the bottles for 10 seconds and analysed by a stability-indicating HPLC method.

No noticeable changes in physical appearance, odour, colour or taste of the suspensions occurred over 14 days. The suspensions had a cloudy white appearance with a faint sweet smell and taste. The authors describe a slight yellow colour, and small changes in odour and taste (less sweet) by day 21, but thereafter these descriptions did not change.

Viscosity and pH were largely unchanged throughout the 91 days (mean pH 3.0 ± 0.16 at 4°C and 2.9 ± 0.20 at 25°C).

At both storage temperatures, the suspensions maintained at least 90% of their initial study concentrations on every study day (days 0, 7, 14, 21, 28, 35, 42, 49, 56, 63, 70, 77, 84 and 91). No obvious degradation trends can be found as any degradation that may have occurred is masked by the natural experimental error associated with the analysis.

The authors comment that the suspending agents used do not contain metabisulfites, and suggest that thiamine suspensions of 100 mg/mL stored at either 4 or 25°C maintain at least 90% of their original concentrations for up to 91 days.

El-Khawas and Boraie (2000) have studied the stability of thiamine hydrochloride in sugar-based liquid dosage forms at various temperatures. Powder was used as the starting material for solutions of 5 mg/mL in vehicles containing 40% sucrose, dextrose 21.1%, fructose 21.1% or dextrose–fructose mixture (21.1% each) at temperatures of 25, 45, 55 and 65°C. The formulations contained 0.1% methyl-hydroxybenzoate as a preservative and were prepared in McIlvain's citric phosphate buffer, with a pH of 3.5.

A sugar-free formulation was also prepared in 30% sorbitol solution in McIlvain's citric-phosphate buffer (pH 3.5), in addition to a control sample using buffer with no sugar added. All formulations were placed in amber glass bottles with a screw-cap. Five samples of each formulation were prepared.

Physical appearance, pH and drug concentration were monitored in addition to drug content, using a stability-indicating reversed-phase HPLC assay. The methyl-hydroxybenzoate was used as an internal standard. Intra-day relative standard deviation (RSD) ranged from 1.2% to 2.3% ($n = 8$) and inter-day RSD ranged from 1.0% to 2.7% ($n = 15$).

The elution times were 2.5 minutes and 6 minutes for the thiamine hydrochloride and methyl-hydroxybenzoate, respectively. After samples were stored, a new peak developed at 2 minutes, accompanied by a decrease in the intensity of the thiamine peak, thus indicating the instability of the drug in the presence of sucrose.

The authors found first-order degradation described as 'quite fast' in the presence of 40% sucrose, 21.1% fructose, 21.1% dextrose and a mixture of dextrose and fructose (21.1% each). Drug losses in the sugar solutions ranged from 5% to 12% after 7 days at 45°C. However, the control solution and the sorbitol solution showed good stability, retaining over 90% of the initial concentration after 136 days at 45°C.

Given the differences between the control and the sugar solutions, the authors suggest that the loss of drug is mainly due to interactions with the sugars rather than any inherent instability of thiamine.

At room temperature, most of the formulations displayed acceptable stability. The control solution, sorbitol 30% solution and dextrose 21.1% solution all maintained over 90% of drug content for 2 years, with the remaining solutions reaching this threshold after approximately 7–8 months.

El-Khawas and Boraie (2000) conclude that thiamine is potentially 'unstable' in liquids that contain sucrose, and that a sugar-free sorbitol 30% solution is recommended as an alternative.

Related preparations – total parenteral nutrition (TPN) formulations

Baumgartner *et al.* (1997) investigated the stability of thiamine and ranitidine in TPN solutions. Each litre of the TPN solution contained 60 g amino acids, 250 g carbohydrate, 35 mEq sodium chloride, 40 mEq potassium acetate, 5 mEq calcium gluconate, 7.5 nM potassium phosphate, 16 mEq of magnesium sulfate, combination trace elements, multivitamins and sterile water.

The solutions were compounded on each study day, stored in a fridge and protected from light until used. Either ranitidine 150 mg/L and/or thiamine 5 mg/L were added to the bags, in the presence or absence of multivitamins.

Each of the study solutions were prepared in triplicate and sampled at 0, 3, 6, 12, 24, 36, 72 and 168 hours. Some solutions were sampled again at 240 and 408 hours. The drug concentrations were assayed using stability-indicating HPLC methods. The thiamine degradation did not show zero-order or simple first-order kinetics, necessitating the use of computer simulation programs. The authors suggest that the time taken for 10% biexponential first-order degradation of thiamine to be 12.9 hours in the solution containing ranitidine and multivitamins, 11.1 hours in the solution containing multivitamins and 33.4 hours in the solution containing thiamine hydrochloride (at 25°C). Ranitidine did not have a deleterious effect on the stability of thiamine in the solutions. No degradation of thiamine could be detected in TPN salt solutions (in the absence of amino acids and carbohydrates).

The stability studies for thiamine at 4°C showed negligible degradation at 24 hours in the presence and absence of ranitidine. After 7 days, 93.5% of the thiamine remained in the presence of ranitidine, and 97.4% remained in the absence of ranitidine. There were no significant changes in pH (typical range 5.6–6.3); those solutions containing multivitamins showed a darkening of the yellow colour after 48 hours at room temperature.

The authors conclude that the degradation of thiamine in TPN involves amino acids and/or carbohydrates, and that thiamine as a component of multivitamin mixtures degrades faster than thiamine alone, regardless of the presence or absence of ranitidine.

Kearney *et al.* (1995) studied the stability of thiamine in TPN mixtures stored in ethylene vinyl acetate (EVA) and multilayered bags, given that the stability of vitamins is a major limiting factor in the shelf-life of TPN regimens. The authors suggested that the use of reduced gas-permeable bags may compromise the stability of thiamine, as reduction appears to be the major degradation mechanism in TPN mixtures.

A stability-indicating HPLC method was used to examine various concentrations of TPN. Stability was highly dependent on the TPN constituents. Thiamine 50 mg/bag (as Multibionta) was degraded in TPN mixtures that contained a metabisulfite-stabilised amino acid infusion (Freamine III 8.5%), but was relatively stable in mixtures containing amino acid infusions either without stabilisers (e.g. Vamin 14) or with weak reducing agents (e.g. Aminoplex 12).

In addition, degradation of thiamine in Freamine-containing mixtures was more rapid in multilayered bags when compared to gas-permeable EVA containers. This may be because metabisulfite will be more stable in bags with minimum oxygen content.

The authors conclude that only reduced gas-permeable containers should be used for storing compounding TPN mixtures with added vitamins.

Related preparations – food products

Nisha Rekha *et al.* (2004) have studied the degradation of thiamine in solution and in red gram splits (*Cajanus cajan* L.), pH 4.5, 5.5 and 6.5 over a temperature range of 50–120°C. The degradation followed first-order kinetics and increased with increasing temperature and increasing pH.

Thiamine is known to be sensitive to processing conditions (especially temperature and pH) in the cooking and baking of meats, vegetables and fruits and during storage (Bender, 1966). Stability is known to be affected by pH, temperature, heating time, water activity or moisture content, ionic strength, buffer type and processing method (Nisha Rekha *et al.*, 2004).

Bioavailability data

No data available.

References

Allwood MC (1984). Compatibility and stability of TPN mixtures in big bags. *J Clin Hosp Pharm* 9: 181–198.

Al-Rashood KAM, Al-Shammary FJ, Mian NAA (1989). Pyridoxine. In: Florey K, ed. *Analytical Profiles of Drug Substances*, Vol. 18. London: Academic Press, pp. 413–458.

Baumgartner TG, Henderson GN, Fox J, Gondi U (1997). Stability of ranitidine and thiamine in parenteral nutrition solutions. *Nutrition* 13: 547–553.

Bender AE (1966). Nutritional effects of food processing. *J Food Sci* 1: 261.

British Pharmacopoeia (2007). *British Pharmacopoeia*, Vol. 2. London: The Stationery Office, pp. 2026–2028.

Connors KA, Amidon GL, Stella ZJ (1986). *Chemical Stability of Pharmaceuticals: A Handbook for Pharmacists*, 2nd edn. New York: Wiley Interscience.

El-Khawas M, Boraie NA (2000). Stability and compatibility of thiamine hydrochloride in liquid dosage forms at various temperatures. *Acta Pharm* 5: 219–228.

Ensom MHH, Decarie D (2005). Stability of thiamine in extemporaneously compounded suspensions. *Can J Hosp Pharm* 58: 26–30.

Kearney MCJ, Allwood MC, Neale T, Hardy G (1995). The stability of thiamine in total parenteral nutrition mixtures stored in EVA and multi-layered bags. *Clin Nutr* 14: 295–301.

Lowey AR, Jackson MN (2008). A survey of extemporaneous dispensing activity in NHS trusts in Yorkshire, the North-East and London. *Hosp Pharmacist* 15: 217–219.

Nisha Rekha P, Singhal S, Pandit AP (2004). A study on degradation kinetics of thiamine in red gram splits (*Cajanus cajan* L.). *Food Chem* 85: 591–598.

Sweetman SC (2009). *Martindale: The Complete Drug Reference*, 36th edn. London: Pharmaceutical Press.

Williams RR, Spies TD (1988). *Vitamin B1 and its Uses in Medicine*. New York: Macmillan, 1988.

Windheuser JT, Higuchi T (1962). Kinetics of thiamine hydrolysis. *J Pharm Sci* 51: 354.

Tranexamic acid oral mouthwash

Risk assessment of parent compound

A risk assessment survey completed by clinical pharmacists suggested that tranexamic acid has a wide therapeutic index (when administered as a mouthwash), with the potential to cause generally mild side-effects if inaccurate doses were to be administered (Lowey and Jackson, 2008). Current formulations used at NHS trusts generally show a low technical risk, based on the number and/or complexity of manipulations and calculations. Tranexamic acid (as a mouthwash) is therefore generally regarded as a low-risk extemporaneous preparation. There may be occasions, however, when therapeutic failure could lead to more serious consequences (Box 14.49).

Summary

Although there is a lack of stability data for oral liquid formulations of tranexamic acid, the molecule is known to be relatively stable. A 7-day expiry is recommended due to the lack of data and the absence of an effective preservative system.

'Specials' are also reported to be available (P. Mulholland, Pharmacist, Southern General Hospital, Glasgow, personal communication, 2007). The supplier should be able to justify their assigned shelf-life. The specification of the product should be assessed before use by a suitably competent pharmacist.

Box 14.49 *Preferred formulae*

Using a typical strength of 50 mg/mL 5% as an example:

- tranexamic acid injection 100 mg/mL × 50 mL
- sterile water for irrigation to 100 mL.

Shelf-life

7 days refrigerated in amber glass (this is an unpreserved solution).

Alternative

Disperse tablets in water immediately before each dose.

Clinical pharmaceutics

Points to consider

- No excipient issues are expected; the injection typically contains the active drug and water only.
- Tranexamic acid liquid preparations are thought to have little or no taste (Ambados, 2003).

Other notes

- Alternative methods of giving tranexamic acid as a 5% mouthwash include crushing a 500 mg tablet and mixing with 10 mL of water immediately prior to administration (E.R. Lowey, Senior Pharmacist (Dispensary), Airedale General Hospital, West Yorkshire, personal communication, 2006). Whether complete dissolution occurs may depend on the brand of tablets used.
- Ambados (2003) has described the dispersion of a 500 mg Cyklokapron tablet in 10–15 mL of water. Ambados suggests that the tablet will dissolve in 3–5 minutes on standing (or faster with intermittent swirling). Despite the solubility of the drug, patients should be warned that insoluble excipient material will remain after adequate agitation.
- Woods (2001) has also suggested the preparation of a liquid by mixing crushed tablets with water, and then filtering the insoluble excipients. When stored in the fridge and protected from light, a 5-day expiry is recommended, due to the lack of any formal testing.
- A commercially available liquid formulation is reported to be available in some countries (Woods, 2001).

Technical information

Structure

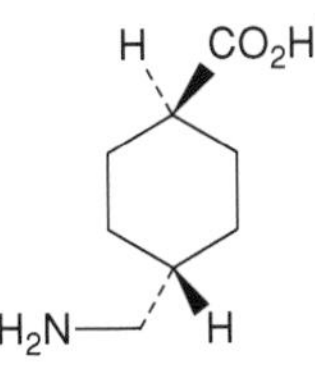

Empirical formula

$C_8H_{15}NO_2$ (Martindale, 2005).

Molecular weight

157.2 (Martindale, 2005).

Solubility

Freely soluble in water (Martindale, 2005). Approximately 167 mg/mL in water (Budavari, 1989).

pK_a

Not known.

Optimum pH stability

Not known.

Stability profile

Physical and chemical stability

Tranexamic acid is a white crystalline powder, freely soluble in water and glacial acetic acid, and practically insoluble in alcohol and ether. A 5% solution has a pH of 6.5–8.0 (Martindale, 2005). The powder is not hygroscopic (Budavari, 1989).

Tranexamic acid has been described as 'chemically stable' (Budavari, 1989). Cyklokapron injection has a shelf-life of 3 years (Summary of Product Characteristics, 2006). It is sterilised by autoclaving.

Degradation products

Not known.

Stability in practice

There are no published stability trials for tranexamic acid mouthwash. Tranexamic acid has been shown to be physically compatible with suspension Diluent A ('Keltrol'), a xanthan gum-based suspending agent (Nova Laboratories Ltd, 2003).

Bioavailability data

No data.

References

Ambados F (2003). Preparing tranexamic acid 4.8% mouthwash. *Aust Prescr* 26: 75.

Budavari S, ed. (1989). *Merck Index: An Encyclopedia of Chemicals, Drugs and Biologicals*, 11th edn. Rahway, NJ: Merck and Co.

Lowey AR, Jackson MN (2008). A survey of extemporaneous dispensing activity in NHS trusts in Yorkshire, the North-East and London. *Hosp Pharmacist* 15: 217–219.

Nova Laboratories Ltd (2003). Information Bulletin – Suspension Diluent Used at Nova Laboratories.

Summary of Product Characteristics (2006). Cyklokapron Injection. www.medicines.org.uk (accessed 13 December 2006).

Sweetman SC (2009). *Martindale: The Complete Drug Reference*, 36th edn. London: Pharmaceutical Press.

Woods DJ (2001). *Formulation in Pharmacy Practice. eMixt, Pharminfotech*, 2nd edn. Dunedin, New Zealand. www.pharminfotech.co.nz/emixt (accessed 18 April 2006).

Vancomycin hydrochloride oral liquid

Risk assessment of parent compound

A risk assessment survey completed by clinical pharmacists suggested that vancomycin hydrochloride has a wide to moderate therapeutic index (when administered orally), with the potential to cause mild side-effects if inaccurate doses were to be administered (Lowey and Jackson, 2008). Current formulations used at NHS trusts generally show a low technical risk, based on the number and/or complexity of manipulations and calculations. Vancomycin hydrochloride is therefore generally regarded as a low- to moderate-risk extemporaneous preparation (Box 14.50).

Summary

Vancomycin injection is licensed for oral use. The Summary for Product Characteristics (2003) states that physical and chemical stability has been demonstrated for 4 days after reconstitution, but several studies have demonstrated that vancomycin is stable for longer periods. The shelf-life is restricted to 7 days as the solution is unpreserved.

Box 14.50 *Preferred formula*

Using a typical strength of 50 mg/mL as an example:

- vancomycin injection 500 mg × 10
- water for irrigation to 100 mL.

Method guidance

Reconstitute each 500 mg vancomycin injection with the appropriate quantity of water for irrigation and shake well to dissolve. Repeat the process as necessary and transfer the contents of the vials to a 100 mL measuring cylinder. Make up to volume with sterile water for irrigation and transfer to an amber glass bottle.

Shelf-life

7 days refrigerated in amber glass.

Clinical pharmaceutics

Points to consider

- At typical strengths (50 mg/mL), vancomycin hydrochloride should be in solution. Dosage uniformity should not be an issue.
- Vancomycin hydrochloride has a bitter taste (Trissel, 2000).
- The injection is licensed for use orally. However, there are serious risk management issues associated with the administration of oral medicines in parenteral syringes (National Patient Safety Agency, 2007). Pharmacy-based manipulation is recommended.

Technical information

Structure

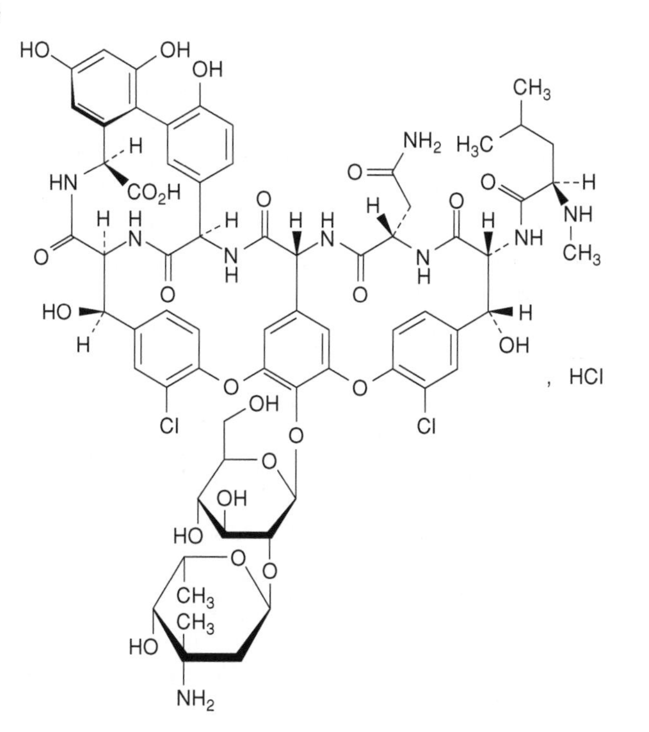

Empirical formula

$C_{66}H_{76}Cl_2N_9O_{24}$,HCl (British Pharmacopoeia, 2010).

Molecular weight

1486 (British Pharmacopoeia, 2010).

Solubility

>100 mg/mL in water (Trissel, 2000).

pK_a

At 25°C: phenols = 12.0, 10.4 and 9.59; amines = 7.75 and 8.89; carboxyl group = 2.18 (Takacs-Novac *et al.*, 1997).

Optimum pH stability

3–5 (Martindale, 2005).

Stability profile

Physical and chemical stability

Vancomycin hydrochloride is the hydrochloride of a mixture of related glycopeptides, consisting principally of the monohydrochloride of vancomycin B. It is a white or almost white powder, hygroscopic, freely soluble in water and slightly soluble in alcohol (British Pharmacopoeia, 2007). It is an amphoteric, free-flowing powder with a bitter taste (Trissel, 2000).

Vancomycin hydrochloride 1.03 mg is approximately equivalent to vancomycin 1 mg (Trissel, 2000). A 5% solution in water has a pH of 2.5–4.5, and solutions are reported to be most stable at pH 3.0–5.0 (Martindale, 2005).

In the pH range 2–10, the principal mode of degradation is thought to be deamidation, following pseudo-first order kinetics (Antipas *et al.*, 1994). This deamidation in aqueous solutions and formulations results in considerable instability, further complicated by the formation of a zwitterion degradation product which is sparingly water soluble (Mallet *et al.*, 1982). Antipas *et al.* (1994) conducted a detailed study of the degradation of vancomycin.

The structure of vancomycin is a heptapeptide backbone with a series of side-chains, one of which is asparagine. Deamidation occurs following hydrolysis of the side-chain amide linkage of the asparagine residue to form a free carboxylic acid. Antipas *et al.* (1994) give further proposed details of the degradation process. The degradation of vancomycin was studied at 50°C, using various buffers and vancomycin appears to be subject to some buffer catalysis in the pH range 3–10. Antipas *et al.* (1994) reported that the maximum stability at 50°C occurred at pH 5.5, slightly higher than previously stated (pH 3–5).

Although vancomycin exists in six ionisable forms, Antipas *et al.* (1994) suggest that in the pH range 2–10 only the carboxyl group and the two amine groups appeared to have a significant impact on the kinetics.

Vancomycin hydrochloride powder should be stored in an airtight container, protected from light (British Pharmacopoeia, 2007).

Degradation products

Deamidation leads to the formation of crystalline degradation product-I (CDP-I), which exists as two rotamers, CDP-I major (CPD-IM) and CDP-I minor (CDP-Im) (Antipas *et al.*, 1994; Khalfi *et al.*, 1996).

Stability in practice

There are several studies describing the stability of vancomycin in various diluents. The stability of the drug appears to provide scope for its formulation as an extemporaneous oral liquid.

In practice, the licensed injection is typically reconstituted and diluted for oral use. Injections are available that are licensed for oral administration. Literature from pharmaceutical companies suggests that solutions of the parenteral powder intended for oral administration may be stored in a refrigerator (2–8°C) for 96 hours (Summary of Product Characteristics, 2003). US literature has historically quoted a shelf-life of 7 days in the refrigerator (PDR, 1984).

The amount of impurities present in the injection was decreased in the late 1980s when a new purified form of vancomycin hydrochloride (Vancocin CP; Eli Lilly) was launched (Walker and Birkhans, 1988). This replaced the Vancocin formulation.

In the early 1980s, research teams led by Mallet published the results of experiments on the older Vancocin formulations. Following reconstitution for oral use, the results suggested that the drug exhibits good stability (Mallet *et al.*, 1982; Mallet and Sesin, 1984). HPLC technology was used to test reconstituted drug at 0, 4 and 25°C for a period of 90 days in amber glass unit-dose vials.

The concentration tested was 83.3 mg/mL (500 mg in 6 mL), with distilled water being used as the diluent. The solution was divided into unit-dose vials and tested over a period of 90 days, including a visual inspection for any signs of colour change and/or precipitation.

Mallet and Sesin (1984) suggest that the results indicate that reconstituted Vancocin was stable for 58 days at room temperature (25°C) and at least 90 days under refrigeration (4°C) and freezing (0°C). However, a precipitate was documented after 6 days at room temperature. From a baseline pH of 3.65, all the solutions showed a gradual increase over 90 days, finishing at 4.32, 3.92 and 3.90 for the vials stored at 25°C, 4°C and 0°C, respectively.

Microbiological activity of the room temperature samples was also documented on days 0, 3, 21 and 35, using *Staphylococcus aureus* penicillinase-producing organisms. The results showed that the four samples tested retained 98% of the theoretical activity.

The authors suggest that reconstituted Vancocin solution remains stable for at least 90 days at 4°C and 0°C. Room temperature storage is not recommended due to the formation of the precipitate. It should be noted that these results refer to unit-dose vials rather than solutions that are used more than once. The unpreserved nature of these solutions would normally limit their in-use expiry to 7 days, unless local validation work has been carried out to extend this limit. It should also be noted that some of the results showed a significant degree of variation, particularly for the solutions stored at room temperature.

Cornu *et al.* (1980) also tested the stability of oral vancomycin solution after reconstitution, using a microbiological assay to analyse the samples. The authors suggested that the solution was stable for 15 days at 4°C and 25°C.

Walker and Birkhans (1988) investigated the stability of the new Vancocin CP formulation in glucose 5% and sodium chloride 0.9%. Assays were conducted using a stability-indicating HPLC method. Polyvinyl chloride (PVC) minibags containing 4 and 5 mg/mL in sodium chloride 0.9% and glucose 5% were tested. Three bags of each solution were stored at 4 or 23°C, and sampled on days 0, 1, 3, 7, 10, 14, 17, 24, 30 and 31. Details of the sampling procedure are given.

All the solutions were reported to remain clear, colourless, and free of visible particles over the 30-day study period. Absorbance did not change and the pH remained between 3.0 and 3.5.

The solutions stored at 23°C degraded significantly faster than those stored at 4°C. The glucose 5% solutions stored at 25°C degraded faster than the sodium chloride solutions at the same temperature. Stability (more than 90% of the initial concentration) was maintained for 14 days and 21 days for the glucose and sodium chloride solutions, respectively. There was no significant difference between the solutions when stored at 4°C.

There were no concentration effects noticeable, with no differences in stability seen between the 4 mg/mL and 5 mg/mL solutions. The refrigerated solutions were stored at 23°C for 24 hours after 30 days. The only solutions to maintain more than 90% of the drug concentration after 31 days were the solutions of vancomycin in sodium chloride. Walker and Birkhans conclude that vancomycin hydrochloride admixtures in normal saline and glucose 5% are stable for 30 and 24 days, respectively, when stored at 4°C. When stored at 23°C, admixtures in normal saline and glucose 5% are stable for 21 and 14 days, respectively.

Das Gupta *et al.* (1986) studied the stability of vancomycin hydrochloride in 5% glucose and 0.9% sodium chloride injections, using pharmaceutical grade powder (Eli Lilly) as the starting material. The two diluents were used to make 5 mg/mL solutions in duplicate. After baseline data were obtained for clarity, pH and drug concentration, the solutions were divided into four parts and stored in clear glass vials at 24, 5 and –10°C (±1°C). Solutions were also stored at 24°C in plastic IV bags (Viaflex). Analysis was carried out using a stability-indicating HPLC assay on samples taken over the 63 days of the study.

The authors suggest that vancomycin hydrochloride was stable for at least 17 days (with a 6% loss of potency) at 24°C in either diluent. Initial pH values were 3.7 and 3.8 for glucose 5% and sodium chloride 0.9%, respectively; these values did not change during the course of the study.

Comparator solutions buffered to pH values of 1.4, 5.6 and 7.1 showed considerable losses after 17 days, as expected given the theoretical pH range of maximum stability (pH 3–5). Degradation in the buffered solutions was lowest at pH 5.6 (89.5% remaining at 17 days), compared with 63.9% at pH 7.1, and 44.9% at pH 1.4. The data suggest that different degradation mechanisms are occurring in these conditions.

When stored at 5 or –10°C, the unbuffered solutions in 0.9% sodium chloride and 5% glucose showed less than 1% loss in potency over 63 days. Visual inspection showed clear solutions, and pH values did not change over the study period. Das Gupta *et al.* (1986) suggest that the 4-day expiry stated by manufacturers is very conservative. Older studies which demonstrated higher losses of vancomycin may have shown different results due to the relative impurity of the vancomycin used. For example, Mann *et al.* (1971) found 45% loss in potency in 60 days at 5°C in both sodium chloride 0.9% and glucose 5%.

Das Gupta *et al.* (1986) also briefly considered the thawing of the samples stored at –10°C. Thawing using a microwave oven for 50 seconds did not appear to degrade the drug, showing 100.3% potency after 63 days at –10°C. The solution was clear, with an unchanged pH value. The authors suggest that thawing the sample in a microwave was not harmful.

In summary, vancomycin hydrochloride 5 mg/mL in 5% glucose or 0.9% sodium chloride was stable for 17 days at 24°C and for 63 days at 5°C and –10°C. The specificity of the assay has been criticised by Walker and Birkhans (1988), as the assay measured total total vancomycin powder rather than the actual concentration of vancomycin B.

Khalfi *et al.* (1996) have studied the stability of vancomycin hydrochloride in various diluents in PVC containers. The study was designed to assess the chemical stability and compatibility of vancomycin hydrochloride 5 mg/mL with PVC bags when admixed in sodium chloride 0.9% or glucose 5% at 22°C for up to 48 hours without protection from light, and at 4°C for up to 7 days with protection from light. The authors also investigated the stability of the drug at 5 and 8 mg/mL during 1- and 24-hour simulated infusions.

A stability-indicating HPLC assay was used for analysis. During development of the assay, vancomycin hydrochloride was found to be chemically stable in acidic or alkaline conditions, but details are not stated. The vancomycin concentration was maintained above 90% of the initial concentration throughout the 24-hour infusion time, demonstrating that there were no problems with adsorption. The vancomycin was chemically stable in both glucose 5% and sodium chloride 0.9% infusions, and no plasticiser was detected in the drug solution.

For the stability study, vancomycin hydrochloride was reconstituted in sterile water and diluted in duplicate in 100 mL of sodium chloride 0.9% or glucose 5% in PVC bags (final concentration 5 mg/mL). These solutions were then stored at 22°C for 48 hours (exposed to light) or 4°C for 7 days and protected from the light. One millilitre samples were removed at intervals up to 48 hours. No significant levels of degradation were detected.

Galanti *et al.* (1997) have studied the long-term stability of vancomycin hydrochloride at 4°C in intravenous infusions in PVC bags. Five bags each of concentrations of 5 and 10 mg/mL were tested over two months using a stability-indicating HPLC assay. The admixtures were visually inspected and pH measurements taken. Assays were carried out in triplicate. For both concentrations tested, more than 97% of the initial concentration was found to be remaining after 58 days. No colour changes were observed, but the authors did note an increase in the pH of the stored solutions (actual figures not stated).

Wood *et al.* (1995) tested the stability of the hydrochloride salt 10 mg/mL in plastic syringes, using water for injection, sodium chloride 0.9% and glucose 5% as diluents. When refrigerated, these solutions were reported to remain stable for 84 days, with 4% or less drug loss. Drug losses were much higher at 25°C, depending on the diluent and brand of syringe used.

Stability in practice – related preparations

Barbault *et al.* (1999) have studied the stability of vancomycin eye drops, as unopened 50 mg/mL vials in 0.9% sodium chloride (protected from light). Over 32 days, the authors assessed pH, osmolality, vancomycin concentration and sterility. Overall results suggested that the vials, when protected from light, are stable for 21 days when stored at 4°C and for 15 days when stored at 25°C.

The impact of deep freezing on ophthalmic solutions has also been studied (Sautou-Miranda *et al.*, 2002). Stability of 25 mg/mL solutions in glucose has been demonstrated for three months at –20°C (±2°C). Thawing of the solutions, either at ambient temperature, or under warm running water from a tap, did not affect the stability of the eye drops.

Bioavailability data

Oral bioavailability of vancomycin is known to be limited. The successful clinical use of oral preparations relies on the drug remaining within the gut lumen.

The absorption of the oral solution was used for comparison with the absorption seen using semi-solid Vancocin Matrigel (Lilly) capsules in six healthy volunteers (Lucas *et al.*, 1987). The cross-over study showed similar levels of faecal availability for the two formulations. The semi-solid gel capsule is a dispersion of a highly water-soluble substance in a water-soluble carrier (polyethylene glycol 6000). The authors commented that the capsule was known to disintegrate in less than 45 minutes and be fully dissolved in 60–120 minutes. Therefore, solution rate was not considered a limiting factor for the capsules in clinical use.

Further data may be available from the pharmaceutical companies that hold product licences that cover the oral use of the reconstituted injection.

References

Antipas AS, Vander Velde D, Stella VJ (1994). Factors affecting the deamidation of vancomycin in aqueous solutions. *Int J Pharm* 109: 261–269.

Barbualt S, Aymard G, Feldman D, Pointereau-Bellanger A, Thuillier A (1999). Stability of vancomycin eye drops (Etude de stabilité d'un collyre a la vancomycine a 50 mg/ml). *J Pharm Clin* 18: 183–189.

British Pharmacopoeia (2007). *British Pharmacopoeia*, Vol. 2. London: The Stationery Office, pp. 2137–2139.

Cornu J, Schorer E, Amacher PA (1980). Etude de la conservation en solution aqueuse de la vancomycine, de la nemycine et de la polymixine B isolées au associées. *Pharm Acta Helv* 10: 253–255.

Das Gupta V, Stewart KR, Nohria S (1986). Stability of vancomycin hydrochloride in 5% dextrose and 0.9% sodium chloride injections. *Am J Hosp Pharm* 43: 1729–1731.

Galanti LM, Hecq J-D, Vanbeckbergen D, Jamart J (1997). Long-term stability of vancomycin hydrochloride in intravenous infusions. *J Clin Pharm Ther* 22: 353–356.

Khalfi F, Dine T, Gressier B, Luyckx M, Brunet C, Ballester L *et al.* (1996). Compatibility and stability of vancomycin hydrochloride with PVC infusion material in various conditions using stability-indicating high-performance liquid chromatographic assay. *Int J Pharm* 139: 243–247.

Lowey AR, Jackson MN (2008). A survey of extemporaneous dispensing activity in NHS trusts in Yorkshire, the North-East and London. *Hosp Pharmacist* 15: 217–219.

Lucas RA, Bowtle WJ, Ryden R (1987). Disposition of vancomycin in healthy volunteers from oral solution and semi-solid matrix capsules. *J Clin Pharm Ther* 12: 27–31.

Mallet L, Sesin GP (1984). Reconstituted oral vancomycin study reveals ninety-day stability and savings potential. *Intravenous Ther News* October 1, 3–5.

Mallet L, Sesin GP, Ericson J, Fraser DG (1982). Storage of vancomycin oral solution. *N Engl J Med* 307: 445.

Mann JM, Coleman DL, Boylan JC (1971). Stability of parentteral solutions of sodium cephalothin, cephaloridine, potassium penicillin G (buffered) and vancomycin hydrochloride. *Am J Hosp Pharm* 28: 760–763.

National Patient Safety Agency (NPSA) (2007). Patient Safety Alert 19: Promoting safer measurement and administration of liquid medicines via oral and other enteral routes. 28 March

2007. http://www.npsa.nhs.uk/site/media/documents/2463_Oral_Liquid_Medicines_PSA_FINAL.pdf (accessed 30 October 2007).

PDR (1984). *Physicians Desk Reference*, 38th edn. Oradell, NJ: Medical Economics Co.

Sautou-Miranda V, Libert F, Grand-Boyer A, Gellis C, Chopineau J (2002). Impact of deep-freezing on the stability of 25 mg/ml vancomycin ophthalmic solutions. *Int J Pharm* 234: 205–212.

Summary of Product Characteristics (2003). Mayne Brand (revised 2003). www.medicines.org.uk (accessed 12 June 2007).

Sweetman SC (2009). *Martindale: The Complete Drug Reference*, 36th edn. London: Pharmaceutical Press.

Takacs-Novak K, Box KJ, Avdeef J (1997). Potentiometric pK_a determination of water-insoluble compounds: validation study in methanol/water mixtures. *Int J Pharm* 151: 235–248.

Trissel LA, ed. (2000). *Stability of Compounded Formulations*, 2nd edn. Washington DC: American Pharmaceutical Association.

Walker SE, Birkhans B (1988). Stability of intravenous vancomycin. *Can J Hosp Pharm* 41: 233–238.

Wood MJ, Lund R, Beaven M (1995). Stability of vancomycin in plastic syringes measured by high performance liquid chromatography. *J Clin Pharm Ther* 20: 319–325.

Warfarin sodium oral liquid

Risk assessment of parent compound

A risk assessment survey completed by clinical pharmacists suggested that warfarin sodium had a narrow therapeutic index, with the potential to cause death if inaccurate doses were to be administered (Lowey and Jackson, 2008). Current formulations used at NHS trusts show a low to moderate risk, based on the number/complexity of manipulations with simple calculations. Due mainly to its high clinical risk, warfarin is regarded as a high-risk extemporaneous preparation (Box 14.51).

Summary

There is a paucity of data to support the use of extemporaneously prepared suspensions of warfarin. No stability work has been carried out to validate the shelf-life. There are optical chemistry and stability issues which require further investigation. Formulation of the drug as a suspension rather than a solution may avoid or mitigate these stability issues. Further analytical work may be required in this area. However, suspensions show poorer uniformity of dose than solutions; this may be critical for warfarin due to the drug's narrow therapeutic index. The suspension should be shaken thoroughly before each dose to attempt to distribute the drug evenly.

However, a 'Special' is now available from a reputable manufacturer. The formulation of 'Specials' should be examined by a suitably competent pharmacist before purchase.

Due to the risks of poor uniformity of dose with suspensions, consideration should also be given to alternative methods of administering warfarin where possible (e.g. dispersing the tablets under water); 500 micrograms tablets are commercially available for all but the smallest dose increments. Where the risk–benefit balance indicates that a suspension is appropriate, a 'Special' should be used where possible. Extemporaneous preparation should only be carried out in extenuating circumstances.

Clinical pharmaceutics

Points to consider

- Whether the drug is in solution or suspension (or a mixture of both) depends on the formulation chosen (see below). As a precaution, shake the bottle thoroughly to improve dosage uniformity.

Box 14.51 *Preferred formulae*

Note: 'Specials' are now available from reputable manufacturers. Extemporaneous preparation should only occur where there is a demonstrable clinical need.

Using a typical strength of 200 micrograms/mL as an example:

- warfarin 5 mg tablets × 4
- Ora-Plus to 50 mL
- Ora-Sweet or Ora-Sweet SF to 100 mL.

Method guidance

Tablets can be ground to a fine powder in a pestle and mortar. A small amount of Ora-Plus may be added to form a paste, before adding further portions of Ora-Plus up to 50% of the final volume. Transfer to a measuring cylinder. The Ora-Sweet can be used to wash out the pestle and mortar before making the suspension up to 100% volume. Transfer to an amber medicine bottle.

Shelf-life

7 days at room temperature in amber glass. **Shake the bottle.**

Alternative

An alternative formula would be:

- warfarin 5 mg tablets × 4
- xanthan gum suspending agent 0.5% ('Keltrol'or similar) to 100 mL.

Shelf-life

7 days at room temperature in amber glass. **Shake the bottle.**

Note: 'Keltrol' xanthan gum formulations vary slightly between manufacturers, and may contain chloroform and ethanol. The individual specification should be checked by the appropriate pharmacist. **The formulation of the suspending agent may be critical.**

- Warfarin sodium tablets are a racemic mixture of *R* and *S* isomers, with different clinical effects.
- Ora-Plus and Ora-Sweet SF are free from chloroform and ethanol.
- Ora-Plus and Ora-Sweet SF contain hydroxybenzoate preservative systems.

- Xanthan gum ('Keltrol') formulations have varying formulations depending on the manufacturer. They may contain chloroform and ethanol. The exact formulation should be checked before use. The pH of the suspending agent is known to be crucial.
- Warfarin is known to have a slightly bitter taste.
- There is no available information on bioavailability of warfarin sodium from oral liquids.

Other notes

- Warfarin is reported to be physically compatible with 'Keltrol' (Nova Laboratories Ltd, 2003) and 1:1 mixtures of Ora-Plus and Ora-Sweet SF (Lowey and Watkinson, 2006).
- 'Keltrol' is made to slightly different specifications by several manufacturers, but typically contains ethanol and chloroform. The exact specification should be assessed before use by a suitably competent pharmacist. Ethanol is a CNS depressant. Chloroform has been classified as a class 3 carcinogen (i.e. substances that cause concern owing to possible carcinogenic effects but for which available information is not adequate to make satisfactory assessments) (CHIP3 Regulations, 2002).
- Although theoretically unlikely, it is possible that the relative amounts of the two isomers may change when warfarin sodium is formulated as a suspension with a pH above 5. At this point, warfarin is ionised and may exhibit increased solubility. Although compatible, care should be taken when using xanthan gum suspending agents such as 'Keltrol', as these preparations are made by a variety of suppliers to different specifications, with varying constituents and pH values (from approximately pH 4 to 6 depending on the manufacturer).
- Care should be taken when changing a patient from tablets to liquids and vice versa.
- Care should also be taken with the possibility of 10-fold errors occurring for ward-based manipulations, given with the existence of 0.5 mg and 5 mg tablets (P. Dale, Paediatric Pharmacist, Royal Cornwall Hospital, personal communication, 2007).

Technical information

Structure

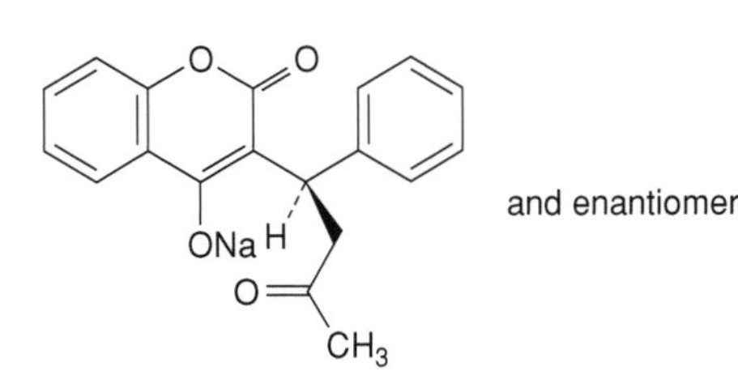

Empirical formula

$C_{19}H_{15}NaO_4$ (British Pharmacopoeia, 2007).

Molecular weight

330.3 (British Pharmacopoeia, 2007).

Solubility

>1 g/mL Warfarin sodium but pH dependent (0.004 mg/mL warfarin base) (Moffat, 1986).

pK_a

5.0 at 20°C (Lund, 1994).

Optimum pH stability

Unknown.

Stability profile

Physical and chemical stability

Warfarin sodium is a white odourless amorphous or crystalline hygroscopic powder with a slightly bitter taste (Goding and West, 1968; Connors *et al.*, 1986). Warfarin sodium solutions (1%) have a pH of 7.2–8.3 (Goding and West, 1968). Although it is very soluble in water (>1 g/mL), the solubility is pH dependent; at pH values below 8.0 the warfarin may precipitate due to formation of the insoluble enol form (Trissel, 2000). At pH 8 or higher, clear stable solutions result as the warfarin is in the soluble form (Hiskey *et al.*, 1962; Trissel, 2005). Adjustment of pH, by adding agents such as sodium hydroxide, may stabilise aqueous solutions of warfarin salts by preventing precipitation of free warfarin (Lund, 1994).

Warfarin is a lactone, an enol and a ketone (Salomies *et al.*, 1994). At pH 9.0 and elevated temperatures, warfarin has been shown to be susceptible to hydrolysis and decarboxylation (Hiskey and Weiner, 1964). Goding and West (1968) reported that during 48 hours at 150°C and pH 9, warfarin underwent hydrolysis and decarboxylation reactions to form 3-(O-hydroxyphenyl)-5-phenyl-2-cyclohexen-1-one. Aqueous solutions of warfarin can theoretically hydrolyse to form *cis*-coumarinic acid, but ionisation of the enol (pK_a 5.0) appears to stabilise the lactone with respect to hydrolysis (Connors *et al.*, 1986; Lund, 1994). Warfarin is discoloured by light and may be prone to oxidation (Salomies *et al.*, 1964; Lund, 1994). Benya and Wagner (1976) suggest that warfarin is sensitive to air and light during thin layer chromatography. The tablets should be stored in light-resistant containers and at controlled room temperature (Martindale, 2005).

Warfarin sodium tablets are a racemic mixture (50:50) of the *R* and *S* enantiomers. *S*-Warfarin is more potent as an anticoagulant. Both *R* and *S* enantiomers are metabolised by the liver, but *S*-warfarin is thought to be excreted more rapidly (Mungall *et al.*, 1985; Holford, 1986). Approximate half-lives are 45 hours for the *R* enantiomer and 30 hours for the *S* enantiomer (Moffat, 1986). Although theory would suggest that racemic mixtures remain in equilibrium (50:50) in solution, it is possible that the relative amounts of these isomers may change.

It is prudent, therefore, to formulate the oral liquid as a suspension rather than a solution, with a pH below that of the pK_a in order to keep the activity as predictable as possible, and as close as possible to the situation that exists in the licensed tablet formulation. This may be achieved using a thixotropic suspending agent such as a Ora-Plus 1:1 Ora-Sweet, Ora-Plus 1:1 Ora-Sweet SF or an alternative xanthan gum suspending agent ('Keltrol'). However, the exact formula of the 'Keltrol' formulations varies between manufacturers, with some exhibiting a pH above 5. This may lead to some of the drug being in solution.

While the pH values of Ora-Plus 1:1 Ora-Sweet SF, methylcellulose and 'Keltrol' are all typically acidic, the 1:1 mixture of Ora-Plus and Ora-Sweet is recommended, as this has the most appropriate and reproducible pH of 4.2 (Paddock Laboratories, 2006). Use of a commercial suspending agent will also suspend any drug which remains bound to tablet excipients.

Degradation products

May hydrolyse in aqueous solution to form *cis*-coumarinic acid (Lund, 1994). No details available on toxicity and/or pharmacological activity.

During forced degradation, warfarin degrades via hydrolysis and decarboxylation reactions to form 3-(O-hydroxyphenyl)-5-phenyl-2-cyclohexen-1-one (Goding and West, 1968).

Stability in practice

Published stability data for warfarin are limited. There is no information related to warfarin sodium as an oral suspension. There is anecdotal evidence only to support the physical compatibility and successful use of warfarin in 'Keltrol' and a 1:1 mixture of Ora-Plus and Ora-Sweet SF (Lowey and Jackson, 2008).

Related preparations – injectable products

Salomies *et al.* (1994) found warfarin sodium 100 micrograms/mL in sodium chloride 0.9% was stable at room temperature for 120 hours in glass, but found significant sorptive losses in PVC containers.

Hiskey *et al.* (1962) and Hiskey and Weiner (1964) have studied the stability of warfarin sodium in injectable solutions (25 mg/mL). Precipitates developed after several days when prepared in sterile water. Solutions buffered with disodium hydrogen phosphate remained clear and colourless for one month, with no change in warfarin concentration (Trissel, 2005; Bahal *et al.*, 1997). Warfarin sodium has been reported to be incompatible with several drugs and solutions, including glucose 5% (Moorhatch and Chiou, 1974). In sodium chloride 0.9% (100 mg in 1 litre), warfarin is visually compatible with no drug loss by UV and HPLC in 24 hours at 21°C (Trissel, 2005). Bahal *et al.* (1997) found that a 1 mg/mL injectable warfarin sodium solution appeared hazy after storage at ambient temperature for 24 hours.

After warfarin sodium lyophilised powder is reconstituted with water, warfarin is reported to be physically and chemically stable for only 4 hours at room temperature (PDR, 2003). The reconstituted solution should not be refrigerated.

Enteral products

Kuhn *et al.* (1989) investigated the recovery of warfarin from various dilutions of enteral feed (Osmolite, Ross), using a stability-indicating HPLC method. Significant reductions were seen of 35–28% at various concentrations of warfarin.

Bioavailability data

No data available.

References

Bahal SM, Lee TJ, McGiees Dobler GL (1997). Visual compatibility of warfarin sodium injection with selected medications and solutions. *Am J Health Syst Pharm* 54: 2599–2600.

Benya TJ, Wagner JG (1976). Warfarin sensitivity to air and light during thin layer chromatography [letter]. *Can J Pharm Sci* 11: 70–71.

British Pharmacopoeia (2007). *British Pharmacopoeia*, Vol. 2. London: The Stationery Office, p. 2160.

CHIP3 Regulations (2002). Chemicals (Hazard Information and Packaging for Supply) Regulations. SI 2002/1689.

Connors KA, Amidon GL, Stella ZJ (1986). *Chemical Stability of Pharmaceuticals: A Handbook for Pharmacists*, 2nd edn. New York: Wiley Interscience.

Goding LA, West BD (1968). The reversible removal of carbon 2 of 3-substituted 4-hydroxycoumarins. *J Org Chem* 33: 437–438.

Hiskey CF, Weiner N (1964). US Patent 3,121,664, February 18, 1964. *Chem Abstr* 60: 15690.

Hiskey CF, Bullock E, Whitman G (1962). Spectrophotometric study of aqueous solutions of warfarin sodium. *J Pharm Sci* 51: 43.

Holford NHG (1986). Clinical pharmacokinetics and pharmacodynamics of warfarin: understanding the dose-effect relationship. *Clin Pharmacokinet* 11: 483–504.

Kuhn TA, Garnett WR, Wells BK, Karnes HT (1989). Recovery of warfarin from an enteral nutrient formula. *Am J Hosp Pharm* 46: 1395–1399.

Lowey AR, Jackson MN (2008). A survey of extemporaneous dispensing activity in NHS trusts in Yorkshire, the North-East and London. *Hosp Pharmacist* 15: 217–219.

Lowey AR, Watkinson A (2006). In-house testing, Leeds Teaching Hospitals NHS Trust, 2006.

Lund W, ed. (1994). *The Pharmaceutical Codex: Principles and Practice of Pharmaceutics*, 12th edn. London: Pharmaceutical Press.

Moffat AC, Osselton MD, Widdop B (2003). *Analysis of Drugs and Poisons*, 3rd edn. London: Pharmaceutical Press. ISBN 978 0 85369 4731.

Moorhatch P, Chiou WL (1974). Interactions between drugs and plastic intravenous fluid bags. I. Sorption studies on 17 drugs. *Am J Hosp Pharm* 31: 72–78.

Mungall DR, Ludden TM, Marshall J, Hawkins DW, Talbert RL, Crawford MH (1985). Population pharmacokinetics of racemic warfarin in adult patients. *J Pharmacokinet Biopharm* 13: 213–227.

Nova Laboratories Ltd (2003). Information Bulletin – Suspension Diluent Used at Nova Laboratories.

Paddock Laboratories (2006). Ora-Plus and Ora-Sweet SF Information Sheets. www.paddocklaboratories.com (accessed 12 April 2006).

PDR (2003). *Physician's Desk Reference*, 57th edn. Montvale, NJ: Medical Economics Company.

Salomies HEM, Heinonen RM, Toppila MIA (1994). Sorbtive loss of diazepam, nitroglycerine and warfarin sodium to polypropylene and warfarin sodium to polypropylene lined infusion bags. *Int J Pharm* 110: 197–201.

Sweetman SC (2009). *Martindale: The Complete Drug Reference*, 36th edn. London: Pharmaceutical Press.

Trissel LA, ed. (2000). *Stability of Compounded Formulations*, 2nd edn. Washington DC: American Pharmaceutical Association.

Trissel LA (2005). *Handbook on Injectable Drugs*, 13th edn. Bethesda, MD: American Society of Health System Pharmacists.

Appendix 1

Change control request form

Section 1. Details of proposed changes – to be completed by the originator		
Date:	Change requested by:	Location:
Description of the proposed change: (Provide a brief description of the proposed change - attach supporting information if necessary)		
Justification for the proposed change: (Provide a full rationale justifying the need for the proposed change and reference supporting information)		
Section 2. Assessment of proposed change		
Impact assessment (consider and identify the impact of the proposed change on other services, processes, equipment, personnel, product safety/efficacy, GMP compliance, etc.)		

Processes: The proposed change involves:			
Equipment	○ Yes ○ No	Training	○ Yes ○ No
Facilities	○ Yes ○ No	SOPs & logs	○ Yes ○ No
Raw materials/Consumables	○ Yes ○ No	Stability	○ Yes ○ No
Worksheets	○ Yes ○ No	Technical agreements/SLAs	○ Yes ○ No
Validation	○ Yes ○ No	Others (give details)	○ Yes ○ No
Specifications	○ Yes ○ No		

Comments

Services: Does the proposed change affect any other service areas? ○ Yes ○ No If yes, give details below					
Section 3: QA assessment of impact of proposed change					
Circulation (identify and contact the appropriate personnel with details of the proposed change.)					
List the personnel contacted below:					
Summary of assessment Give a detailed summary of all processes, documentation, etc. affected by the change.					
Proposed action plan: Give details of the proposed course of action to implement the change.					
Action			Lead person	Timescale	Outcome
Section 4. Approval					
Approval for the change	Yes	No	Name	Signature	Date
Comments					

Appendix 2

Deviation reporting form

<table>
<tr><td colspan="3">Section 1. Details of deviation</td><td>Reference:</td></tr>
<tr><td>Date:</td><td colspan="2">Deviation raised by:</td><td>Location:</td></tr>
<tr><td></td><td colspan="2"></td><td></td></tr>
<tr><td colspan="4">Category of deviation (Tick appropriate category)</td></tr>
<tr><td colspan="2">Planned event</td><td colspan="2">Unplanned event</td></tr>
<tr><td colspan="4">Details of deviation: Give a brief description of the deviation and the planned course of corrective action. Refer to specific instructions or specific pieces of equipment (please append any supporting information).</td></tr>
<tr><td colspan="4"></td></tr>
<tr><td colspan="4">Scope: Indicate what is affected by the deviation (e.g. product details [description, batch no.], equipment, facilities, procedures, personnel, etc.).</td></tr>
<tr><td colspan="4"></td></tr>
<tr><td colspan="4">Justification: Indicate why the deviation/corrective action was necessary.</td></tr>
<tr><td colspan="4"></td></tr>
</table>

Section 2. Approval to proceed	**Yes**	**No**	**Name**	**Signature**	**Date**
Supervisor					
Comments					

<table>
<tr><th colspan="4">Section 3. Management review</th></tr>
<tr><td colspan="4">Comments: Include risk assessments or root cause analysis where appropriate.</td></tr>
<tr><td colspan="4"></td></tr>
<tr><td colspan="4">Action plan: (if appropriate)</td></tr>
<tr><td>Action</td><td>Lead person</td><td>Timescale</td><td>Outcome</td></tr>
<tr><td></td><td></td><td></td><td></td></tr>
<tr><td colspan="4">Closure</td></tr>
<tr><td>Name</td><td colspan="2">Signature</td><td>Date completed</td></tr>
<tr><td></td><td colspan="2"></td><td></td></tr>
</table>

Appendix 3

Example raw material specification

<table>
<tr><td rowspan="3">Citric acid monohydrate raw material specification</td><td colspan="3">SOP ref.</td></tr>
<tr><td>Page x of y</td><td>Issue:
Supersedes:</td><td>Review</td></tr>
<tr><td>Date of issue</td><td colspan="2">Copy reference no.</td></tr>
</table>

General information

Molecular formula	$C_6H_8O_7,H_2O$
Molecular weight	210
Description	White, crystalline powder, colourless crystals or granules
Solubility	Very soluble in water, freely soluble in alcohol, sparingly soluble in ether
Storage conditions	Store in well-closed containers
Retest date	2 years if no manufacturer's expiry
Maximum expiry	5 years or manufacturer's expiry if shorter
Hazard rating	Low

Analytical tests

Analytical tests	Acceptance criteria
Identification (*First identification: B. Second identification: A, C, D*)	
A. Dissolve 1 g in 10 mL of water *R*. The solution is strongly acidic (ref. BP)	Complies
B. Infrared absorption spectrophotometry – comparison with reference spectrum after drying both the substance being examined and the reference substance at 100°C to 105°C for 24 hours (ref. BP)	Complies
C. Add about 5 mg to a mixture of 1 mL of acetic anhydride R and 3 mL of pyridine R. A red colour develops (ref. BP)	Complies
D. Dissolve 0.5 g in 5 mL of water R, neutralise using 1 M sodium hydroxide (about 7 mL), add 10 mL of calcium chloride solution R and heat to boiling. A white precipitate is formed (ref. BP)	Complies
Limit tests	
A. Appearance of solution (ref. BP) Dissolve 2.0 g in water R and dilute to 10 mL with the same solvent. The solution is clear (2.2.1) and not more intensely coloured than the reference solution (ref. BP).	Complies
B. Readily carbonisable substances (ref. BP) To 1.0 g in a cleaned test tube add 10 mL of sulfuric acid R and immediately heat the mixture in a water-bath at 90 ± 1°C for 60 min. Immediately cool rapidly. The solution is not more intensely coloured than a mixture of 1 mL of red primary solution and 9 mL of yellow primary solution	Complies
C. Oxalic acid (ref. BP) Dissolve 0.80 g in 4 mL of water R. Add 3 mL of hydrochloric acid R and 1 g of zinc R in granules. Boil for 1 min. Allow to stand for 2 min. Transfer the supernatant liquid to a test-tube containing 0.25 mL of a 10 g/L solution of phenylhydrazine hydrochloride R and heat to boiling. Cool rapidly, transfer to a graduated cylinder and add an equal volume of hydrochloric acid R and 0.25 mL of a 50 g/L solution of potassium ferricyanide R. Shake and allow to stand for 30 min. Any pink colour in the solution is not more intense than that in a standard prepared at the same time in the same manner using 4 mL of a 0.1 g/L solution of oxalic acid R (350 ppm, calculated as anhydrous oxalic acid)	Complies
D. Sulfates (ref. BP) Dissolve 1.0 g in distilled water R and dilute to 15 mL with the same solvent. The solution complies with the limit test for sulfates (150 ppm)	Complies
E. Heavy metals (ref. BP) Dissolve 5.0 g in several portions in 39 mL of dilute sodium hydroxide solution R and dilute to 50 mL with distilled water R. 12 mL complies with limit test A for heavy metals (10 ppm). Prepare the standard using lead standard solution (1 ppm Pb) R	Complies
F. Water (ref. BP) 7.5–9.0%, determined on 0.500 g by the semi-micro determination of water	Complies
G. Sulfated ash (ref. BP) – not more than 0.1%, determined on 1.0 g	Complies
Assay Dissolve 0.550 g in 50 mL of water R. Titrate with 1 M sodium hydroxide, using 0.5 mL of phenolphthalein solution R as indicator. 1 mL of 1 M sodium hydroxide is equivalent to 64.03 mg of $C_6H_8O_7$	99.5–101.0% of stated contents

Appendix 4

Example worksheet

<table>
<tr><td colspan="8">XXXX NHS Trust - Pharmacy</td><td colspan="2">Site:</td></tr>
<tr><td colspan="8">Extemporaneous dispensing worksheet</td><td colspan="2">Batch no:</td></tr>
<tr><td rowspan="2">Product:</td><td colspan="7" rowspan="2">Lorazepam suspension 100 micrograms in 1 mL</td><td colspan="2">Dispensing date:</td></tr>
<tr><td colspan="2">Expiry: 7 days
Storage: Fridge</td></tr>
<tr><td>Batch size:</td><td colspan="7">50 mL</td><td colspan="2">Expiry date:</td></tr>
<tr><td colspan="4">Patient name:</td><td colspan="6">Worksheet prepared by:</td></tr>
<tr><td colspan="4">Ward/Department:</td><td colspan="6">Worksheet checked by:</td></tr>
<tr><td>Ingredients</td><td>Amount</td><td>Manu-facturer</td><td>Batch no./QC ref. no.</td><td>Expiry date</td><td>Assem-bled by</td><td colspan="2">Booked out by (labels on reverse)</td><td colspan="2">Accuracy checked by</td></tr>
<tr><td>Lorazepam tablets 1 mg</td><td>5</td><td></td><td></td><td></td><td></td><td colspan="2"></td><td colspan="2"></td></tr>
<tr><td>Ora-Plus</td><td>25 mL</td><td></td><td></td><td></td><td></td><td colspan="2"></td><td colspan="2"></td></tr>
<tr><td>Ora-Sweet (Sucrose free)</td><td>To 50 mL</td><td></td><td></td><td></td><td></td><td colspan="2"></td><td colspan="2"></td></tr>
<tr><td colspan="10">Equipment required: L.E.V. cabinet or waysafe, mortar, pestle, stirring rod, one 50 mL measuring cylinder, 50 mL amber glass bottle and cap</td></tr>
<tr><td colspan="3">Method</td><td colspan="4">Compounded by:</td><td colspan="3">Checked by:</td></tr>
<tr><td colspan="7">1. Count out five lorazepam 1 mg tablets and place in a mortar</td><td colspan="3"></td></tr>
<tr><td colspan="7">2. Measure out 25 mL of Ora-Plus in a measuring cylinder</td><td colspan="3"></td></tr>
<tr><td colspan="7">3. Place the mortar in a cabinet and grind the tablets to a fine uniform powder using a pestle</td><td colspan="3"></td></tr>
</table>

4. Add a small volume of Ora-Plus (~10 mL) to wet the powder and mix with the pestle to form a smooth paste. Perform further stepwise additions of Ora-Plus (~5 mL at a time) and mix in the paste upon each addition until homogeneous. Continue until all the Ora-Plus has been added. Transfer the suspension to one of the 50 mL measuring cylinders	
5. Rinse the mortar with approximately 10 mL of the Ora-Sweet and add the residue to the suspension. Repeat the rinsing stage and add the residue to the suspension	
6. Make up to 50 mL with the Ora-Sweet. Mix thoroughly.	
7. Transfer the suspension to a 50 mL amber glass bottle and cap. Clean and label the bottle. Attach a fridge sticker	
8. Clean all equipment and working surfaces	

Labelling	
Sample label	**Label used** Labels printed by Labels checked by
Comments	
Final product and worksheet check by: ______________	Date: ______
Final product release by: ______________	Date: ______
Health and safety precautions	
Activity type: Crushing tablets, mixing or reconstituting powders.	
Hazard category: 3	
Environmental control: Local containment, e.g. Waysafe, fume cupboard, micro isolator.	
Personal protective equipment: Disposable latex/vinyl gloves + disposable apron or disposable coat + mobcap.	
Stability and storage information	
See QC monograph.	

Appendix 5

Technical agreement for commissioning of extemporaneous product preparation service

Between
(The "CONTRACT GIVER")
Name and address
And

(The "CONTRACT ACCEPTOR")
Registered pharmacy
Effective date:

Table of contents

1. Objective

The objective is to define the technical agreement for ensuring that products ordered by the Contract Giver and prepared by the Contract Acceptor are made in accordance with Good Preparation Practice in a Registered Pharmacy and that any changes or deviations are reported by the Contract Acceptor to the Contract Giver and vice versa.

This agreement is made between:
Organisation name
Whose facilities are situated at:
Address

(Hereinafter referred to as the CONTRACT GIVER) and
Organisation name
Whose facilities are situated at:
Address
(Hereinafter referred to as the CONTRACT ACCEPTOR)

2. Definitions

Good Preparation Practice as set out in the Guide to Extemporaneous Products and the PIC/s Guide to Good Preparation Practices for the Preparation of Medicinal Products in Healthcare Establishments

3. Responsibilities of the Contract Giver

- Shall supply the Contract Acceptor with a valid prescription for the product ordered.
- Shall report any defects, Adverse Drug Reactions (ADRs) and other problems to the Contract Acceptor.
- Shall ensure all prescriptions are clear and unambiguous.
- Shall notify the Contract Acceptor of any changes and give reasonable notice for the termination of the contract.
- Shall give the Contract Acceptor at least two weeks notice prior to an audit or any request for SOPs or calibration certificates.

4. Responsibilities of the Contract Acceptor

- Shall work in compliance with GPP and not subcontract work without prior written agreement of the Contract Giver. Shall prepare all products under the supervision of an Accountable Pharmacist.
- Shall prepare all products in accordance with the supplied prescription to validated methods using calibrated equipment.
- Shall ensure that all products are released for issue by the supervising pharmacist.
- On request shall submit all formulations methods and procedures to the Contract Giver for approval.
- Shall notify the Contract Giver of any changes to formulations, site of preparation, the Accountable Pharmacist and other major changes.
- Shall notify the Contract Giver of any changes in price.

- Shall supply products ordered within one week of receipt of the prescription and order. Shall notify the Contract Giver if any problems will prevent this.
- Shall archive all documentation as appropriate.
- Shall inform the Contract Giver of any major deviations or failures including the results of any investigations.
- Shall inform the Contact Giver of any known product defects and product recalls in good time.
- Shall allow the Contract Giver to audit the site upon request.

5. Contracts

The Primary contacts for this agreement are set out below:

Contract Giver	Contract Acceptor
Name	Name
Job Title	Job title
Address	Address

6. Miscellaneous

This agreement will be interpreted in accordance with the Laws of England and Wales. The Product remains the property of the Contract Giver.

Signed by

On behalf of

Signed by

On behalf of

Appendix 6

Audit tool for extemporaneous preparation

Location:	Date:

This audit tool should be used in conjunction with the *Handbook of Extemporaneous Preparation: A Guide to Pharmaceutical Compounding*, which gives detailed standards for each section. Audit results are categorised as critical (C), major (M), minor (O) or satisfactory (S) for each section.

Risk management

Acceptance criteria	**Comments**
Follow-up issues from previous audit:	
Suitable risk management systems in place. Procedure available for carrying out risk assessment (2.1)	
Records maintained of risk assessments (2.1)	
No clinically equivalent licensed product available (2.2)	
Prescribers are made aware of unlicensed status of extemporaneous preparations (2.2)	
Alternative options considered and evaluated (2.3)	
Appropriate clinical risk reduction measures taken (2.4.7, 2.4.8, 2.5.1)	
Appropriate technical risk reduction measures taken (2.4.1, 2.4.2, 2.4.3, 2.4.4, 2.4.5, 2.5.2)	
Appropriate health and safety risk reduction measures taken (2.4.6)	
Systems in place to review effectiveness of treatment (2.4.8, 2.5.1)	
Summary comments	
Audit result (CMOS)	

Quality management

Acceptance criteria	Comments
Follow-up issues from previous audit:	
Comprehensive pharmaceutical quality system (PQS) in place (3.1) PQS is fully documented (3.1) Recorded monitoring of effectiveness of PQS in place (3.1) Documented change control procedure and records in place and effective (3.1, 3.5) Documented deviation management procedure and records in place and effective (3.1, 3.6)	
Summary comments	
Audit result (CMOS)	

Personnel and training

Acceptance criteria	Comments
Follow-up issues from previous audit:	
Nominated accountable pharmacist (AP) and appropriate deputising arrangements in place (4.1.1) AP conversant with the necessary standards (4.1.2, 4.1.3) All extemporaneous preparation supervised and released by a pharmacist (4.1.4) Adequate number of competent personnel at all times (4.17) Documented duties and responsibilities of all personnel (4.19) Documented comprehensive training programme and records for all personnel (4.1.10, 4.1.11) Documented assessment of competence for all personnel (4.1.12) Documented procedures in place for hygiene behaviour (4.2.1) Effective control of contamination in place (4.2.2) Appropriate clothing procedures in place (4.2.3)	

Documented procedures in place for staff infections/ skin lesions (4.2.5) Appropriate handwash facilities and procedures in place (4.2.6)	
Summary comments	
Audit result (CMOS)	

Premises and equipment

Acceptance criteria	**Comments**
Follow-up issues from previous audit:	
Suitable premises and equipment (5.1.1, 5.1.2, 5.1.11, 5.1.12) Adequate product segregation. Layout and systems designed to reduce risk of cross-contamination (e.g. only one product handled at a time) (5.1.3, 5.1.5) Satisfactory environmental conditions – temperature, humidity and lighting (5.1.4) Appropriate cold storage and adequate monitoring (5.1.7) Suitable and dedicated clothing in use (5.1.10) Equipment segregated for certain product types. Policy in place. Adequate health and safety control measures in place (5.2.1, 5.1.8) Equipment calibrated, maintained and used within suitable range (5.2.2) Balances suitable and calibration checked at regular intervals (5.2.5, 5.2.6) Balances sited appropriately (e.g. away from draughts) (5.1.6) Glassware appropriate and in suitable condition (5.2.3, 5.2.4) Fume cupboards and containment devices subject to maintenance and monitoring (5.2.7) Effective cleaning (5.3.1) Availability and location of sinks (5.3.2) Containers and lids clean and dry before use (5.3.4) Adequate pest control measures (5.3.5)	
Summary comments	
Audit result (CMOS)	

Documentation

Acceptance criteria	Comments
Follow-up issues from previous audit:	
Comprehensive, QA approved system (6.1) Standardised style (6.1) Regular review (6.1)	
Superseded documents (6.1) Risk assessments documented (6.2.1) Specifications in place for starting materials (including packaging) and finished products (6.3) Approved product-specific instructions (master documents), including specimen labels available for all products (6.4) Worksheets correctly completed – checks, calculations and release recorded. Audit trail available (6.5.1) Systems in place for error-free reproduction of worksheets (e.g. dedicated worksheets from approved masters) (6.5.1, 6.5.2) Suitable worksheet design – content (6.5.3) Suitable label design (6.5.4) Suitable worksheet and label design – medication error risk (6.5.5) Computerised systems (labels and worksheets) password protected (6.5.6) All necessary SOPs in place, comprehensive and current (6.6) Suitable additional documentation in place to support extemporaneous preparation process (6.7)	
Summary comments	
Audit result (CMOS)	

Preparation

Acceptance criteria	Comments
Follow-up issues from previous audit:	
Prescription available and verified (7.2.1, 7.2.2, 7.2.3)	

Prescription verification procedure available and comprehensive (7.2.4) Records maintained (7.2.5, 7.2.6) Appropriate checks of worksheets. Labels and calculations carried out (7.3) Appropriate assembly checks carried out on starting materials – correct, approved and in date (7.4.1, 7.4.2, 7.4.3, 7.4.4) Equipment and containers clean and fit for use (7.4.5) Evidence is available that checks and calibrations of balances have been performed regularly – records kept (7.5) Systems for measuring liquids appropriate for purpose (7.6.1, 7.6.2) Measures and vessels regularly inspected for damage (7.7.3) Independent volume checks (7.6.4) Powder uniformity checks following tablet crushing (7.7) Powder homogeneity checks (7.8.1) Suitable systems for mixing liquids (7.8.2) Suitable systems for mixing semi-solids. Mix time validated and documented when using mechanical mixers (7.8.3) Primary containers fit for purpose – environment and physico-chemical properties considered (7.9.1, 7.9.2) Labels are clear, legible, and contain all the information required to meet statutory and professional requirements (see page 7.9.4) Waste disposal appropriate (see page 7.10)	
Summary comments	
Audit result (CMOS)	

Formulation and stability

Acceptance criteria	**Comments**
Follow-up issues from previous audit:	
All preparations supported by published or in-house stability data (8.1)	

Reference to stability data on worksheet (or by other suitable method) (8.1) Data critically assessed for local conditions (8.2) In absence of supportive data, appropriate risk assessments carried out and documented (8.3) Risk assessments consider pharmaceutical, clinical and patient aspects (8.4) Shelf-lives reflect microbiological and chemical aspects. Shelf-lives assigned are minimum required when not supported by stability data. Max 28 days – preserved, max 7 days – unpreserved (8.3, 8.4.1)	
Summary comments	
Audit result (CMOS)	

Quality control

Acceptance criteria	Comments
Follow-up issues from previous audit:	
All starting materials are from reputable sources and stored in their original containers (9.2.1) Identity of starting materials confirmed and quality assured. Risk assessment carried out for non-pharmacopoeial starting materials without MA (9.2.2, 9.2.3, 9.2.4) Certificates of analysis obtained and appropriate (9.2.4) Date of first opening marked on containers. Max expiry of 2 years (9.2.6) Appropriate storage of starting materials. Storage area monitored (9.2.7, 9.2.8) Water used for preparation should be potable or a sterile product with a maximum in-use shelf-life of one working day after opening (9.2.9) COSHH assessments carried out on materials and appropriate controls in place (9.2.11) All ingredients TSE-free – evidence available (9.2.12) Expired material destroyed appropriately and recorded (9.2.13) Release in line with current legislation (9.4.1)	

Release independent of preparation and considers all aspects (9.4.2, 9.4.3, 9.4.4) Procedure available and includes action to be taken in event of failure (9.4.4, 6.6.2)	
Summary comments	
Audit result (CMOS)	

Complaints, product recalls and adverse events

Acceptance criteria	**Comments**
Follow-up issues from previous audit:	
Policy and procedures in place for handling complaints, recalls and adverse events (10.1) Records maintained – logs etc. (10.1.1, 10.1.4) Trends reviewed (10.1.4) Regular review and audit (10.1) Corrective and preventative action taken following quality problems and documented (10.2) Complaints investigated and documented. Appropriate follow up action taken. Procedure in place (10.1.2, 10.3) Recall procedures in place for initiating recalls and responding to external recall affecting preparations (10.4) Appropriate procedures in place for handling returned and rejected products (10.5)	
Summary comments	
Audit result (CMOS)	

Procurement and quality assessment

Acceptance criteria	**Comments**
Follow-up issues from previous audit:	
Product made under a specials manufacturing licence (11.1) Supplier has current and appropriate licence (11.2.1)	

Supplier audit carried out (11.2.1) Valid formulation and shelf-life – supportive data available (11.2.2) C of A/C of C available and appropriate (11.2.2) TSE statement available (11.2.2) Medication error potential assessment carried out on labelling and packaging (11.2.2) Suitable technical agreement available when outsourcing extemporaneous preparation services (11.3) Technical agreements in place for Other services contracted out (11.4)	
Summary comments	
Audit result (CMOS)	

Audit and monitoring

Acceptance criteria	**Comments**
Follow-up issues from previous audit:	
Regular planned self-audits covering all aspects of process. SOP available (12.1.1, 12.1.2) Independent audit 12–18 monthly (12.1.3) Observations recorded and actions identified. Timescale for corrective action agreed (12.1.6) Supplier audits carried out for outsourced services (12.2.1) Level of environmental monitoring relevant to the scale of operation/types of products made (12.3.1) Records of monitoring activities maintained. Trend analysis carried out (12.3.2) Medicine storage areas temperature controlled and monitored. Records kept (12.3.3) Monitoring programme documented and limits set. Appropriate action taken when limits exceeded (12.3.4, 12.3.6) Monitoring focused in risk areas (12.3.5) Testing laboratories compliant with good laboratory practice (12.3.9)	
Summary comments	
Audit result (CMOS)	

Appendix 7

Suspending agents

Ora-Plus, Ora-Sweet and Ora-Sweet SF

(Paddock Laboratories http://www.paddocklabs.com)

- **Ora-Plus** contains: purified water, microcrystalline cellulose, carboxymethylcellulose sodium, xanthan gum, flavouring, citric acid, sodium phosphate, simethicone, methylparaben, and potassium sorbate. pH 4.2.
- **Ora-Sweet** contains: purified water, sucrose, glycerin, sorbitol, flavouring, citric acid, sodium phosphate, methylparaben and potassium sorbate. pH 4.2.
- **Ora-Sweet Sugar-Free (SF)** contains: purified water, glycerin, sorbitol, sodium saccharin, xanthan gum, flavouring, citric acid, sodium citrate, methylparaben, propylparaben, potassium sorbate. pH 4.2.

Xanthan gum

Synonyms: Keltrol, corn sugar gum.

Xanthan gum is a high molecular weight polysaccharide manufactured by a fermentation process and is widely used within the pharmaceutical industry as a suspending and thickening agent.

Keltrol is the trade name for the xanthan gum products produced by CP Kelco.

Xanthan gum is commonly used as a suspending agent at concentration between 0.4 and 1.0% and is stable over a wide pH range (pH 3–12). It has been successfully used for the suspension of many medicines. However, it is an anionic material and is not usually compatible with cationic materials, which can cause precipitation. There have been reports of incompatibility with some tablet film coatings and some drugs (e.g. tamoxifen, verapamil and amitriptyline) (Rowe *et al.*, 2009).

Reference

Rowe RC, Sheskey PJ, Quinn ME (2009). *Handbook of Pharmaceutical Excipients*, 6th edn. London: Pharmaceutical Press.

Glossary

accountable pharmacist (AP)	the named pharmacist locally responsible and accountable for all extemporaneous activity on a specific hospital site
BNF	British National Formulary
ADR	adverse drug reaction
COSHH	Control of Substances Hazardous to Health
cross-contamination	the accidental contamination of a particular medicine or starting material with traces of another
detergent	a cleaning agent with wetting or emulsifying properties used to remove poorly soluble residues from hard surfaces such as worktops
disinfection	the process of reducing the number of viable microorganisms in or on a surface by washing or wiping with solution with proven antimicrobial activity
EMEA	European Medicines Agency
excipients	ingredients with no inherent therapeutic activity added to medicines to achieve desirable characteristics such as chemical stability of the active ingredients or patient acceptability. Examples include agents to adjust pH, sweeteners, colouring agents
extemporaneous preparation	preparation to the order or prescription from a doctor or dentist, from individually selected starting materials. The medicine is dispensed immediately after preparation and not kept in stock
external audit	an external audit is undertaken by staff who are not managerially accountable within the corporate structure in which the extemporaneous preparation activities occur, and are independent of any service provision to the facility in which these activities occur
GMP	Good Manufacturing Practice
GPP	Good Preparation Practice
ISO	International Standards Organisation
master label	a pre-formatted and pre-authorised label template to which the operator must adhere with the addition only of patient-specific details, date of preparation and 'expiry'

MA	MHRA marketing authorisation (previously known as a product licence)
MHRA	(UK) Medicines and Healthcare products Regulatory Agency
ML	MHRA manufacturer's licence
MS	MHRA manufacturer's 'Specials' licence
NHS	National Health Service (UK)
organoleptic properties	sensory properties of a medicine such as taste, smell, texture, etc.
PQS	pharmaceutical quality system
primary packaging	the container such as the bottle or ointment pot in which the medicine is contained and with which it comes into direct contact
Pro-File	database of NHS hospital-manufactured 'Specials': www.pro-file.nhs.uk
QA	quality assurance
QC	quality control
RPSGB	The Royal Pharmaceutical Society of Great Britain
secondary packaging	the outer container of a medicine such as a carton which has no direct product contact
Standard Operating Procedure (SOP)	a detailed written document covering all aspects of extemporaneous preparation, formally authorised by the accountable pharmacist and, if appropriate, quality assurance
semi-solid	substance such as a cream or ointment with properties intermediate between those of a sold and a liquid
SI	(UK) Statutory Instrument
SLA	service level agreement
'Specials'	unlicensed medicinal products manufactured under an MHRA manufacturer's 'Specials' licence (MS)
starting materials	all the pharmaceutical ingredients used for the preparation of a medicinal product, excluding packaging
SPC	Summary of Product Characteristics
trituration	the process of reducing the particle size of a powder by grinding (e.g. in a mortar and pestle) or of mixing two or more powders or semi-solids together
TSE	transmissible spongiform encephalopathy
vehicle	therapeutically inactive base or diluent such as aqueous cream or water
potable water	water freshly drawn directly from the public 'mains' supply and suitable for drinking
WDL	MHRA wholesaler dealer licence
WDLI	MHRA wholesaler dealer (import) licence

Index